Mineral
Economics

Fourth Edition

W0080979

Mineral Economics

Fourth Edition

RK Sinha

Formerly of Indian Bureau of Mines

NL Sharma

Formerly of Indian School of Mines

Oxford & IBH Publishing Co. Pvt. Ltd.

New Delhi

(A Unit of CBS Publishers & Distributors Pvt Ltd *)*

CBS Publishers & Distributors Pvt Ltd

New Delhi • Bengaluru • Chennai • Kochi • Kolkata • Mumbai

Bhopal • Bhubaneswar • Hyderabad • Jharkhand • Nagpur
• Patna • Pune • Uttarakhand • Dhaka (Bangladesh)

Mineral Economics
Fourth Edition
ISBN-13: 978-81-204-0331-4
ISBN-10: 81-204-0331-2

© 1970, 1976, 1980, 1988 RK Sinha and NL Sharma
Reprint: 1993, 1998
CBS Reprint: 2019, 2022

OXFORD & IBH
New Delhi
(A Unit of CBS Publishers & Distributors Pvt Ltd)

Published by Satish Kumar Jain and Produced by Varun Jain for

CBS Publishers & Distributors Pvt Ltd
4819/XI Prahlad Street, 24 Ansari Road, Daryaganj, New Delhi 110 002, India.
Ph: 23289259, 23266861, 23266867   Fax: 011-23243014   Website: www.cbspd.com
e-mail: delhi@cbspd.com;
cbspubs@airtelmail.in.
Corporate Office: 204 FIE, Industrial Area, Patparganj, Delhi 110 092, India
Ph: 4934 4934   Fax: 4934 4935   e-mail: publishing@cbspd.com;
publicity@cbspd.com

Branches

- **Bengaluru:** Seema House 2975, 17th Cross, KR Road, Banasankari 2nd Stage, Bengaluru 560 070, Karnataka, India
  Ph: +91-80-26771678/79   Fax: +91-80-26771680   e-mail: bangalore@cbspd.com
- **Chennai:** 7, Subbaraya Street, Shenoy Nagar, Chennai 600 030, Tamil Nadu, India
  Ph: +91-44-26680620, 26681266   Fax: +91-44-42032115   e-mail: chennai@cbspd.com
- **Kochi:** 42/1325, 1326, Power House Road, Opp KSEB, Power House, Ernakulam Kochi 682 018, Kerala, India
  Ph: +91-484-4059061-65,67   Fax: +91-484-4059065   e-mail: kochi@cbspd.com
- **Kolkata:** 147, Hind Ceramics Compound, 1st Floor, Nilgunj Road, Belghoria, Kolkata-700056, West Bengal, India
  Ph: +033-25633055, 033-25633056   e-mail: kolkata@cbspd.com
- **Lucknow:** Basement, Khushnuma Complex, 7 Meerabai Marg (Behind Jawahar Bhawan), Lucknow-226001, UP, India
  Ph: +0522-4000032   e-mail: tiwari.lucknow@cbspd.com
- **Mumbai:** PWD Shed, Gala no 25/26, Ramchandra Bhatt Marg, Next to JJ Hospital Gate no. 2, Opp. Union Bank of India, Noorbaug, Mumbai-400009, Maharashtra, India
  Ph: 022-66661880/89   e-mail: mumbai@cbspd.com

Representatives

Hyderabad	0-9885175004	Jharkhand	0-9811541605	Nagpur	0-9421945513
Patna	0-9334159340	Pune	0-9623451994	Uttarakhand	0-9716462459

Printed at Chaman Enterprises, Daryaganj, New Delhi, India

Publisher's Note

This book is dedicated to the memory of late Dr. R.K. Sinha who expired shortly after revising the text for this fourth edition.

Preface to the Third Edition

It gives us a great pleasure in bringing out the third edition in a span of a short time after printing of the second edition in December, 1976. Usefulness of this text is natural because it is the only book available providing the basic knowledge in mineral economics.

In preparing the third edition due care has been taken in incorporating all the useful suggestions and comments received at appropriate places in different chapters. All the statistical tables have been updated providing the latest data.

R.K. SINHA
N.L. SHARMA

Preface to the First Edition

This textbook on mineral economics has been designed to serve the needs of students of geology and mining. Although the idea of mineral economics was mooted for quite some time the subject has now assumed great importance and is regarded as a special branch of geological science, dealing essentially with technology, and related political science, geography and economics. A regular teaching of mineral economics is being gradually introduced in universities in India and abroad that have a postgraduate course in geology. It is also taught as one of the optional subjects to students of mining engineering in many universities.

At present there is no complete book on mineral economics. The first book on this subject was attempted by the American Institute of Mining and Metallurgical Engineers which organised lectures at the Brookings Institution, Washington. These were later brought out in a book entitled *Mineral Economics* (AIME series) edited by F.G. Tryon and E.C. Eckel and published by McGraw-Hill Book Company Inc. in 1932. The Brookings Institution was also responsible for publishing in 1943 *World Minerals and World Peace* by C.K. Leith, J.W. Furness and C. Lewis, experts in the field of mineral resources and economics. Besides these two books, there are number of publications available, but they deal with only one or two aspects of the field. Of these, mention may be made of *World Minerals and World Politics* by C.K. Leith; *Minerals in World Affairs* by T.S. Lovering; *Minerals in World Industry* by W.H. Voskuil and some others. In recent years, great need for a comprehensive book was felt which could provide basic knowledge in the field or Mineral Economics. In writing this book, our endeavour has been to fulfil this need and to provide a deep insight into the vast facets of mineral economics and to prepare students for the responsibility of mineral economists in steering the development of mineral wealth.

Besides attempting to meet the requirements of students of mineral economics, the subject matter dealt with in this book is of considerable interest. The topics on mineral legislation in India and abroad, grading and marketing of minerals and ores, methods of estimating mineral reserves including petroleum, principles and techniques of mineral dressing, etc., have been discussed in such details as to make them useful to all concerned with Mineral Industry. A bibliography has been provided at the end of the book for those who seek more detailed information on individual topics.

The authors are very much indebted to Prof. Ronald D. Parks, Massachusetts Institute of Technology and also associated with the Federal Government, Washington, in work related to mineral industry, taxation and other phases of mineral economics for his keen interest in the book from the very beginning. We acknowledge with deep gratitude the help received from Shri B.K. Dhruva Rao and Shri N.K. Mukherjee, formerly of the Indian Bureau of Mines and now with the Geological Survey of India; Prof. N.V.R. Subrahmanyam and Prof. S.D. Singh of Indian School of Mines, and Shri P.N. Jagtap, formerly of O.N.G. Commission, now with the Directorate of Geology and Mining, Govt. of Maharashtra in the preparation of the chapter 'Methods of Estimation of Reserves'. We greatly acknowledge the help given by Shri D.V. Kulkarni, and Shri N.N. Subrahmanyam, of the Indian Bureau of Mines, in the preparation of the chapter 'Principles and Methods of Mineral Dressing', To Sarvashri S.L. Rai and H.G. Goel, of Indian Bureau of Mines, our thanks are due for giving material help for writing of 'Mineral Taxation' and in preparation of statistical tables and other related matters; and to Shri S.N. Virnave of the Department of Atomic Energy for providing necessary help in collating some useful data. Our special thanks are due to Sarvashri K.S. Mahapatra, Controller, Indian Bureau of Mines, Dr. S.C. Chatterjee, Head of Department of Geology, Ujjain University; Shri S.S. Prasad, Director, Mines Safety, Dr. A.K. Dey, former Senior Specialist, Planning Commission and Dr. P.P. Agarwal, Deputy Director of Industries, Government of Bihar, for their keen interest in bringing out this book. We acknowledge with gratitude help offered by the Embassy of the Netherlands in India for promptly lending us photographs of the diamond cutting and polishing industry in Holland and to the editor of *World Mining* for allowing us to incorporate several location maps appearing in the various issues of the journal. We wish to record our sincere appreciation to Shri R.M. Soundarkar, who typed the entire manuscript and showed extraordinary promptitude in delivering the material in time. Our sincere thanks are due to Mrs. Vasudha Deshmukh for readily agreeing to go through the typed script and make corrections wherever necessary, to Shri P.W. Mankar, for going through the proofs and to Amitabh Sinha for preparation of the index.

R.K. SINHA
N.L. SHARMA

Contents

1

Mineral Economics
and Its Concept

THE USE of minerals has been instrumental in raising the standard of living of mankind. The names of minerals and their products have been used to christen various eras of civilisation, such as the Stone Age, the Bronze Age, the Iron Age and the Nuclear Age. The sophisticated world of today is largely the result of the enlarged use of minerals, whether it be as fertiliser for food; coal, petroleum, natural gas and atomic energy as sources of power; or countless other necessities of life, like automobiles, aeroplanes, ships, modern communications and a host of chemicals which are derived from the use of minerals. All engineering and structural material, machinery, plants, equipments and anything from pins to planes are manufactured from metals and their innumerable alloys. The properties to withstand extreme temperature, pressure and corrosive actions possessed by refractory minerals like chromite, magnesite, fireclay, quartzite, dolomite and several alumino-silicate minerals which are required for lining the furnaces and smelting tanks, have made it possible to smelt and treat minerals and ores for obtaining metals, alloys and chemicals. Iron and steel and their special alloys are the most common metals which largely enter into the fabrication industry. Other metals which are required most commonly are aluminium, copper, lead, zinc and tin. Alloying metals like manganese, silicon, tungsten, chromium, cobalt, nickel, vanadium, etc., find wide applications in the manufacture of engineering goods, plants, equipments and other components. Chemical industry depends largely on the use of sulphur. There is not a single industry which can do without minerals or their products. Minerals thus form a part and parcel of our daily life. Since the beginning of this century the use of minerals has been greatly diversified and expanded. Their consumption has shown an unprecedented increase, year after year. It has been estimated that the quantity of minerals consumed in the last 50 years even exceeds the aggregate quantity consumed in previous human history. The sharp rise in consumption has accelerated attempts in continuous search for locating new deposits and even deeper probe into the womb of the earth and the ocean beds.

Minerals do not occur where we want them to be nor deposits become assets unless explored and developed. Experience shows that no country possesses adequate resources of all minerals. Several countries are practically devoid of mineral wealth, many have inadequate resources, some possess resources of important minerals adequate for their own needs only, and few have enormous reserves of certain minerals by virtue of which they hold monopoly in the supply to the world market.

Even advanced countries like the U.S.A., the U.S.S.R., the U.K., West Germany and many others lack in several minerals. The entire economy of Japan is based on imported ores. Because of the uneven distribution of minerals bestowed by nature, the international trade in this commodity is unavoidable. At the same time it has led to several regional groupings, monopolistic tendencies, cartels and imposition of preferential tariffs and other taxes and even causes of conflicts. The formation of European Common Market, European Free Trade Association, European Coal and Steel Community, Latin America Free Trade Association, Inter-Governmental Council of Exporting Copper Countries, International Tin Council, Organisation of Petroleum Exporting Countries, etc., is the result of such inter-regional groupings. Wars increase the consumption of minerals. Fear of interruption in supplies of strategic minerals compels some countries like U.S.A. to maintain stockpiles. Even in normal times, most industrialised countries invest in exploration and mining in other countries to ensure regular supplies of minerals for their industries.

Concept

The concept of mineral economics has emerged from these economic problems rooted in the peculiar character of mineral deposits; the basic fact being their localisation and exhaustibility. The pressing needs of augmenting mineral production, the intricacies of assuring future supplies, economic evaluation of the deposits, the economics of location of plants in relation to domestic and imported mineral raw material are the subject matters of a special study. These form a broad spectrum of mineral economics under the geological science requiring in-depth study of the world-wide distribution of mineral resources, reserves, utilisation, ore dressing practices, research, planning and development of mineral based industries, trade in minerals, the part played by minerals in economic productivity, their influence on economics and politics—both national and international—and such other related matters. Mineral economics can, thus, be defined as the synthesis of such theories and practices of geological science, mineral engineering, political science, law and economics that are involved in or attracted to the planned development and management of the country's mineral resources.

Scope of Study

The purpose of giving special treatment to the subject is to stimulate study in providing proper background of mineral administration and to prepare students for holding key posts in industry and government departments responsible for mineral development. The subject provides an intimate knowledge of the country's mineral wealth in relation to its industrial applicability. A mere knowledge of mineral occurrences will not be of much help to the students of mineral economics. Rather an intelligent study of various deposits in relation to their economic evaluation is required to be made. In mineral industry usually a number of minerals go to form a single product. The economics of selecting the site for a particular industry in proximity to mines and certain deposits, besides other considerations, require a great deal of scrutiny. It is first to be decided whether it is essentially a power based or raw material based industry. Power based industries like aluminium, ferromanganese, electrolytic copper and zinc and several others need not necessarily be anchored near the deposits but near the source of cheap electricity because of the high consumption of electricity required. The following data will indicate power requirements in the main plants in principal electro-metallurgical and electro-chemical industries.

In the case of power based industries a careful study of the cost analysis of power consumption in relation to overall cost involved in transportation of raw material to the proposed site is required to be made. Cost analysis can be worked out in a manner given on

Industry	Unit of power consumption per tonne
Metals & Alloys	
Aluminium	18 500 kwh
Electrolytic cobalt	6,000 ,,
Electrolytic zinc	4,500 ,,
Ferro-manganese	3,400 .,
Electrolytic copper	2,500 ,,
Chemicals	
Caustic soda	4,000 kwh
Calcium carbide	4,000 ,,
Phosphoric acid	3,900 ,,
Calcium ammonium nitrate	3,200 ,,
Element	
Elemental phosphorus	18,000 kwh

page 5. To give an example aluminium industry has been chosen. It is clear from the cost data that power consumption alone accounts for about 41 % in the reduction plant (potroom), the largest single item in relation to other raw material.

The total cost of metal production in most of the cases works out to be about three times the cost of raw material required.

On the other hand primarily raw material based industries like iron and steel and cement are necessarily to be located near the deposits because in such industries the bulk cost involved is in the transport of raw material. In a steel plant, for every tonne of steel production nearly five tonnes of raw material in terms of iron ore, coking coal, limestone, dolomite, manganese ore, and several other fluxing and refractory material are required to be mined, hauled and put into the furnace, besides ferro-manganese. Raw material requirements per tonne of steel production are approximately 1.7 tonnes of iron ore of 60% Fe, nearly same quantity of coking coal, 0.5 tonnes and 0.25 tonnes of flux grade limestone and dolomite respectively, 40 kg of manganese ore, 10 kg of ferro-manganese and 65 kg of refractories of which 40 kg, 20 kg and 5 kg are of fireclay, basic and silica bricks respectively. The cement industry requires for every tonne of cement nearly 1.3 tonnes of limestone, 400 to 500 kg

Raw Material	Qty. required per tonne of ingot production	Average per tonne price at the factory site	Total cost of raw material
Bauxite (Average 50% Al_2O_3)	5.5 tonnes	Rs. 65	Rs. 358
Caustic soda	200 kg	Rs. 1700	Rs. 240
Cryolite	33.5 kg	Rs. 7500	Rs. 251
Aluminium fluoride	33 kg	Rs. 9400	Rs. 310
Fluorspar	3.5 kg	Rs. 2250	Rs. 8
Petroleum coke	445 kg	Rs. 2100	Rs. 934
Fuel oil	320 kg	Rs. 851	Rs. 43
Anthracite and coke	25 kg	Rs. 3900	Rs. 97
Electricity (for reduction plant only)	18,500 kwh	Rs. 0.12	Rs. 2220

Total: Rs. 4461

of clay, a small quantity of laterite, bauxite or silica to balance the alumina and silica proportion and 40 to 50 kg of gypsum of 75 to 80% purity. Obviously for these reasons the raw material based industries are generally located in proximity to the source of principal mineral deposits.

The most crucial study required is about the investment which baffles planners. It is rather difficult to say off hand the likely investment or amount required to be spent in prospecting, mining and milling and related metallurgical plant, as these depend on numerous variant factors. In each case a detailed study is required. However, a simple formula works well with good proximation in estimating likely investment. In case of producing mine an investment at the rate of 5% of the total value (selling price at the mines head) of mineral produced can safely be earmarked each year for prospecting and improving further reserves. If a mine has rated capacity to produce say, 100,000 tonnes of chromite, the total sale proceeds in a year will come to Rs. 10,000,000 calculated on the selling price of Rs. 100 per tonne. The 5% of this sale proceeds Rs. 500,000 i.e. at the rate of Rs. 5 per tonne can be invested each year for prospecting. For the same capacity of iron ore mine fetching price Rs. 20 per tonne at the pit's head an investment of Re. 1 per tonne can be made on prospecting. No ready-made criterion can be evolved for initial prospecting especially of hidden deposits. Total investment in mining can be safely made up to one and half times the value of one year's production at targetted capacity. This ratio is workable provided the reserves available are sufficient to last for 20 years. In exceptional cases, the investment in mining may be increased not exceeding twice the value. In case of metallurgical plants the total investments twice the ex-plant value of the envisaged annual installed productions of the metal is found reasonable. The norm of investment indicated above also takes into account the expenditure to be incurred on townships. Based on these norms investment per tonne on setting up alumina aluminium, steel, electrolytic copper and cement plants work out to be Rs. 3350; Rs. 25,000, Rs. 4000; Rs. 7000 and Rs. 800 respectively. It should be carefully realised that the norm of investment is closely linked with the optimum level of production planned and to be maintained below which capacity it may not be economically viable. For example, in the present day economy installation of a new unit of an integrated aluminium smelter having less than one lakh capacity a year and that of iron and steel plant less than 2.5 million tonnes capacity a year is considered not economical. Similarly, in the case of a new cement plant and asbestos-cement plant a minimum of 6 lakh tonnes and 36,000 tonnes respectively annual capacity is suggested.

Related studies arise in acquiring a thorough knowledge of the laws governing mines and minerals, the method and procedure of obtaining mining rights, mineral taxation and royalties, tariff and other taxes, mining finance, grading and marketing. These among many others, like production and the study of consumption patterns of minerals in different industries, projecting future requirements, stock-piling and procurement programmes, methods and prospects of utilising low-grade minerals and ores, conservation, substitution, trade in minerals and formulation of mineral policy, are the special features of mineral economics.

The subject matter of mineral economics is so varied and complex, and at the same time so fascinating that one has to make an exploratory trip through the entire range of the mineral industry. In fact, mineral economics is mineral intelligence. The study should impart to students, the idea of becoming a 'spying agent' for the search, mobilisation and utilisation of minerals. A mineral economist has necessarily to acquire this professional acumen. He must maintain an inventory of world mineral resources, the sources from which other countries are obtaining their supplies and keep a vigil on any development in this respect; and depending upon his own country's deficiency or inadequacy he should frame policy and advise his government appropriately in such matters. His responsibility does not cease here. He must study and work out plans to ensure the inflow and maintenance of regular supplies both during peace and war. He has to study and work out such strategy. A mineral economist can easily predict a war, and also from which quarters it is emanating. The outbreak of the Second World War was predicted by the American mineral economists about seven years before it actually took place. It was possible to predict by making an intelligent probe into the stock-piling programme which the Germans had then undertaken. On the same analogy he can prevent a war by studying the procurement programme of mineral raw material of hostile countries and getting their supplies cut by diplomatic means or even by force.

A mineral economist is regarded as a custodian of mineral wealth of his country and the repository of mineral information. He is required to be concerned with the management, conservation and development of nature's non-renewable mineral resource. Realising the basic importance of minerals, many countries have well organised mineral economics departments to foster the development of mineral resources.

2

Peculiarities Inherent in Mineral Industry

MINERALS were formed predominantly in weak zones like mountains, faulted and folded regions, lakes, troughs and continental shelves by geological processes continuously going on since billions of years. They inherited, since their origin, some distinct characteristics which distinguish them from other natural resources. Out of many peculiar characteristics inherent in minerals, the one that they are non-renewable natural assest is most significant and distinctive. While the flora and fauna undergo seasonal or annual regeneration, the minerals are not affected. It takes epoch making time for geological processes, sedimentary, igneous or metamorphic to form any useful mineral.

It has been often spoken that the vast resource of the manganese nodules containing over 20 metals found on the oceanic floors are renewable. It is said that manganese nodules being formed are dispersed all over the oceanic floors at the rate of about 10 million tonnes every year. This statement has no scientific bearing. The process of forming manganese nodules is slowest possible. Four alternative theories are given for the genesis of nodules: (i) hydrogenous origin—produced from the sea water solution, (ii) volcanic origin—could have been produced by changes in basalt, (iii) biogenous origin—could have been produced by micro action, or (iv) diagenetic origin—could have been deposited from the pore solution held within the oceanic sediments. The growth of the nodule is found facilitated by marine organism like radiolarian ooze whose skeleton remains have been found attached to the nodule surface. Concentrations of the nodules are found at depths of 4-5 km on abyssal sea-floor, far from the land and that is also in few localities.

Although the age of the nodules has not been precisely ascertained but their origin can be traced back from the time rifting and drifting of the continents took place. Limited work done so far indicates that the nodules are found resting on Tertiary sediments.

The beginning of the process of forming nodules, therefore, may be considered related to mountain building epochs during Tertiary time (13–60 million years ago) and possible wide spread volcanism extruding oceanic floors. The average size of the nodules has been found to vary between 2 and 8 cm. The largest nodule found weighed 860 kg.

By and large minerals are depleting assets and once mined out the deposits get depleted without any chance of replenishment. Many other peculiarities inherent in minerals as enumerated below, are not shared by any other commodity on the earth.

(1) The occurrence of minerals is localised and limited and the total area occupied by them is far less than that by populated cities or agricultural lands. Majority of mineral deposits is very sporadic in distribution and rich deposits are distinctly rare. Compared to metalliferous deposits, the formations containing coal and petroleum are bigger. The coal-bearing formations of India, for instance, occupy about 4% of the total area of the country, though the coal itself seems to occupy only a small fraction of the coal formations. The Rand Gold deposits of the Republic of South Africa, which are responsible for half the gold output of the world occur in an area 80 km by 32 km only. Similarly, the Kolar gold fields of India occupy an area of only 8 km by 1.6 km. Over 60% of the world's nickel production comes from mines near the small town of Sudbury in Canada. Most of the world's production of molybdenum comes from a single property, the Climax mine in Colorado, U.S.A. The Cerro de Pasco Corporation's property in Peru is the world's largest bismuth producer. An area of 10 sq km near Butte in Montana, U.S.A. is responsible alone for 12% of world's production of copper. The Butte Hill has therefore been called "The richest hill on Earth".

The depth to which metalliferous mines have been worked in exceptional circumstances, is up to 3 km or even slightly deeper e.g. the Rand gold field mines and the Kolar gold field mines. The crude petroleum is being produced from oil wells down to the depths of 6 km (20,000 ft). The production of oil at a depth of 3 km is not uncommon.

(2) Mineral wealth is exhaustible. It cannot be cropped every

year like agricultural products. Once it is mined and consumed, it is lost for ever, except for the thrown out material as refuse or waste which can be consumed afterwards by improved technological methods of benefication and utilisation. This exhaustibility gives a characteristic life cycle to a mine through which every mining district has to pass. There are three distinct stages in the life cycle of a mine, namely youth—development stage; maturity—firmly established; and old age—declining ore reserves. Some of the deposits which were originally considered to be of enormous size, have lost their importance due to depletion of the reserves. The cryolite mine at Ivigtut, Greenland, which was under production for a hundred years was closed down in 1962 as the reserves exhausted. This was the only natural deposit of cryolite known to the world. The Cornish tin (cassiterite) mines, which were the world's most important source for about 1000 years are not very active at present. There is almost a complete record of production of tin from this area dating back to 1156. The copper ore deposits of Michigan, U.S.A., the silver mines of Potosi in Mexico and the once fabulously rich Comstock lode in Nevada, U.S.A. have lost their importance. The most productive petroleum well so far known to the world, viz. Cerro Azul, 4, at Tampico, Mexico, after having produced about 60 million barrels of crude, suddenly began giving salt water. It started as a gusher in February, 1916, when the oil shot up from the borehole to a height of about 180 metres above the ground and was giving a daily production of over a quarter of a million barrels for some months. The closure of the North Skelton mine, the last of the Cleveland mines, brought to an end hundred years or more of ironstone mining on the North-East coast of the U.K.

(3) The unpredictable character of mineral occurrences, mineralisation, considerable variation in grade in a single deposit, uncertainties of extent and persistence at depth, and the increasing cost of mining with the increase in its depth make it quite distinct from any other industry. Mining has its own problems. Depending on the modes of occurrence of the mineral deposits, the operations have to be either opencut or underground. In case of underground mining, the sinking of shafts, development, drifting, haulage, ventilation, drainage, pumping, safety and several other factors are to be taken into account to run a mine efficiently, economically and with safety. The other important points that have to be kept in view in case of

deep mining beyond 150 metres of depth, are the temperature gradient and the rock burst. The temperature of the earth's crust is found to increase by 1° F for every increase in depth varying from 30 to 60 m. Thus at a depth of about 3 km (10,000 ft) the rock temperature in the Kolar gold mines is in the range of 135-150° F. In other words, everything is burning hot if air cooling is not provided at that depth.

(4) Rock burst is a phenomenon met frequently in deep mines caused by the rock pressure of the superimposed load along with uncertain structural features and residual stresses from the previous deformations. Assuming the specific gravity of earth as 2.7, one cubic metre of rock will weigh 2704.8 kg, and on the basis of superimposed load alone the pressure at 300 metres depth will be about 84 kg/sq cm. These abnormal peculiarities of temperature and pressure variations coupled with rock burst, inundation, fire and gas hazards, collapse of roof and side walls are encountered only in mining industry.

(5) Production of minerals is affected by the changing needs of manufacturing industries, accumulation of stock, fluctuations in prices, and re-use of metallic scrap. Discovery of new deposits, nature of ore bodies and the related advancement in recovery and metallurgy of new finds also affect the other mining activities adversely. The refining of less complex and high grade ores involves less expenditure than the complex ores. Evidently, the country possessing higher grade ores can market the products at far cheaper rates, than is possible by other nations. It depends also upon the technological advancement made by individual countries. A country less advanced in technology possessing even richer grades of ores may suffer keen competition in international market by the advanced countries possessing poorer grades because of improved methods of mining, automation and processing adopted, thereby reducing the cost of production considerably. Take for example the low grade beach sand deposits of Florida. They contain only 2-3% ilmenite. Technology of processing Florida's beach deposits has practically squeezed the market of Indian ilmenite which is found in rich concentration as high as 80% in beaches of Kerala and Tamil Nadu. Discovery of new deposits in any country immediately reflects in pattern of international trade so far as pricing is concerned. It gives in effect to buyers' market from sellers' market. To quote few examples, mention can be made of iron and manganese ores which enter into world

market in bulk quantities. The emergence of large number of countries in Latin America and Africa besides Australia as iron and manganese ores producers has affected India's competitive position in the export trade. The steep fall in the f.o.b. prices of the manganese ores during June, 1954 after the cessation of Korean War, from Rs. 190 and Rs. 100 per tonne of ore with 48% Mn and 38% Mn respectively to Rs. 85 and Rs. 45 for the same grades of ore in July 1954, led to the sudden closure of hundreds of manganese ore mines, in India.

The manufacturing industries usually keep a big reserve stock of minerals in normal times; and during a period of recession when the output of manufactured goods goes down, mineral production is the first to be cut down or reduced, whereas in boom period, this production is the last to be revived or increased because of the sufficient quantities of minerals already stored in the big reserve stock.

(6) Mineral occurrences know no political boundary. There is no nation which has within its own borders adequate production and resources of all the minerals needed for its industrial development. Minerals have international character of occurrence. There is unequal distribution of minerals which makes some countries acquire monopolistic rights in respect of certain deposits resulting in geo-politics and rivalries. The unequal distribution also makes the trade in mineral inevitable. Minerals, therefore, acquire international importance and figure prominently in political and economic discussion between countries.

(7) Trade in minerals is quite different from that in any other commodity because considerable variations take place in the quality of the same deposit. Selective mining, blending and processing for grading and marketing pose a special problem in mining industry. Fixing of price is done on the basis of tenor and commercial grades for which minerals mined are required to be prepared according to specifications. Premium or penalty is imposed for any deleterious constituents present which affects the manufacturing process.

Some 40 to 50 minerals enter prominently into the international mineral trade besides metals.

(8) Lastly, the search for minerals, their exploration and development rest upon sound geological and technical experience. Much exploration has to be done and many trial pits have to be developed before deposits of economic value are found. All these require investment of huge capital. The ore bodies buried deep underground are unearthed or discovered by geological as well as some specialised

methods or prospecting like geochemical and geophysical followed by deep drilling. A systematic geological work needs to be followed to find a deposit and even then, there may be certain unforeseen geological factors like faulting, pinching out of the ore bodies, quality deteriorating at depth, etc., which may ultimately affect the life of a mineral deposit.

The investment in minerals, namely, by way of prospecting, drilling, development, mechanisation and exploitation is quite different in nature and involves a good deal of uncertainties, risk and hazard resulting from the depletion of reserves, unpredictable character of mineralisation, practically negligible salvage value of machinery employed and variant working conditions underground. A number of companies in the past are reported to have undergone liquidation because of the nonavailability of possible ore bodies, sudden pinching of the veins, deterioration in quality at depth, etc. All these investments prelude to possible hope of profit, require risking of large capital. Mineral industry for these reasons is placed on altogether different footing.

3

World Resources of Minerals

To ACQUIRE and possess worldwide knowledge of mineral distribution is of vital interest for the development of trade and study of their strategic importance. For studying the regional distribution and concentration of mineral deposits, it may be convenient to examine them individually by the major continents, namely, Africa, Asia, Australasia, Europe, North America and South America, besides the Island countries of Pacific and Caribbean sea. The U.S.S.R. falls both in Asia and Europe. A portion of the land approximately west of Ural Mountains falls in Europe and that of east in Asia. For convenience of description purpose, the mineral potentialities of the U.S.S.R. and China are dealt separately and their account of mineral production and resources has not been included with the countries of Europe or Asia.

Studies reveal, as we shall find later, that even if we consider all the countries falling in a continent as one group, they may be self-sufficien in most of minerals but still lack in some of the useful minerals. This has made international trade in minerals imperative.

AFRICA

It is one of the most emergent continents in the world with greatest future hope of new finds. This continent is a major supplier of a large number of minerals to the European, American and Asian markets. Except for nine countries, namely, Somalia, Burundi, Chad, Gambia, Mali, Malawi, Niger, Dahomey and Mozambique, which record insignificant mineral production, all the rest yield minerals substantially.

The mining and allied mineral industries are controlled by about 800 organised companies of substance and repute. This continent as a whole makes substantial contribution to the world's supplies in some important minerals, viz. diamond, gold, cobalt, platinum, manganese ore, chromite, phosphate, vermiculite, vanadium ore, beryl, copper, antimony, asbestos, uranium, tin, zinc, lead, bauxite, iron ore, and cadmium. Africa can make sufficient supply of ilmenite, rutile, graphite, lithium minerals, cobalt, columbite, tantalite, wolfram and uranium minerals. New reserves are coming up for nickel and mica. It lacks in supply of gypsum, sulphur, and coal required for her own consumption while molybdenum and mercury are non-existent. Nickel, copper and tin are undoubtedly the coming mineral resources of the continent. Niger has emerged as an important source of uranium. Reserves in terms of metal content is placed at 160,000 tonnes. It produces minor quantity of tin ore. Oil producing countries are Algeria, Libya, Tunisia, Nigeria, Angola, Gabon, Congo, Ethiopia and Egypt. The Central and North Sahara region is regarded as potential source of oil. Nigeria and Angola have shown impressive improvement in oil output. Some of the African countries lead in the world's production in certain minerals and make substantial supplies to export market. Discovery of new deposits in Africa is being made with a great spurt and speed by virtue of which it may become a leading source of practically all minerals.

Libya has largest reserves of petroleum in the continent. Proved reserves of crude oil and natural gas are placed at 26,000 million tonnes and 821 billion m^3 respectively. Production of crude has multiplied several times over last few years. Now it ranks amongst the first eight oil producing countries in the world including U.S.S.R. Hafra, Ora and Beda are the main producing centres. The crude is transported through the pipelines to Ras Lanuf port for export. Gypsum is the only other mineral, the production of which was started in mid-1964 near Tripoli. It is a Government-owned quarry operated by a British firm under contract. It has low grade iron ore deposits estimated at 795 million tonnes with 52% Fe.

Morocco possesses enormous reserves of phosphate estimated at 40,000 million tonnes and is second largest producer in the world after the U.S.A. Youssoufia (Safi), and Oued-Zem (Khouribga) phosphates are guaranteed to contain 70% minimum tri-calcium phosphate of lime. Annual output records over 19 million tonnes and is world's largest exporter of phosphate rock. Entire mining and marketing of phosphate rock is controlled by office Cherifen des Phosphate, a State mining company. Export is made mainly through Safi and Casablanca ports. It has gained world distinction as having the only primary producing cobalt mine in the world—80 m in the north. It is emerging as an important producer of iron ore, manganese ore, antimony ore, barytes and fuller's earth.

Tunisia stands fourth amongst world phosphate exporters. Tebessa and Gafsa are the two areas where phosphate is mined. The TPL content varies from 61 to 62%. The ore is exported through Sfax port. There is likely to be substantial increase in the production of iron and steel, lead and zinc. Since mid-1964, this country has become potentially self-sufficient in oil. A rich oil belt has been discovered in the region of EL Borma along the Algeria-Tunisia border.

Algeria is the second largest producer of barytes in the continent. Production comes from Affensou deposits. Djebel Onk area is the important producer of phosphate rock. Reserves of phosphate rock are estimated over 500 million tonnes. Mines de l'Ouenza Bou Khadra has emerged as an important iron ore producing mine with annual production of about 5 million tonnes. It has one of the principal producers of antimony ore after the Republic of South Africa and Morocco and the third largest producer of oil in the continent after Nigeria and Libya. For the transport of crude oil a pipeline from the central Sahara to Mediterranean has been laid. A new pipeline from Hassi Mc Saoud to Arzew has been laid for the transport of the North Sahara crude. The production is well over 53 million tonnes a year. It has a massive reserves of natural gas estimated at 2800 billion m^3. The world's first commercial methane liquefaction plant (three units) was set up in this country which went into production during August, September 1964 and March 1965. Most of the crude oil production goes to France, the U.K. and the U.S.A.

Western Sahara is the second important phosphate producing country in North Africa. Discovery was made only during 1964 and thereafter. Phosphate is mined by a State controlled company Fosfatas de Bu Craa which operates Bu Craa deposits estimated to contain over 1300 million tonnes with 70 to 72% TPL. It is mined by open pit. It is planned to raise production to 10 million tonnes a year comparable to Morocco level of production. The mined ore is conveyed by belt conveyor system to the port of El Aaiun on the Atlantic coast for a distance of 100 km from Bu Craa. It is the longest belt conveyor system in the world carrying some 2000 tonnes of phosphate rock per hour. Belt conveyor system was considered a better choice in comparison to rail, road and pipeline transport. The conveyor system was installed by Fried, Krupp of West Germany.

Nigeria is an important producer of tin ore containing columbite. Tin is smelted at Makari. Since 1962, most of the tin ore produced in Nigeria is smelted locally and exported as tin ingot. It exports sizeable quantity of columbite concentrate also. Most of the coal output comes from coal field near Enugu. The first production of crude oil was started only in 1964. A new oil field in Okan offshore has been established. Nigeria's first refinery began operating late in 1955. It now ranks first in the production of crude in the continent. Nigerian Petroleum Refinery Company is owned 50% by Government and 25% each by the Shell and British Petroleum.

Sierra Leone is an important producer of alluvial diamond. Diamond export accounts for over 60% of the export earning. Prior to 1970, Sierra Leone was rated to be second largest producer of rutile after Australia. Technical and market problem forced them to practically abandon the mining activities. Rutile mining was again revived during 1978 by Sierra Rutile field with 100,000 tonnes annual production target from Gbanghama alluvial. Besides diamond, iron ore and bauxite are the two minerals which are actively mined for export. Sierra Leone Development Company is engaged in mining iron ore from its Masaboin and Ghafal deposits which have been developed for a rated capacity of 3 miilion tonnes annually. Sierra Leone Ore and Metal Company works bauxite deposits.

Liberia has emerged as a leading exporter of iron ore with export of about 22 million tonnes a year including two million tonnes of pellets. The iron ore is exported mainly to the U.S.A. Japan, W. Germany and other European countries. The Nimba mine of Lamco (The Liberian American-Swedish Mineral Co.) alone contributes about 10 million tonnes a year. Production of industrial and gem varieties diamond is reported from Nimba, Lofa and Grand Cape Mountains.

Ivory Coast contains substantial deposits of manganese and iron ores with limited known diamond fields. *Senegal* and *Togo* are big producers and suppliers of phosphate rock.

Gabon is estimated to contain about 300 million tonnes of manganese ore. It is one of the principal sources of manganese ore to the U.S.A. and European countries. The Moanda mine worked by

Comilog produces annually nearly 2 million tonnes of metallurgical grade manganese ore. The main shareholder of Comilog is U.S. Steel Corporation. Mouana is an important uranium mine held by French interests. Several new uranium deposits have been discovered at Oklo, Boyindzi, Okelonbondo and Kaya-Kaya. Established reserves are placed at 30,000 tonnes of contained uranium metal. There are 12 fields of oil but all are of small potentialities.

Zaire is a big producer of industrial diamond from Bakwanga area and gem diamond from Kasai area. The black (industrial) diamond found here is regarded as the hardest, and extensively used in the manufacture of cutting tools of highest quality. Zaire alone accounts for 75 % of the world output of industrial diamond. It is the second largest producer of copper in the continent followed by Zambia. It is the largest producer of cobalt in the world with average annual production of over 10,000 tonnes. Copper mines are centred round three towns of Katanga Province, Lubumbashi (Elisabethville), Jadotville and Kolwezi. The mining zones are referred to as the southern, the central and the western groups of mines respectively. All are non-porphyritic copper ore deposits. The principal copper ore is chalcopyrite. Cobalt is invariably found to occur in economic quantity in copper deposits of Zaire; and in fact copper ores of African continent is found to contain cobalt as co-metal unlike molybdenum as found in the European and North and South American countries. Cobalt, therefore, forms an important product of this country. Zinc is also found associated with copper ores. Gecamines are the largest producers of these metals in the country and owns 7 concentration plants and 3 smelting plants. The Prince Leopold mine in the southern groups is famous for its high grade containing 10 per cent copper. Rest two groups contain on an average 4 % copper. The Tenke area in Shaba Province contains one of the world's richest grade of copper ore with ore reserves of 45.7 million tonnes with 5.5% copper and 0.34% cobalt. The Kamoto copper and cobalt mine in the western group is the largest mine in Katanga which mines three million tonnes of ore annually. It also produces gold and tin in substantial quantity.

Congo situated adjoining to western boundary of the Zaire has good resources of manganese ore. Potash (KCl) deposit of big magnitude occurs near Pointe Noice and mined at Holle. It is a small producer of copper, lead and zinc ore.

Zambia contributes substantially in the production of copper, lead, zinc and cobalt. It is the fourth largest producer of copper in the world after the U.S.A , Canada, Chile. The big copper mines of this country such as Nchanga, Mufulira and Chambishi are well known. There is another side of Zambian copper mining—the small scale operations. Small mines are worked which are situated away from the main copper belts. Allies mine situated 400 km east of Lusaka started production in 1970 with 5 tonnes a month of electroytic copper which was raised to 10 tonnes subsequently and eventually planned to raise production to 30 tonnes a month. Against this background the source of the ore is rich grade ore bodies but having small reserves. The ore occurs in the shear zone in quartz-biotite schists over a strike length of 500 m with 120 m of barren gap. Total reserves are placed at 250,000 tonnes with 3.5% Cu. The zone being mined lies between 24 and 45 m below the surface containing 8% Cu. All the copper ores in this zone is in oxidised form malachite with secondary cuprite and chrysocolla. Copper is extracted using solvent extraction and Torco processes and then by electrowinning. Copper export provides over 90% of foreign exchange earning of Zambia. Production of coal less than one million tonnes a year comes from Lake Kariba area. It produces significant quantities of gem stones amethysts and emerald and have good reserves of limestone and talc. The latest addition to Zambia is Munali nickel deposit south of Lusaka where drilling was completed in 1972.

Angola is an important exporter of iron ore (taconite), diamond and to small extent in manganese ore. About 70 % of the total diamond production is of gem variety. The export of minerals is the main foreign exchange earner in Angola economy. Production of oil comes from Tobias field.

Uganda is famous for its copper mine—Kilembe Mines Ltd.

Tanzania is famous for diamond production from the properties held by Williamson Diamond Ltd. It produces both gem and industrial varieties of diamond.

Kenya is an important source of soda from unique Magadi Soda Lake.

Ghana is an important producer of manganese ore, diamond and gold; the Ashanti gold fields are one of the richest gold mines in the world. The only bauxite mine is worked actively at Awaso in the western region.

The *Republic of South Africa,* by and large, is the major producer in the continent and accounts for 40 % of the African mineral yield.

It contains the world's largest deposits of gold, silver, diamond, platinum and its group metals (palladium, rhodium, ruthenium, osmium and iridium) and vanadium; and possesses large reserves of uranium, chromite, asbestos, ores of manganese, iron, antimony and copper, vermiculite and coal. It has emerged as one of the major countries in the continent in the export of iron ore to Japan and West European countries on long term contracts. Major producing areas are North West of Capes. Production of iron ore is over 26,000,000 million tonnes per annum. The Bushveld Complex is the largest source of metallurgical grade chromite. Platinum is found associated with chromite. The Merensky Reef in the Bushveld Complex in Transvaal where the Rustenburg Platinum Mines Ltd. operates is probably the only place in the world where the ore is rich enough to justify mining directly for platinum. The Rustenburg operates three mines, Rustenburg, Union and Amandelbult with total annual production capacity of about 1,515,000 tonnes. The ore contains about 0.25 troy oz of platinum group of metals per tonne, of which 72 % of platinum metal. The platinum minerals are found in about a 30 cm thick gently dipping band in the lower part of norite zone within the Merensky Reef. The production of fluorspar, magnesite and phosphate rock is of substantial magnitude. Cape Province and Transvaal are the largest, rather the only producers of crocidolite (blue asbestos) followed by a small production in Australia and Bolivia. Similarly, the north-eastern Transvaal is the largest source of amosite seconded by comparatively much smaller resources in Russia. Palabora copper open pit located at Phalaborwa in the north-eastern Transvaal is the largest mine on daily tonnage basis in whole of Africa and third in the world outside the U.S.A. The estimated reserves are 315,019,500 tonnes with cut-off grade of 0.3% down to depth of 360 metres. Presently, it is being mined with grade at 0.79% Cu with 0.4 % cut-off grade. In the production of antimony ore it ranks first in the world excluding China. Rooiberg Minerals Development Co. Ltd. operating near Warmbaths, Transvaal has expended the production capacity of tin concentrate to 4000 tonnes annually thus becoming a leading tin producer in African continent. Finsch open pit diamond mine in Cape Province, owned and operated by De Beers Consolidated Mines Ltd. is the country's largest diamond mine producing 4 million carats annually out of which 25% is of gem varietey. Finsch mine contains the highest tenor 66.65 carats per 100 tonnes. The lowest grade of diamond mine mined in S. Africa is Koffiefontein mine containing 11.04 carats/100 tonnes, which is also very high, much above the world average. The development of Trojan nickel mine containing 1% nickel in ore by Anglo-American Corporation of South Africa has become Africa's largest nickel mine. One of the largest deposits of andalusite is found in the metamorphosed rock of Pretoria series, on the north-eastern rim of the Bushveld igneous complex, east of Chuniespoort, northern Transvaal. It has developed Richards Basy in Natal which will produce sizeable quantities of rutile, zircon and ilmenite. The Republic of South Africa is probably the only country in the world producing synthetic petrol utilising low grade coal. The plant is located at Salol in Johannesburg. Daily 14,000 tonnes of locally mined coal is treated to make 66 gallons of liquid per tonne.

Mozambique has in very recent years started vigorous attempts to discover minerals in its territory. Modest production of coal and bauxite is reported. Large deposits of fluorspar and iron ore have found.

Zimbabwe is a major world's source of chromite, corundum, lithium minerals, asbestos both of chrysotile and crocidolite and gold. Important asbestos mines of world fame are Shabani, Mashaba, Pangani, Boss, Vanguard and Rex. It is an important producer of copper also. The major quantity of Rhodesian chromite goes to U.S.A. The production of lithium minerals is entirely geared to overseas market demand as there are substantial reserves. Bikitia area is probably the only pegmatite zone in the world containing rich deposits of lithium ores (amblygonite, lepidolite, potalite, eucryptite and spodumene). Beryl is found in good quantity which is hand sorted. This country possesses large reserves of coking coal and exports it to many countries of the continent. Zimbabwe's chromite deposits are under United States control. The reserves of chromite are estimated to be more than 600 million tonnes. The great asbestos mines are British owned. Tin mining is controlled by Holland. Nickel is the newest mineral resource added to this country worked by Anglo-American Corporation (Rhodesia) Ltd. Corundium occurs at O'Briens some 56 km north of Salisbury.

Namibia (*South West Africa*) is famous for the gem quality diamond and its production from the off sea-shore. Tsumeb Corporation Ltd. mines complex ores, smelts and recovers copper, lead, zinc, cadmium, silver and sulphuric acid. This country is also one of the important sources of lithium, minerals, uranium and tin. Uranium mine has been developed in the Langer Heinrich mountain containing 60 million tonnes grading 2 pound/tonne of ore.

Egypt possesses sufficient reserves of iron ore, manganese ore, gypsum, talc and limestone required for domestic consumption. Phosphate production is in surplus, Kossier, Safaga and Sabaiya are three important phosphate mining centres. The TPL content varies between 55 and 60%. All phosphate mining and export are under State control. Phosphate is exported mainly to India, Sri Lanka and European countries. A new phosphate deposit has been discovered between the oasis of El-Dakhale and El-Kharga. New oil field has been discovered in extension to Balayim Marine field in Gulf of Suez. The current production is sufficient for its normal consumption. More new oil fields have been discovered in the western part of Egypt stretching from El Alamin to the Libyan border. The Sinai desert containing number of oil fields which was under occupation of Israel was reverted back to Egypt on 25th April 1982.

Guinea in West Africa is estimated to have the largest bauxite reserves after Australia, in the world, more than 2400 million tonnes. Compagnie des Bauxites de Guinee's Boke (CBG) is the largest bauxite mining concern in Guinea and raises about 5 million tonnes of bauxite annually. Guinea now ranks amongst leading bauxite producers in the world and can be rated with Surinam and Guyana (about 5 million tonnes annual production each) but behind Jamaica and Australia with about 12 million tonnes annual production each. CBG is 51% owned by Halco Mining Inc of the U.S.A. and 49% by the Government of Guinea.

Mauritania has sufficient reserves of iron ore. Miferma mine has been developed by the European finance which now exports nearly 12 million tonnes of high grade iron ore per year. New iron ore reserves containing 37% Fe in quartz-magnetite ore bodies have been established at Guelbs. Copper ore is worked at Akjoujt and huge gypsum reserves is mined at Nouak chott.

Botswana has singularly become important for highest incidence of diamond known so far in the world. The Orapa open pit diamond mine of De Beers Consolidated Ltd. which was brought into production during 1971 gives on an average one carat per tonne of ore mined. It produces both industrial and gem diamonds. Annual production from this pit alone averages 4.5 million carats. Another giant Jwaneng mine has been opened which is heading one of the biggest diamond mines in the world.

Development of Pikwe-Selebi nickel-copper mines by Bamang-wato Concessions Ltd. has brought this country into international limelight. Production has started in 1973. The mine has been developed with initial capacity of 2.2 million tonnes of ore; producing some 46,000 tonnes of nickel-copper matte which is shipped to Amax nickel refinery plant at Brathwaite, Louisiana, U.S.A. Bamangwato Concessions Ltd. is owned 15 % by the Bostwana Government and remaining 15 % by the American interest. Pikwe contain proven reserves of 24 million tonnes with 1.45% Ni and 1 14 % Cu; and Selebi 11 million tonnes with 0.7 % Ni and 1.56 % Cu. One of the important features of nickel-copper matte production is that SO_2 gas emanating from the furnace is passed through columns of pulverised coal to recover elemental sulphur. Nearly 127,000 tonnes of sulphur is obtained as by-product every year. Bostwana has also started production of coal with a rate of 2 million tonnes per annum. Proved reserves of coal are placed at 500 million tonnes.

Sudan and *Ethiopia* are reported to contain several useful minerals but the production is relatively very small. *Basutoland* and *Lesotho* have small reserves of diamond. *Swaziland* is predominantly important for chrysotile asbestos and recently for export of iron ore.

Malagasy possesses the largest reserves of flake graphite and phlogopite mica in the world. The graphite rich zones are found in Tamatave, Tanarive and Amapanihy close to South-central coast. Phlogopite is mined north-west of Fort Dauphin in beds of pyroxenites. The dune deposits near the Fort Dauphin are reported to contain heavy minerals including uranium and thorium minerals. In the lake Alaotra region large deposits of chromite have been discovered. Petroleum reserves are adequate.

ASIA

All countries in Asia with exception of Nepal, Lebanon, Jordan and Cambodia are significant from mineral point of view. This region forms a major source of many useful minerals, especially manganese ore, mica, iron ore, ilmenite, graphite, kyanite, sillimanite, gypsum, barytes, petroleum, tin ore, tungsten ore, chromite, bauxite, soapstone, magnesite, fluorspar and coal. In supply of fluorspar Thailand has come out singularly as a leading source in Asia. India has become a powerful source of barytes in the world market. It alone possesses one-fourth of the known world's reserves. Since late seventies, Iran and Indonesia have become a potential source of copper metal and the latter also in nickel. Burma and Ceylon are the world's largest sources of gems—jade, ruby, sapphire, aquamarine, topaz, chrysoberyl and moonstone. Ceylon is also the largest producer of crystalline graphite of high purity. The resources of mercury are confined to Turkey and Philippines only. Japan, however, produces a small quantity of mercury. Turkey is the second largest producer of borax in the world and third largest producer of chromite closely followed by Philippines. Asia is the largest producer of tin ore in the world, the monopolistic possession is however maintained by Malaysia, Thailand and Indonesia only. India possesses world monopoly in kyanite, sillimanite and sheet mica. In petroleum, Asia now ranks largest producing continent, the second being North America and third the U.S.S.R. The West Asian countries possess nearly 61% of the total known world reserves of petroleum.

Asia as a whole excluding China and U.S.S.R. lacks in resources of asbestos, sulphur, nickel, mercury, cobalt, bismuth, platinum group of metals, copper, lead, zinc, antimony and anthracite and coking coal. The occurrence of anthracite is confined to Southern Vietnam, South Korea and Japan. This region can, however, be a large producer of iron ore, fluorite, ilmenite, rutile, gypsum and bauxite. The largest source of nickel in Asia is found in Indonesia.

Discussing about the countries from West of East, *Turkey* holds great importance in production of refractory grade chromite, boron minerals, and in some extent mercury and magnesite for which it has exportable surplus. Chromite is mainly shipped to the U.S.A. and Japan. Possibilities of finding substantial reserves of phosphate rock are indicated. It is the only source of boron minerals in Asia. Colemanite (hydrated calcium borate) mixed with ulexite (hydrated sodium calcium borate) are the principal boron minerals mined. They are shipped to European countries for converting them into borax. The production of mercury is small, of the order of 100 tonnes a year only. It is also an important source of antimony ore and barytes. The crude oil production is small, less than a million tonnes per year. The present production meets hardly 25% of the domestic requirements.

Cyprus possesses good reserves of pyrite and cuprous pyrite which are mined for the extraction of copper, sulphur and sulphuric acid.

The Cyprus Mines Corpn. of U.S.A. is the largest producer of above minerals from the Movrovoumi underground mine and the Skouriotissa open pit.

Israel is one of the two major producers of phosphate rock in Asia. The reserves of phosphate rock at Oron are estimated to be 40-50 million tonnes with 30 % P_2O_5; 10 million tonnes at Hor Hohar and 120 million tonnes at Arad. It has a good source of potassium chloride from the Dead Sea works. Copper ore is worked at Timna in the extreme southern part of the country. Oil production is small. New fields have been discovered in Mir Am-Helatz area. For the present Israel meets 90 % of her requirements of petroleum from Iran. It has a well developed gem cutting and polishing industry especially of small diamond crystals.

Iran, Iraq, Saudi Arabia, Kuwait, Neutral Zone between Kuwait and Saudi-Arabia, *Qatar, Bahrain island*, and a tiny island *Abu Dhabi* in Persian Gulf hold strategic positions for supply of petroleum and is a very sensitive region of geo-politics. Sixty-one per cent of the proved reserves of oil in the world are in West Asia, 10 % in U.S.A., and 9% in U.S.S.R. and remaining elsewhere. Nearly one-fourth of the world's supply of petroleum comes from West Asia and it controls 50% of the world trade in petroleum. The petroleum industry in this region is mostly controlled by the companies of America, British, France, Dutch and Japan or their combine with subsidiaries from Italy, W. Germany and Spain. But all these foreign owned companies are facing threat of nationalisation or a greater control in the equity shares by the host countries.

Iran is a major source of zinc and lead ores, barytes, chromite and gypsum and to a lesser extent of manganese ore. Iran is steadily becoming an important producer of copper. Sar Cheshmeh copper deposits are now government owned and being developed with a technical assistance provided by the Anaconda Company of U.S.A. Minak Company with Japanese partners, has erected a 400 daily tonne mill for copper ore at Qaleh-Zari, south of Birdjan. The concentrates containing 30% copper also contains 15 grammes gold and 240 grammes silver per tonne. Besides the Iranian firms, the Rio Tinto Zinc Corporation Ltd., American Metal Climax, Inc. and Cerro Corporation of America and Societe Generale des Minerals of Belgium are associated with the mining of lead-zinc ores. Japan is the regular buyer of Iranian lead and zinc concentrates. This country can also form a source of supply of lead-zinc concentrates to

India. It is one of the richest producers and exporters of oil in West Asia. Rich oil wells are found in the inland and off-shore areas of Iran. Agha Jari, Marun, Gachsaran, Alborg, Nafti-Shah, Bharegan Sar Sasan and Darius are important oil fields.

Kuwait, a tiny country with a population of about 322,000 is the rich oil country in West Asia having an average annual revenue of over Rs. 25,000 million. It ranks amongst first ten largest producers in the world. *Saudi Arabia* is the third largest producer of oil in the world. There are more than 10 producing fields of which Ghawar, Abquaiq, Safaniya and Khursaniya are important. The recent geological survey has revealed prospect of finding good copper ore reserves at Jabal Sajid, 320 km north-east of Jeddah. The *Neutral Zone* between Kuwait and Saudi Arabia is held by Arabian Oil Company, a Japanese firm and crude produced from the area is mainly exported to Japan.

Iraq has good deposits of gypsum and limestone and is oil rich country. The greater part of the production comes from the Iraq Petroleum Company's field at Kirkuh where oil was struck in 1972.

Jordan has practically no oil but it is the largest producer of phosphate in Asia. The principal mine is at Ruseifa about 16 km north of Amman. Small deposits of iron ore exist in two localities, namely, at Ajloun and Wadi-Sabra. The Jordan side of Dead Sea is exploited for potassium chloride and other salts. Medium-sized industry exists including a cement factory, an oil refinery, a marble industry and an iron works.

Syria is a small producer of oil. All oil fields are situated in the north-east region of Karatchouk, Hamzah and Rumailan worked under the control of the National General Petroleum Authority since 1958.

Lebanon is poor in mineral resources, except for small production of phosphate, iron ore, limestone and salt. No other mineral is reported to have been produced.

Afghanistan, a land-locked country, has good resources of beryl chromite and rock salt. Beryl carrying veins about 200 in number varying from 1 to 20 metres in width have been recorded in Darra Pech in Kurar Province. The beryl content of the vein is 0.2 to 0.6%. Lithium and tantalum are found associated with the ore. Several beryl pegmatites contain muscovite mica. Production of coal is small. Several oil and gas discoveries have been made in the Sar-i-pal area of the north-west, at Sultan Kot, Khwaya-Gougwer and Yatin Dagh.

TABLE 1

Production of Crude Oil during 1978-82

(in million tonnes)

Country	1978	1979	1980	1981	1982
World	3,097.3	3,189.1	3,059.1	2,765.6	2,643.2
Abu Dhabi	70.0	71.1	64.5	54.5	42.5
Algeria	58.8	53.2	51.5	46.3	45.8
Argentina	22.9	23.9	52.2	25.2	25.6
Australia	20.8	20.6	18.0	19.0	19.5
Brazil	8.3	8.5	9.4	11.0	13.4
Canada	74.7	83.2	83.0	72.9	70.0
China	104.0	106.2	106.07	101.2	101.7
Dubai	18.5	17.7	17.5	18.0	17.8
Ecuador	9.7	10.5	10.8	10.7	10.4
Egypt	24.4	26.0	30.1	34.1	34.9
India	11.2	12.8	9.4	14.9	19.7
Indonesia	82.4	79.1	78.5	82.2	69.0
Iran	260.8	151.4	76.6	65.2	119.8
Iraq	128.9	168.0	130.0	44.9	49.6
Kuwait	108.9	127.2	81.4	48.2	34.9
Libya	96.2	98.9	85.9	58.7	54.7
Malaysia	10.8	13.4	13.2	12.6	14.6
Mexico	66.4	80.8	106.8	128.3	149.4
Nigeria	94.0	113.5	101.8	71.2	63.5
Norway	16.8	18.3	24.0	24.9	24.6
Oman	15.5	14.6	14.0	16.4	16.2
Qatar	23.4	24.4	22.8	20.2	16.3
Saudi Arabia	421.9	475.2	496.4	491.3	327.9
UK	54.0	77.9	80.5	89.4	103.4
USA	481.5	478.6	482.2	482.8	484.5
USSR	572.5	586.0	603.0	609.0	612.2
Venezuela	115.73	122.8	112.9	112.5	101.3

Source: Indian Petroleum and Petro-Chemicals Statistics, 1982-83 & 1984-85.

Pakistan except for gypsum and rock salt and to some extent anti-mony (stibnite) and chromite is deficient in almost all other econo-mic minerals. The domestic production of coal, limestone, silica sand, chromite and barytes is of small magnitude. There is no coking coal deposit. Copper ore was being prospected at Saindak in Chagai district, Baluchistan Province. Iron ore deposit with 55% Fe content is reported in Dalbandeen in Chagai district extending up to Afghan border. No further work on the deposit has been reported. New deposits of phosphorite and magnesite have been reported from Abdotabad area in Hazara district. Phosphorite is reported to contain 25 to 35% P_2O_5. Chromite is mined at Hindubagh at an annual rate of 25,000 tonnes. Gypsum is mined in the Salt Range and Sulaiman Range areas. Production of coal is reported to be small of the order of about 1 milion tonne per year. Attock oil field and Sui gas field are the main source of oil and gas in Pakistan. Production of crude is also very small, roughly about 1/2 million tonne every year. Pro-duction of natural gas is about 126,000 million cubic feet per year. Sui gas is transported through 16-inch pipe from Multan to Lyallpur viz. Kabirwala, Darkhana, Jarkala, Toba Tek Singh and Gojra.

Bangla Desh is a new independent nation which came into exist-ence on 17th Dec. 1971. Formerly it was the eastern wing of Pakistan. The mineral sources of Bangla Desh include natural gas, coal, lime-stone, china clay, glass sand, gypsum, iron ore and barytes but the resource developed hitherto has been very poor. Natural gas is re-ported to be in plentiful. Coal of Gondwana range is found to occur at Semutang near Chittagong. New deposits have been discovered in Paharpur-Jamalgang area, Bogra district where about 700 million tonnes of reserves up to a depth of 1110 metres are reported. Exten-sive deposits of limestone are reported to occur at Takerghat and Jaipurhat. Presently India is supplying most of the limestone require-ments for its Chattak cement plant of two million tonnes capacity. It is also reported that the Government of Bangla Desh have opened their coastal areas for prospecting of natural oil and gas.

Nepal, although unknown for mineral occurrence, has since 1953 made significant efforts in search of deposits. The Nepal Bureau of Mines has discovered several mineral deposits of economic impor-tance. The most important discovery is of cement grade limestone at Chobbar, Bjainse and several other places of southern Nepal. The total reserves are estimated at 120 million tonnes. Chobbar limestone containing reserves of 8 millions (49% CaO, 1% MgO) has been developed by Himal Cement Company Pvt. Ltd. for 175 tonne a day of cement plant.

Other recent discoveries of economic importance are: 8 million tonnes of hematite iron ore (56% Fe) near Phulchoki; 500,000 tonnes of talc; 100 million tonnes of magnesite; 60,000 tonnes of ochre at Kharidhunga; 16 million tonnes of crystalline magnesite (44% MgO) at Kampur Ghat; and 3 million tonnes of pyrite (25 per cent sulphur and assaying 0.2 to 0.7 per cent copper) at Bering Khola.

Copper deposits have been discovered at Didhi Khani, Waspa Khani, Bhut Khola, and Bag-lung. Lead and zinc deposits are known at Phulchoki, Nawakot-Kandrangarhi, and Sallmor valley. Good quality mica deposits are at Kathmandu, Dhankuta Bhojpur, and Bajhang. Beryl at Kathmandu, semiprecious stones at Taplejung-Sakhuwa Sabha, and phosphate rock at Dang. Deposits of antimony, barytes, cobalt, gypsum, kaolin, kyanite, marble and nickel are also known.

India is a leading supplier of sheet mica in the world and is the only country (except a small quantity in Ceylon) in Asia possessing this mineral. It has also the largest reserves of high grade iron ore in the world and possesses large and extensive deposits of ilmemite, chromite, manganese ore, limestone, dolomite, soapstone, china clay, gypsum, barytes, magnesite, kyanite, sillimanite, bauxite, rutile, monazite, zircon sands, uranium minerals and coal. Several large size phosphorite deposits have been discovered. The supply position of chrysotile asbestos, industrial diamond, tungsten, crystalline and flake graphite and non-ferrous metals except aluminium is limited. The deposits of mercury, tin and antimony ores, platinum and its group of metals, elemental sulphur, borax are non-existent.

The exploratory works carried out since 1955 have established several prospects in Rajasthan, Bihar, and Madhya Pradesh for copper ore. Sizable quantity of lead-zinc-silver ore are confined to Rajasthan. The petroleum resources are fairly good. Digboi, Nahorkatiya, Moran and Sibsagar Lakwa and Rudrasagar in Assam and Ankleshwar, Cambay and Kalol in Gujarat and the Bombay high in the offshore are important producing centres. The present production capacity nearly meets less than one-third of the total petroleum requirements of the country. India's mineral resources have been dealt with separately in this book.

Burma is famous for production of lead-zinc and silver, tin and tungsten ores, precious stones like jade and ruby and to some extent sapphire. Precious stone of Burma is of world fame. Lack of finance and capital equipment are reported to be the main cause for decline in production. All the Burmese mines are in the hands of five State owned corporations. Also the entire export-import trade is controlled by the Govt. The production of lead-zinc and silver comes from once world famous Bawdwin mine in Northern Shan State worked by Myanma Bawdwin Corporation. Formerly, the ore mined contained 30 to 35% of lead. At the end of 1976 ore reserves were estimated at about 6 million tonnes with 11.2% lead; 5.6% zinc; 0.3% copper and 7.8 ounce of silver per tonne. Zinc concentrate contains 0.3 % cadmium. Three more deposits of lead-zinc and silver ores situated at Longh Keng, Bawsaing mine and Yadana Theingi mine are expected to be developed. The large Mawchitin-tungsten mine is the largest single mine producing these two minerals in Burma. These two minerals occur in lode and detrital deposits; the alluvials carry cassiterite only. The Burma's tin belt is an extension of the Banka, Billiton, West Malaysia and Thailand belt and extends from the extreme south of the country northward to near Mandalay city for a distance of 1500 km. The production of tungsten comes from Mawchi. Production of tin ore has centred mainly in the region of Tavoy, Mergui, Tenasserim area in Southern Burma. Production of copper comes chiefly as a by-product of Bawdwin lead-zinc-silver mine. It has become self-sufficient, rather surplus, in production of industrial minerals especially gypsum and barytes.

Burma continues to be a major exporter of jade, ruby and sapphire in the world. Jade occurs in the vicinity of Kamaing in northern Burma. Mogok rubies are the world famous for their quality and colour. Oil and natural gas is mined by Myanma Oil Corporation. The domestic production is sufficient for home demand. The natural gas is mostly utilised for the manufacture of nitrogenous fertilisers.

Srilanka (Ceylon) is important for its highest quality crystalline graphite (98 % C min.) and exports annually 10,000 to 11,000 tonnes of it to many countries. It is a major producer of gemstone, sapphire, topaz, zircon, chrysoberyl and moonstone. It has also become a major producer of ilmenite and rutile from the beach sand.

Indonesia is the third largest producer of tin in Asia after Malaysia and Thailand and fourth in the world excluding U.S.S.R. and China. This country has become an important producer and exporter of copper, nickel, bauxite and petroleum in addition to tin. Several multi-national mining companies are operating in Indonesia.

The production of tin ore comes chiefly from three islands Bangka, Billiton and Singkep, off the north coast of Sumatra. The copper ore is worked at Ertsbery in West Irian owned by Freeport Indonesia, Inc. It has started production of copper metal in 1972. It also exports sizeable quantity of copper concentrate of Japan and elsewhere. Ertsberg copper deposit contains high gold and silver values. In this deposit a minimum of 33 million tonnes of copper ore containing 2.5 Cu has been estimated. Nickel ore deposits are located in the Pomala Tandjoeng Pakar areas of South Kolaka, Celebes (Sulawesi) island. Extensive deposits of nickel ore are found also in Central Celebes. The Sulawesi Nickel Development Company Ltd., a Japanese firm, mines nickeliferous chromite containing on an average 3.2% Ni. A rich nickel-cobalt deposit estimated at 76 million tonnes is reported on the island of Waigeo, in West Irian. The largest deposit of nickel in Asia are located near Tjilabjag on the south coast of Java. Consortium of eight Japanese firm have reported discoveries of nickeliferous laterite in two small islands Obi and Gobe about 200 km apart located south and south-east of the large Halmaheva island. The total reserves have been estimated at 60 million tonnes with 1.6% nickel. Bauxite deposits are found on the islands of Bintan and Kojang in the Riouw Archipelago. Deposits occur also on three other islands — Bangka, Batam and Singkep. Bintan deposits are regarded as the largest. One of the largest sulphur deposits is located in West Java with one million tonnes of proved reserves with 50% sulphur content. There are also sulphur deposits in the province of Kedu, Central Java and a volcanic sulphur deposit near Bandung. Occurrences of copper ore are reported on the islands of Borneo, — Ceram, Timor, the western tip of New Guinea and the Moluccas islands. These deposits are believed to be an extension of the Pacific volcanic copper belt running through Japan, Farmosa and the Philippines. Indonesian coal reserves are large, the bulk production coming from Bukit Assam coal field. Central Sumatra and Borneo are richly endowed with oil. Minas, Duri in Central Sumatra and Tandjung in Kalimantan in Borneo are the important producing fields Petroliferous bed has also been found in the offshore area west of Java. Indonesia has substantially improved its oil production from the number of new oil fields brought into production and by reactivation of a number of older fields. Oil is mined both by State owned company and a few foreign companies. Japan has a great interest in the Sumatra oil field.

Thailand is an important producer of cassiterite, fluorite, barytes, wolfram, antimony and lead. It ranks second in tin production in the world. Large deposits of tin ore occur off the west coast. Important centre of tin mining is located in the sea off the Phuket island. In fluorspar production also it stands amongst first three fluorspar producing countries in the world including U.S.S R. The production comes chiefly from northern province of Lampoon. It has a good source of gypsum after India, Pakistan and Burma. India imports fluorspar from Thailand. Production of manganese ore is small. Though it has no natural resources of oil, but has an estimated petroliferous shale reserves of 300 million tonnes in Mae Sod district. The plan is to produce 1.5 million tonnes of shale per year to produce 200,000 barrels of oil.

Malaysia is the largest producer of tin and accounts for about one-third of the world production. There are about 875 mining locations for tin ore. Sawarak produces large quantity of bauxite which finds export to Japan. Ulu Rompin and Dungun in Pahang and Trenagganu are the largest iron ore producing mines in the country. Malaysia is strong competitor of Indian ilmenite exporting over one lakh tonnes a year. Swarak, North Borneo and Brune (British Protectorate) all are a good source of oil. Miri oil field is the largest producer.

Laos possesses good resources of tin and iron of which only tin ore is mined. Tin mining is carried on mostly at the nationalised mine Phou Tiou near Thakhek. The tin is smelted in Malaysia before export. Extensive deposits of iron ore have been found near Xien Khouang estimated to contain 1000 million tonnes with 60% Fe. Oil fields have been discovered between Thakhek and Pakse and in the Vientiane valley. Coal is also reported to occur in Vientiane valley. The political difficulties in this country held these deposits unexploited so far.

South Vietnam has not progressed satisfactorily in mineral production owing to the continued hostilities between north and south. The Aa Hoa area about 40 km to the south-west of Tourane is reported to contain a large variety of minerals including ores of iron, copper, lead, zinc, nickel, gold, mica and graphite. This area is an important source of anthracite. The Laokay area is said to contain about 1000 million tonnes of apatite covering both North and South Vietnam.

North Vietnam contains appreciable reserves of iron ore at the

Trai-Cau mine area which supplies ore to Thai Nguyen Iron and Steel Work. It possesses substantial reserves of chromite, anthracite, coal and tin. The production of tin ore is probably supplied to Russia and apatite to China.

Formosa (Taiwan) — Known resources of coal reserves are estimated at 200 million tonnes. Small production of sulphur comes from the Tatun volcanic region. It also produces pyrite, over 10,000 tonnes of electrolytic copper and 20,000 tonnes of aluminium metal from imported alumina per year. Silver and gold is recovered as by-products of copper refinery. This island has reserves of copper ore estimated at about one million tonnes in the northern coastal water, with copper content varying from 0.7% to 2%. For the first time a production of about 6000 tonnes of mica was reported during 1965 from the island which was all exported. Thereafter, no production of mica has been reported. It has good resources of talc and marble. It is the major supplier of cement for local consumption in the South Vietnam.

Republic of South Korea is the largest producer of tungsten ore (scheelite) containing bismuth, in the world excluding China and U.S.S.R. Production of scheelite largely comes from Sangdong mine. The reserves of amorphous graphite and anthracite are also very large. The production of copper, lead, zinc, molybdenum, nickel and bismuth is adequate. It has become an important source of fluorite and has expanded its cement industry expeditiously recently. Small production of iron ore comes from two major mines, the Mulkum mine about 40 km north-west of Pusan and the Kyongin mine about 40 km west of Seoul. Korea has a growing steel industry with annual capacity of 5.5 million tonnes during 1978. For the steel plants iron ore is imported from Australia and South America. The Kyongsang Pukdo area in the east central region is the main lead and zinc producing centre. The Yonhwa mine in this area produces about one half the total lead and zinc ores mined in the country. The lead and zinc concentrates contain 65% Pb and 45-50% Zn, respectively.

North Korea is also an important producer of tungsten ore and gold besides lead and zinc with average annual production of 50,000 tonnes of lead and 65,000 tonnes of zinc. Magnesite production is of the order of one million tonnes yearly. The steel production is two million tonnes annually using domestic iron ore and coal.

Japan is the largest smelting and refining country in the world using imported ores and concentrates, so much so that the economy of many countries is dependent on the long-term contract for export to Japan. It has set a brilliant example of industrialisation based on imported raw material. Although the level of production of mineral and ores are much below the demand of its industry, still it meets 40 % of copper, 65% of zinc and nearly 100 % of coal from domestic resources. At the same time it is an important exporter of metals and alloys produced from the imported ores and concentrates. Domestic resources of antimony, mercury, molybdenum, iron and tungsten ores are small; and nickel, chromite, manganese ore and bauxite practically nil. Sulphur of volcanic origin is found in many islands stretching from the north down to the south near Formosa. Japan is very systematically developing all its deposits. The systematic prospecting has discovered a new type of deposit Kuroko (Kuro—black, Ko—ore) in the Hokuroku province. Kuroko contains galena, chalcopyrite, sphalerite, pyrite with gold and silver and recoverable quantities of gypsum and barytes. The discovery of these deposits has brought a great boom in mining activities. None of the ore body crops out on the surface. The total reserves of Kuro ore have been estimated at 100 million tonnes. The ore now mined contains 2.5% Cu, 1.5% Pb, 2.75% Zn and 20 % pyrite and recoverable quantities of gypsum and barytes. The by-product recovery of gypsum has brought Japan nearly self-sufficiency in this mineral which was mostly imported for the cement manufacture. Domestic production of petroleum is negligible. Over 98% of her requirements of petroleum are met by imports from several countries in Middle East, Far East and the U.S.S.R. Japan controls the production of Arabian Oil Company's offshore Khafji field in the Neutral zone between Saudi Arabia and Kuwait. The production from this field is 50 million barrels a year.

Philippines is the second largest producer of refractory and metallurgical grades of chromite in Asia after Turkey and the fourth largest producer in the world after Republic of South Africa and the U.S.S.R. Most of the production of chromite is exported to the U.S.A. Philippines is also a dominant source of copper, nickel, and mercury. Annual production of copper metal is well over 260,000 tonnes. All the copper ores of Philippines are of porphyritic. They are being mined by over a dozen important companies of international repute. The copper ores are also an important source of gold and silver as co-products. Biga pit being worked by Atlas

Consolidated Mining and Development Company contains 0.42% copper which is the lowest grade of copper deposits worked in this country. Several new lateritic nickel deposits have been found. Sherritt-Gordon Mines Limited have started producing nickel since 1970. The refinery is located at Nonoc Island.

AUSTRALASIA

Australia, New Zealand, New Guinea and Papua have been included under continent Australasia for description.

Australia having an area approximately $2\frac{1}{2}$ times and population 1/50th of India is a singularly fortunate country today having enormous quantities of bauxite, iron ore, coal and rutile and ilmenite already proved. There are vast areas as yet incompletely unexplored for development with potential hope of finding rich deposits to supply to increasing world demand. More than 90 overseas companies jointly with Australian capital are engaged in mining and exploring new deposits. It is the largest exporter of rutile in the world market and is an important exporter and producer of iron ore, coal, bauxite, lead-zinc-copper, ilmenite, tin and zircon sand. The production of tin and tungsten has been greatly expanded. New discoveries of uranium ore have been reported from Jim-Jim and Nabarlak areas, associated with Archaean rocks in Northern Territory. Several deposits of manganese ore have been located in Ripon hills in Western Australia. The reserves are placed at 60 million tonnes. The grade is, however, low containing 30% manganese and 16% iron. It is worked by Australian Anglo-American Ltd. The deposits of nickel sulphide ore have been reported at Agnew, north of Kalgoorlie containing an average 1.9% nickel. Proved reserves are placed at 33 million tonnes with 2% nickel. Some of the world's largest workable bauxite deposit occur at Weipa in North Queensland; at Gove, Northern Territory; and at Jarrahdale in Western Australia. Australia now ranks number one in the production of bauxite. The Weipa deposit alone is estimated to contain over 3000 million tonnes of bauxite with 50% Al_2O_3 content. The deposits are being exploited by a number of firms of international repute. The annual production capacity of aluminium metal had reached nearly 3 lakh tonnes by the end of 1973. There are three primary aluminium metal producers viz. Comalco Ltd. having annual smelter capacity of 94,000 tonnes at Bell-Bay, Tasmania; Alcan Mining

Company having 100,000 tonnes of annual capacity at Kurri-Kurri, New South Wales; and Alcoa of Australia Ltd. having smelter capacity of 90,000 tonnes annually at point Henry, Victoria. Two giant firms, namely, Queensland Alumina Ltd. Comalco Ltd. have set up alumina plant at Gladstone, Queensland and at Weipa, Northern Territory. The former has two million tonnes and latter 1.6 million tonnes of alumina production capacity per year. Australia is now the world's largest producer of sapphire from Swanbrook and Frazer Creek gravel in the northern part of NSW.

Mount Isa and Mount Morgan areas in Queensland are the important sources of copper-gold, whereas the Broken Hill properties in New South Wales are the sources of copper, lead and zinc. A number of new lodes have been found. Active mining operations continue for copper at Queenstown and for lead-zinc-copper at Rosebury in Tasmania. Victoria is famous for brown coal, which occurs in Yallourn area. Metallurgical coal comes from Wollongong area, New South Wales. Baralaba area in Queensland is famous for anthracite coal worked by Mount Morgan Ltd. Western Australia has a vast reserves of iron ore. About 90 million tonnes of iron ore are exported from Western Australia every year.

The production of tin comes from Tasmania only. There are two important producers, namely, Renison Ltd. and Cleveland Tin. The reported reserves in the Cleveland area are 3.19 million tonnes with 1.91% Sn and 0.38% Cu. In Renison area, the estimated reserves of tin ore are placed at 12 million tonnes with 1.1% Sn. All the production of tin ore from Tasmania is from hard rock. The exciting development in Australian area has taken place from early sixties as a result of interest shown by many countries in getting assured supply of minerals from the vast potentialities available in Australia. Besides U.S.A., Canada, U.K., Sweden, W. Germany, Italy and other European nations Japanese have also leaned heavily on Australia for import of iron ore, coal and bauxite, and have entered into long-term contract with several mining concerns in Australia for import of these raw materials in addition to whatever supply of copper, lead-zinc and nickel concentrates are made available to them. Australia has come into the world market for export of chrysotile asbestos from newly located huge deposit, near Barraba, in northern New South Wales. Production is well over 70,000 tonnes annually. The mine produces only shorter fibres. It is worked by Wood Reef Mines Ltd. The fibre content in the mine ranks from

0.3% to 6.94%. From nil production in petroleum, a major exploration effort has resulted in two onshore production oil fields of Barrow island and Moonie and in the much larger off-shore fields in Bass Strait, in the South east coast. In addition, natural gas fields of commercial significance have been discovered and the exploitation of them started in four States. Australia is now meeting 65% of its oil requirements from domestic resources.

New Zealand with high mountains, with the geysers of iceland dotted by beaches, numerous lakes and many river flows presents very beautiful topographical features which give it a look of Switzerland of Europe. It comprises of two islands, North Island of about 112,640 sq km and South Island comprising 148,480 sq km separated by Cook Strait about 32 km across its narrowest. The production of mineral in the country is of small magnitude. Over 90% of the total value of mineral production comes from coal, limestone and road metals. Yearly production of coal is of the order of about 2.8 million tonnes from about 100 underground mines and 50 open cast mines, worked both under private and public sectors. The coal reserves are likely to last for 100 years with the present rate of production. There are large reserves of titaniferous iron sands in the west coasts of both the islands, and a start has been made towards their exploitation. A steel industry of a half million tonnes a year capacity has been set up utilising the titanomagnetite sands at Waiket Heads in North island and using coal, limestone and electric power which are available at low coast. Non-metallic minerals chiefly bentonite, diatomite, dolomite, perlite, pumice, salt, serpentine, wollastonite and clays are reported to occur in small quantities During 1973, Rennecott Copper Corporation announced a big discovery of chrysotile asbestos in South Island. Reserves are placed at 100 million tonnes in terms of fibre. Subsequently Cassair Asbestos Corporation has undertaken detailed prospecting of the entire area. No potential source of oil has been found so far. Kapandi oil and gas field in Taranaki produces 60 million cu ft of gas and 4500 barrels of condensate a day. New Zealand has, however, abundant supply of cheap electricity from hydropowers. It is one of the most electrified countries in the world which opens good prospects of setting up metallurgical industry based on imported concentrates. An aluminium smelter has been set up in 1971 at Bluff based on imported alumina from Australia.

Papua and *New Guinea*, once world famous for the gold workings, have closed down the dredging operation since 1966 reportedly due to exhaustion of the deposits. Bougainville island has made a big name in copper porphyry. It is regarded to be the largest discovery of the second half on the 20th century. Copper mineralisation is associated with diorite and granodiorite rocks which have been intruded by a series of andesitic volcanic rocks. The mine is worked by open pit which went into production in April, 1972. Proved reserves are placed at 900 million tonnes assaying 0.48 % copper and 0.018 % gold. The gold content is probably the highest in the copper-gold deposits of the world. The open pit owned by Bougainville Copper Ltd. has been developed for rated production of two lakh tonnes of ore per day with related crushing and milling facilities. The con-

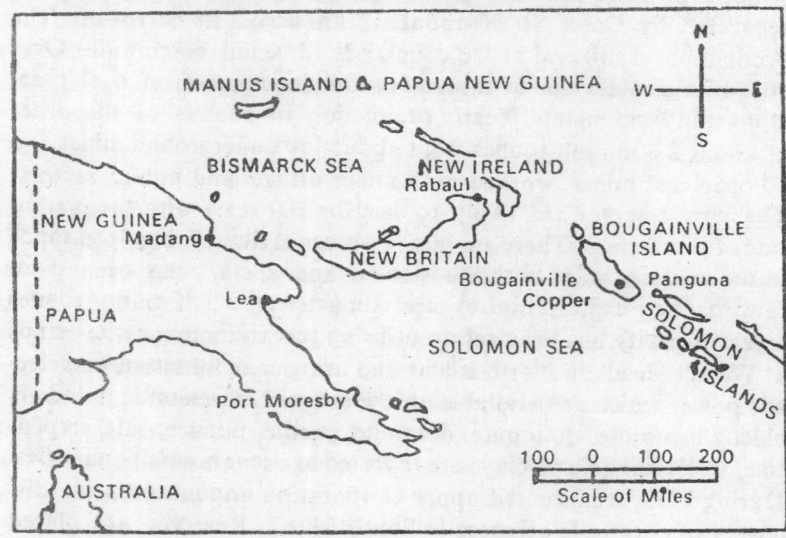

Fig. 1

centrates are chiefly shipped to Japan and European countries. Bougainville copper deposit has been developed at a total investment of 436 million dollars by Conzinc Rio Tinto of Australia Ltd. and New Broken Hill Consolidated Ltd. and Govt. of Papua and New Guinea. Chances of finding copper deposits in other island namely New Britain, New Ireland and in the mainland in the Morobe, Madang, Eastern Highlands and Sepik districts are quite favourable. A number of companies of international repute are prospecting for copper.

EUROPE

This continent is a potential source of many minerals but it is also a great consumer of minerals. The mining industry of the continent is very old. Most of the mines have reached the stage of maturity, therefore, the trend is now on the utilisation of leaner ores and recovery of all associated metals as co-product by highly improved technology in mining and metallurgy. Extraction of sulphur and sulphuric acid from pyrite and of the by-product sulphuric acid from the smelting of sulphide ores is the common feature of all European countries. The Orkala process developed in Norway and Outokumpu process developed in Finland have taken a worldwide lead in the extraction of sulphur from pyrite. In Europe, except for Denmark, Luxemburg, Switzerland and Belgium, all other countries produce significant quantity of minerals. Europe is the largest producer of mercury in the world. Mercury production mainly comes from Italy and Spain. Portugal produces also much of tungsten. The U.K., Ireland, France and West Germany are the major producers and exporters of anthracite coal. Some countries like Italy, Sweden, Switzerland and Norway, which have inadequate or no coal, have substituted it by hydro-electricity to a great extent. The continent as a whole has adequate supply position of fluorspar, lead and zinc. The availability of copper has substantially improved from Bulgaria and Poland. Sources of ilmenite are confined to rock type deposits occurring in Finland and Norway only. Europe is the leading continent in the world utilising phosphoric iron ore for steel production. Europe lacks in supply position of chromite, manganese ore and tin. The deposits of mica and phosphorites are practically absent. Europe has suddenly improved its position in petroleum resources by discovering huge reservoirs of petrol in the North Sea by the U.K. and the Norway. It is the most spectacular and significant discovery for Europe. In the North Sea bordering U.K. and Norway, the U.K. have reported discoveries of petroleum in the Forties and Auk fields and that Norwegien in the Ekofisk field. The prospect of discovering further oil fields around Shetland island in the North Sea is rated very high by the U.K. By this discovery, more than 100 million tonnes of crude petroleum is expected to be produced annually. The U.K. is going to be more than self-sufficient in oil resources by this discovery and so will be the case with Norway. In recent years, West Germany has also stepped up its production of crude petroleum. Domestic resources of petroleum in Austria, Hungary and Rumania are adequate. Rest of the European countries depend largely on imports from Middle East and Latin American countries.

The United Kingdom is famous for its Cornwall and Goovean China clay which finds good export market. She is a major producer and exporter of bituminous coal and anthracite in Europe, with an average annual production of 170 million tonnes. The anhydrite mine at Billingham-in-Tees, Durham County with a production of over 5000 tonnes a day is the largest in the world. Billingham is one of the few places in the world where anhydrite is used as a source of sulphur and sulphuric acid. It is worked by the Imperial Chemical Industries. Production of tin ore comes from Wheal Jane and South Crofty mines, Cornwall area. South Crofty is one of the world's largest underground mine for tin. The fluorspar milling plant in Derbyshire which treats the Glebe mine deposit is the largest milling plant in Europe. Large tonnage of barytes and lead concentrates are recovered during fluorite washing. The U.K. lacks very much in domestic resources of other minerals and depends on import and has made heavy investments in several African, Latin American, Asian countries, and in Australia in the mining and mineral industry. The domestic iron ore (iron stone) worked near Colsterworth (Northamptonshire), Lincolnshire contains 34% Fe only with 3% Ca, 12-18% SiO_2 and with 1 to 3% P. It is rather a sandy deposit occurring in a bed 3 to 4 metres thick underlying the Jurassic formation. The United Steel Companies Ltd. operates three open pits of iron stone in the areas to produce 34 to 38 thousand tonnes per week. The same company operates underground mines at Scunthrope and Frodingham. The ore of these areas contains 22%, CaO 22% and SiO_2 8.10% and 0.2% P. By carefully blending of 35-40% Northampton sand with 60-65% Frodingham, an ideal self-fluxing blast furnace feed is prepared with a lime silica ratio of 1 to 1:5.

Ireland has become an important source of lead-zinc concentrates for the European smelters since December, 1965, by the start of production in Tynagh mine of Northgate Exploration Ltd. The ore occurs in the limestone country 4.8% Pb; 4.3% Zn; 0.6% Cu and two ounces of silver per tonne. About 2000 tonnes of ores are treated to produce concentrates for shipment. A number of new lead-zinc mines have been brought into production. Barytes is recovered as by-product and with a daily production of 1000 tonnes, this country has emerged as one of the world's leading producers and suppliers of mineral

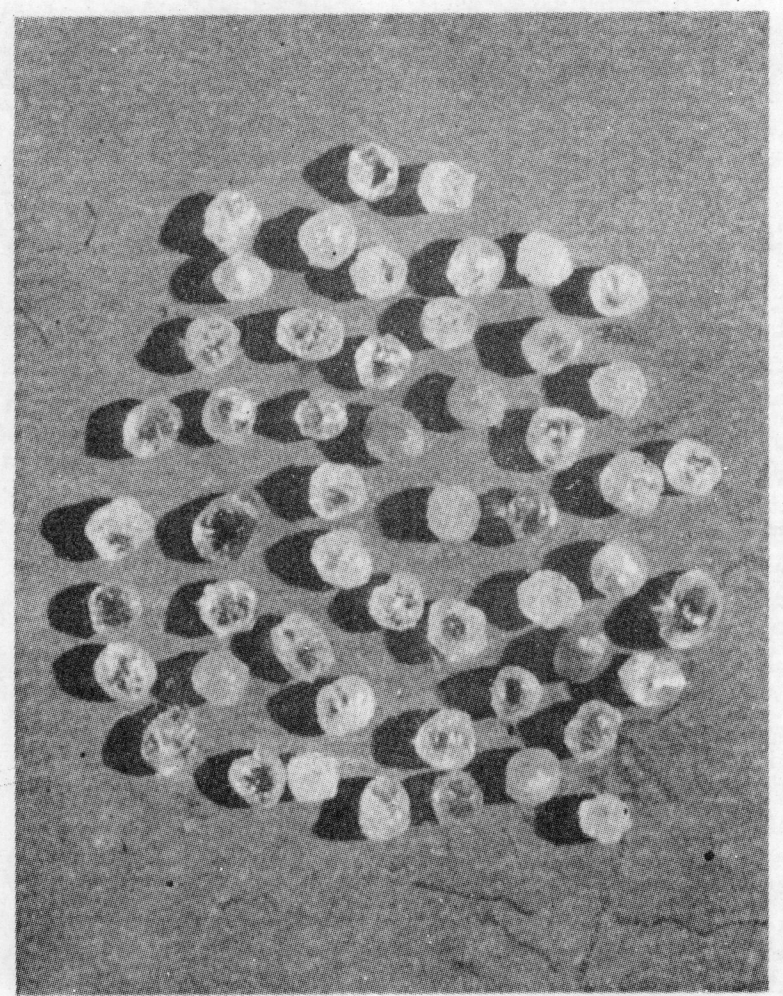

Fig. 2. Uncut diamonds

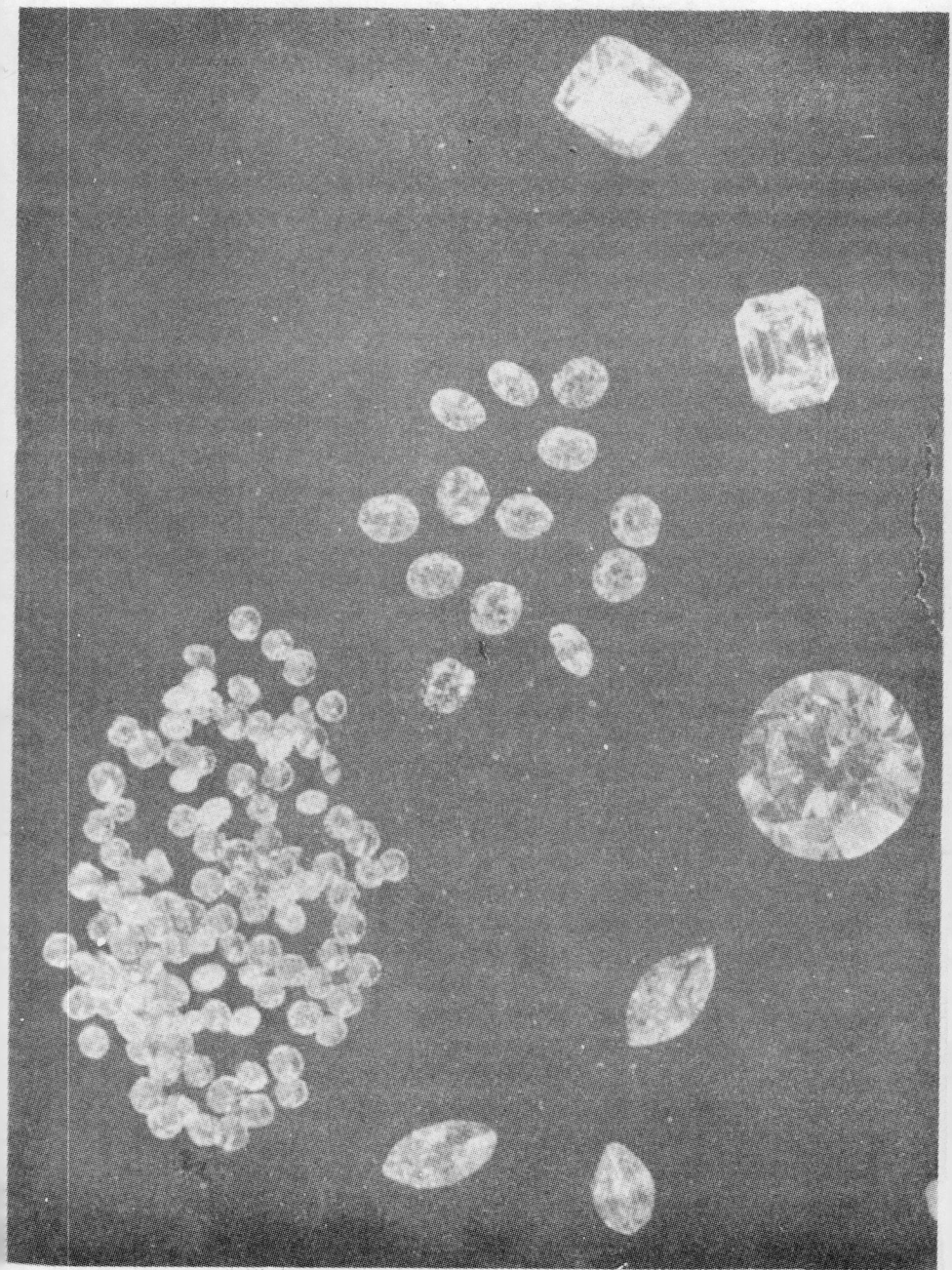

Fig. 3. Cut diamonds

Fig. 4. Splitting of diamonds in Amsterdam

Fig. 5. Polishing of diamond in Amsterdam

barytes. It has also large deposits of copper-silver ores in the Vale of Avoca, Wicklow County and Tipperary. It is emerging as a potential source of mercury and tin ores.

Portugal is the largest producer of tungsten ore outside the U.S.S.R. Beralt Tin and Wolfram Ltd. and Metallium Corporation are the chief producers and exporters of tungsten ore. There is no production of oil.

Spain is famous for mercury. The Minas de Almanden mine is one of the largest producing mercury mines in Europe. This area has been under continuous production for more than 2000 years and is still one of the world's greatest mineral treasures. Cinnabar is found deposited in the hairlines and fissures of the quartzite crossing the planes of fracture and penetrating between the quartz grains of quartzite wherever there has been easy circulation of the dissolved sulphide. This country is also an important producer of copper, zinc, lead and aluminium metals from indigenous ores. The production of sulphur and sulphuric acid comes from abundantly available pyrite deposits. Production of tin is small. There are adequate reserves of coal, iron ore, fluorspar and potash salt. It is largest producer of fluorspar in Europe. The estimated reserves of iron ore are 1500 million tonnes. The ore is phosphoric i.e. containing more than 2% P. There is no commercial oil production. The only other hydrocarbon source is three small gas wells in the foothill of the Pyrenees at Castillo, near Victoria. Spain has rich deposits of uranium ores. There is a concentration plant at Andujar in South Spain. Another concentration plant is located in Salamanca Province.

France has large reserves of boron, bituminous and anthracite coal. With nearly 3 million tonnes of annual production of bauxite, this country is the largest producer in Europe. The natural gas in Lacq in lower-Pyrenees has become a great source of sulphur with annual recovery of over 1.5 million tonnes. The source of sulphur from natural gas in France can be regarded second only to Alberta and Saskatchewan in Canada. This recovery of sulphur has become a strong competitor to pyrites mining. Output of iron ore containing 30 to 36% Fe and 0.5 to 0.7% P comes from Lorraine area. The iron ore reserves in the Lorranie basin are estimated at 4500 million tonnes comprising some 2500 million tonnes of calcareous ore and 2000 million tonnes of siliceous ore. The annual production is about 50 million tonnes. It has rich deposits of potash and uranium ores. France holds potential areas in Algeria, Sahara, Gabon and Zaire for petroleum which form important sources of supply to mainland. France also has a stake in the Middle East through State owned Cie Francaise des Petroles (CFP). The domestic production of crude is approximately two million tonnes against the annual requirements of 40 million tonnes. Oil fields are located in central area, mostly south-east and east of Paris.

Luxembourg has small resources of minerals except iron ore. The reserves of iron ore are placed at 225 million tonnes. It is of the same type as found in France. Steel industry is the backbone of Luxembourg economy.

Belgium has few known occurrence of minerals. Coal and iron ore of Lorraine region are produced in small dimensions. It has sound metallurgical industry based on important raw materials. Cement, polished diamond and steel form the major items of export. In diamond cutting and polishing it leads in the world.

The Netherlands possesses large reserves of natural gas in the North, East and West Holland. The production of iron and steel and aluminium is based entirely on imports. Coal production is of the order of 12 million tonnes per annum. Bauxite is imported from Surinam, mined by the Dutch Billiton concern. The first aluminium smelter was set up in 1965 near the port of Delfzijl. It was started with a capacity of 32,000 tonnes which was subsequently increased to 15,000 tonnes. Second smelter has been set up at Flushing having capacity of 85,000 tonnes a year. Holland is now rated as a major producer in aluminium in Europe. Amsterdam is world-famous for cutting and polishing of gem diamonds.

Italy is the leading producer of mercury in Europe. Monte Amiata is Italy's foremost mercury producer and exporter. The ore cinnabar in the deposits contain 0.5 to 0.7 per cent mercury. This country is the fourth largest producer of pyrite in the world, and also an important producer of elemental sulphur found as bedded deposits occurring with gypsum around Slicily. It possesses large reserves of antimony and cadmium-bearing argentiferous lead-zinc ores and fluorspar which are under exploitation. The central part of the island near Orani (Barkagia) contains one of the finest qualities of talc. The estimated reserves are 2.5 million tonnes. It is an important producer for aluminium metal from domestic bauxite and iron and steel from imported ore and scraps obtained from France, West Germany, Belgium and Luxembourg. It is, however, embarrassingly deficient in raw materials like iron ore, metallurgical coal and also manganese ore. Coking coal is entirely imported. The bauxite reserves of San Giovanni deposits having depleted. Italy's dependence on imports of bauxite for aluminium production is imminent. Ragusa oil field, south-east of Sicily, is the only oil field under production for a considerable long period but the production is far insufficient for domestic requirements.

Finland is a most refreshing and challenging country to the mineral enterprisers. There are four important mining firms, namely (i) Outokumpu, (ii) Otanmaki, (iii) Vuoksenniska, and (iv) Rautaruukki, who are responsible for the major production of minerals and metals. The county has achieved a great metallurgical feat in treating complex ores and produces good amount of cobalt, cadmium, copper, lead, selenium, zinc, nickel, titanium, vanadium, iron and sulphur from pyrite. Iron ore mines are located in Karvasvaara, Jussaro and Raajarvi in northern Finland, north of Arctic Circle. The ore of the Metsamonttu mine of Outokump company contains gold, silver, lead, zinc and copper. The pyrite of Pyhasalmi open pit of the same company contains 0.85 % Cu, 2.8 % Zn, 33 % S and 37 % Fe. The reserves of pyrite are estimated to be seven million tonnes. A new cuperiferous nickel ore mine Vuonos has been developed by the same company. The ore contains 0.2 % nickel, 0.03 to 0.09 % copper and 0.02 to 0.07 % cobalt. This company has developed in 1961 a famous process itself known as Outokumpu, for the recovery of sulphur and sulphuric acid from pyrite. The company has such a plant at Kokkola where iron is also recovered from pyrite. The recovery of sulphur from the Kokkola plant was however discontinued from May, 1978. Two mainly copper mines, Outokumpu and Ylojarvi; one Vihanti zinc mine, one Raajarvi iron ore mine, and one Kotalahti nickel mine are also worked by the above company. This country is the major producer of mercury and nickel in Europe from domestic resources. It is a third country after India and the U.S.A. which produces wollastonite, entirely for domestic consumption. Cadmium and mercury are recovered at Kokkola plant. The Otanmaki Company mines well-known for iron-titanium-vanadium ores at Otanmaki and recovers concentrates of all the three metals for smelting. The ore contains 40 % magnetite, 30 % ilmenite and 0.25 to 0.3 % vanadium. Vuoksenniska Company mines mainly Jussaro iron ore deposit. Rautaruukki is the largest producer of vanadium in Europe rather in the world. Annual production is about 5300 tonnes in terms of V_2O_5 which accounts for 10–15 % of the world's production. There is no crude oil production. Russia is the main supplier of crude to the only refinery at Naantali owned by the State Government.

Sweden is famous for its Bergaslagen and Kiruna iron ore deposits in the northern Lapland. The iron ore reserves amount to 4000 million tonnes of which more than 3000 million tonnes are situated in Lapland and some 1000 million tonnes in central Sweden. The Lapland ores have a phosphorus content ranging from 0.05% up to 5% phosphorus and more than half of the total reserves in this area is accounted as phosphoric. The LKAB (Luossavaara Kiirunavaara Aktieboleg) or commonly known as Kiruna, from the name of town, itself operates three underground mines, namely, Kiirunavaara, Luossausaara, Malmberget, and two open pit mines Zenobia and Svappavaara. Kiirunavaara is the largest underground iron ore mine in the world with computerised traffic control of the highest order of efficiency. The ore is worked at 420 metres level. The ore is magnetite and martite, assaying from 51% to 58% Fe with phosphorus varying from 0.1 to 0.7%. LKAB prepares several grades of marketable ores after magnetic concentration and pelletisation, and exports over 20 million tonnes yearly to European countries. The pellets assay 67% Fe and 0.06% P. Since Swedish ore is magnetic removal of phosphorus content poses no problem. It is easily reduced to low content during magnetic separation. Kiruna's new pelletisation plant at Malmberget which went into production on 29th June, 1965, is the largest and the most modern one in the world. The designed capacity exceeds 15 million tonnes a year. In this plant, ferrous sulphate ($FeSO_4.7H_2O$) is used as a binder instead of conventional bentonite. This country has also good resources of sphalerite, pyrite zinc, tungsten and vanadium. The abundance of hydro-electricity is responsible for the development of electrometallurgical industry which is well developed. Swedish steel is world-famous for making razors, blades, tools and other provisions instruments. The new open pit Aitkit copper mine opened in 1966 by Boliden Gruvaktiebolag in Lapland is world's most northerly copper mine and the lowest known copper content (0.6%) deposit in the world. The ore contains silver and gold besides copper. Boliden Gruvaktiebolag is the largest producer of base metals in Sweden, having 15 underground mines and 3 open pit Aitik, Rakkejaur and Kankberb of which Aitik with the production of 3.2 million tonnes of ore per year is the second largest pit in Europe. A new tungsten ore deposit has been discovered in 1967 in Bergslagen province. Shale containing on an average 304 gm of uranium oxide is mined in large quantity near Skovde. Oil shale is an important industry of this country. It is a major source of petrol, petroleum gas, sulphur and ammonia, besides electricity.

Norway is famous for the recovery of sulphur from pyrites by 'Orkala Process' developed in this country by the Orkala Company. It is the second largest producer of aluminium metal in the continent based on imported bauxite. Norway offers the cheapest electricity in the world. Nickel is produced at Kristiansand, south Norway. The production is based on Canadian nickel matte supplied by Falconbridge Nickel Company of Canada. The smelter is working under the name of Falconbridge Nikkelverk Aktiaselskap. The matte contains 50 % Ni, 25 % Cu and 20 % S in addition to small percentage of cobalt, selenium, platinum, silver and gold which are recovered. Copper ore is mined at Repparfjord, the most northerly important town in the world. The ore contains 0.7 % copper and is being produced at the rate of six lakh tonnes a year. The ore is mined from a series of lenticular bodies on a mountain top. The ore is crushed and floated at the site and the concentrate is shipped to West Germany. The massive ilmenite-norite deposit occurring between Plekkefjord and Egersund is the major source of ilmenite in the continent. The deposit outcrops for a distance of 2.56 km with average width of 180 metres.

The estimated reserves are 350 million tonnes with 39% ilmenite (18% TiO_2), 2% magnetite with minor amounts of cobalt, nickel, copper and apatite. It is worked by A/S Titania, a subsidiary of the National Lead Company, New York. Iron ore and pyrites are the other important minerals mined. The principal pyrites deposit at Bjorkaaset at Ofotenfjord having exhausted, new sources have been developed at Tverfjeilet, close to the Sate Railway in the Dovre mountain. Norwegian pyrites contains usually over 40% sulphur and 1% Cu. Norway is one of the three important countries in the world, the other two being Canada and U.S.S.R., which produces nepheline syenite. It is mined on an Island in the Alta Fjord in the Norwegian Artic at a latitude of 70° 15′ by a subsidiary company of Elkem-Spiger-verket A/S. They mine and process two grades—glass and ceramic grades. Glass grade consists of 100% minus 0.5 mm particles whereas ceramic grade contains at least 50 % minus 0.01 mm size fines. In both the cases iron content is kept low as far as possible to satisfy market specifications. It is done by passing the crushed material from secondary and tertiary crusher through several stages low to high intensity magnetic separators.

German Federal Republic (West Germany), a well-advanced country in Europe, possesses good reserves of lead-zinc ores, pyrite, fluorite, barytes, potash and rock salt and anthracite, bituminous and brown coal. The chief centre of coal production is Ruhr. There are several small fields scattered along the northern border of the central upland linked with the chain of fields that extends west from upper Silesia, through Ruhr to Central Belgium and Northern France. This country is a major exporter of coal. After U.S.S.R., West Germany with over 43 million tonnes annual production, is the largest producer of steel in Europe and the fourth largest producer in the world, the first, second and third being U.S.A., U.S.S.R. and Japan. The domestic resources of iron ore is comparatively small. The production has been going down every year since 1960. It is the Europe's largest producer of copper, lead and zinc and aluminium. However, the domestic resources of manganese ore, mica, chromite, magnesite, nickel, titanium, phosphate are practically nil. The crude output is about seven million tonnes per annum. The refining capacity is about 60 million tonnes. New oil fields are Bramberge, Hankensbuttel, Ruhlermoor and Sinstorf.

German Democratic Republic (East Germany) is the leading producer of lignite accounting for one-third of the world production and has largest potash deposits in Europe. The potash deposits occur in Werra, Sudhartz, and Mansfeld districts and has been under production for over 70 years. These potash mines have always been subjected to intense and frequent outburst of rock and gas, and mining engineers have trying time to work the deposits. The outburst consists of sylvanite, carnallite, halite and hartz salts, the accompanying gas consists of a mixture of carbon dioxide, methane, hydrogen sulphide and nitrogen. More than 10,000 outbursts have taken place. Potash salts after processing find a good export market. Copper ore with 1.5 % Cu is mined in the Mansfelder and Sangerhauser Mulde; tin-tungsten ores in the Erz-Mountains and in the Vogtland; lead-zinc ores in Freiberg (Saxonia) and nickel ore in West Saxonia. The nickel plant was open in 1961 at St. Egidien. During the same year, an electrolytic zinc refinery with capacity of 14,000 tonnes a year was erected on the eastern outskirts of Freiberg. The resources of iron ore is small. The ore is of low grade and mostly consist of chamoisite.

Most of the processing units operate on imported raw materials. East Germany has practically no oil. It depends entirely on the supply from the U.S.S.R. Natural gas occurs in Salzweded field bordering W. Germany. The recent discovery of oil in East Prussia by Russians has given this country and Poland a new impetus for prospecting in the coastal areas.

Austria is a leading source of graphite and magnesite. It ranks first in the production of magnesite in Europe closely followed by Czechoslovakia and Greece. Graphite is mostly of amorphous type. It has been used as a substitute for coke in blast furnaces. Except for bauxite, this country is self-sufficient in almost all useful minerals including natural oil and gas. Austria is the inventor of top blown oxygen process for steel-making commonly known as Linz Donawitz process (L.D. process). This process is most effectively replacing the conventional O.H. process in steel making.

Greece has immense reserves of bauxite existing in the central part of the country. Aluminium production is also substantial. Russia imports considerable quantity of bauxite from this country. Greece is third major producer of magnesite in the continent and exports it after calcining The magnesite is of massive amorphous type and is found in altered serpentine masses occurring in Euboae island. Iron ore of local origin, mainly from Thasos island, contains only 30% Fe. Nickel ore is found at Laryma on the east coast to Euboae. Annual production of nickel is 4000 tonnes, and about 8000 tonnes to 10,000 tonnes of steel are produced utilising the by-product iron. Zinc is mined and smelted near Laurium, South of Athens, by a French Company. Other lead and zinc deposits are located in Kirki. Nearly six tonnes of lignite is raised from Ptolemais area. Asbestos is mined from Korani area. The Nippon Mining Company of Japan has established copper deposits in Skouries area, Chalkidiki Peninsula.

Poland has a good source of coking coal, the important mines being Porabka, Szezyglowice and Zoflowka. It is a leading producer of sulphur, aluminium, lead, zinc and copper. Sulphur is produced in the Tarnoprzeg district. Annual production of elemental sulphur now exceeds two million tonnes. Production of copper metal witnessed 100% expansion. Aluminium production is of the order of 100,000 tonnes. Steel mills are located at near Cracow and Miociny. Production of oil mainly comes from Crobla in Bochina district. Prospect of finding oil in the northern part of Poland bordering

Lithuania and Kalinui districts of U.S.S.R. is said to be good. Poland gets substantial supplies of oil from U.S.S.R.

Czechoslovakia has good resources of coal followed by magnesite, graphite, kaolin, opal and uranium ores. Small production of tin-tungsten ore and fluorite is reported. The major coal mines are situated in Ostrava-Karvina basin in Silesia. This country produces yearly, on an average, 27 million tonnes of bituminous coal, including coking coal and over 70 million tonnes of lignite. The reserves of high grade coal in unworked ground are estimated at 20,000 million tonnes. The Czech iron ore is mostly of siderite (iron carbonate). It is mined in the region between Kladero and Plezn in Bohemia 'and from the Slovakian Mountain. Production is small. Major requirements are met by imports from India, U.S.S.R. and also from Sweden, Brazil and Finland. The occurrence of manganese ore is located at few localities but the tenor is low, between 10 and 17% Mn with high sulphur. Pyrite and magnesite are produced from the Ore Mountain. Czechoslovakia is well advanced in the metallurgical industry and foundry plants, and is one of the most important gem garnet cutting centre in Europe. The upper Miocene beds at Hodonin and Gbely are oil producing centres which meet a partial requirement of the country. Rumania is the chief supplier of oil. Czech is also world famous in the production of cut garnet stones.

Hungary is one of the leading and richest producers of bauxite, oil and natural gas in Europe. In bauxite production, it is second to only France in the continent. Production of coal comes mainly from Zobak area, Mecsek, South Hungary. Coal mines are worked below 720 metres level. Unusual deep seated porphyry copper deposit, the top ore body 600 m below surface has been discovered and is under development at Recsk. It is the most northerly representative of the porphyries occurring in the great Bulkan – Carpathian copper-molybdenum belt. Petrochemical industry is highly developed in this country. There are more than 750 wells under production in the Zala County petroleum belt.

Rumania in Europe appears to be an important source of some important minerals. It has sufficient reserves of bauxite, chrome ore, manganese ore, mercury, lead, zinc, gold, silver and molybdenum and coal. The reserves of limestone, marble, oil, gas and rock salt are regarded as immense. There are about 240 salt domes producing high purity sodium chloride. The iron ores are of low grade with 16 to 40% Fe. Reserves of barytes are also adequate. Copper sulphide ores are mined in Moldova-Nova area in Temeserve. It has surplus production of petroleum. The entire crude output is processed in the country's eleven major oil refineries. About half of the product is exported. It imports nearly its entire requirements of apatite and sulphur from the U.S.S.R. and Poland. Imports of aluminium and other ferro-alloys from the U.S.S.R. are considerable.

Bulgaria has good deposits of lead, zinc and copper ores and pyrites. Non-ferrous metals mining and metallurgy hold important position in the national economy. There are four principal zones of mineralisation. These are (1) Rhodope, (2) Sakar, (3) Strandia, and (4) Ossogovo. The Rhodope zone comprises the southern part of the country up to the Greek border. There are 90 lead-zinc deposits in this zone and they constitute the country's main reserves of these metals. The Strandia zone comprises the eastern part of the country and extends from the town of Burgas south to Turkish border. This is the region where major reserves of copper in the country are concentrated. There are 300 main veins running almost parallel. All copper ore bodies are associated with molybdenite. Medet in the central area is the major centre of copper production by open pit. Copper-molybdenum mineralisation is confined in granite-monzonite porphyry intrusions. This area accounts for 75% of the copper production in the country. On an average eight million tonnes of copper ore is mined annually from the Medet pit containing on an average 0.36% Cu and 0.008% Mo. Bulgaria on an average produced more than one lakh tonnes of lead, 80,000 tonnes of zinc and 55,000 tonnes of copper per year.

Yugoslavia is rich in several important minerals which include iron ore, lead-zinc ores, copper ore, nickel bauxite, chromite, manganese, pyrites, barytes coal and lignite. Now copper ore reserves have been proved at Veliki in Serbia. An open pit mine is being opened for production of 20,000 tonnes of electrolytic copper annually with by-product gold, silver and molybdenum. New reserves have also been established for nickel in Kosovo-Mitrovica where reserves are estimated at 38,000,000 tonnes with 1.6% Ni and some cobalt. The most important iron ore mines which account for about 90% of the total output are Ljubija in north-west Bosnia and Vares near Sarajero. The reserve in Ljubija field are estimated at 130 million tonnes with 52% Fe. The two other important iron ore mining centres are in Macedonia and Tomasica. The total iron ore reserves are placed at 444,286,000 tonnes with 34% Fe by the

Federal Geological Institute. The known magnetite deposits are
under production. Magnetite mines are worked near Skopje, in the
Kapaonik massif and near Prokuplie in Southern Serbia. A wollastonite deposit and an asbestos deposit of big magnitude are reported to have been discovered near village Jaram in Serbia. Domestic
production of crude oil is small—of the order of three million tonnes
per annum only.

Albania is rich in chrome ore containing nickel. Small production
of lignite, copper ore and iron-nickel ore is reported. The iron-
nickel ore is mainly hematite with subordinate amount of magnetite
containing 1% nickel. It is worked in Pishkash, Cervenake, Bushtrice
and Prenjas. This country possesses Europe's largest asphalt deposits. Oil and chrome account for half the Albania total export.
Export is made mainly to China, East Germany, Italy and Bulgaria.
Most of the mines and metallurgical industries have been developed
with Chinese assistance.

SOUTH AMERICA AND CENTRAL AMERICA—LATIN AMERICA

The South America is joined to North America through Panama,
a narrow land link between them, each of them forming a continent.
The Central America constitutes the countries south of the U.S.A.
up to Panama. The Latin America is the word given to all the countries lying south of the U.S.A. comprising Central America, South
America and the adjoining islands excluding those of Caribbean Sea.
Here South and Central Americas have been described separately.

The South America accounts for the world's total output of nitrate
from Chile and nearly the total output of electrical grade quartz
crystals from Brazil. Chile and Peru have become one of the main
hubs of copper production in the world and also in the export of
iron ore. Big discoveries of porphyry copper deposits have been
made in Colombia. Peru is the largest producer of bismuth in the
world. Bolivia ranks fourth in production of bismuth. Argentina,
Brazil, Chile and Peru offer greatest diversity of mineral resources
like iron ore, manganese ore, bauxite, copper, lead, zinc, nickel, tin,
bismuth, sheet mica, mercury, borax, potash, lithium minerals, beryl,
barytes, fluorspar and crude petroleum. Except Uruguay, Paraguay,
Ecuador and French Guiana, all other countries of the South
America contribute significantly to the world supplies. In Ecuador

the Japanese Overseas Mineral Development Corporation and the Prospection Limited, a Canadian firm had carried out deep drilling explorative work in the southern part of the country for possible location of copper-molybdenum porphyry deposits, but both the firms met with disappointing results. In French Guiana, substantial reserves of bauxite are indicated where mining has already commenced by Suriname Aluminium Company, a subsidiary of the Aluminium Company of America. The South America continent, as a whole, is a growing source of iron ore. Except in Columbia where small production of iron ore is consumed in domestic furnace, the ore is primarily mined for export, mainly to the U.S.A., Europe and Japan. Estimated reserves of iron ore in Latin America (South American countries plus Mexico of Central America only) are estimated at 4800 million tonnes and possible reserves at 76,800 million tonnes. Pelletisation plants have been erected in Brazil, Venezuela and Peru. Venezuela with vast potentialities of oil and natural gas has emerged as a second largest exporter of oil in the world outside West Asia. The U.S.A. is the largest buyer of all Latin America's mineral output. The U.S.A. has invested large sums in mining industry in Latin America which reflects its need for basic mineral raw materials found in abundance nearer home. The South America is, however, in short supply of coal and coking coal, sulphur, chromite, magnesite, phosphate and asbestos. The known occurrences of chromite and magnesite are confined to Brazil, and Guano and phosphate to Chile, Venezuela and Peru only.

SOUTH AMERICA

Venezuela has become the biggest exporter of iron ore in the continent exporting nearly 20 million tonnes per annum from the eastern region. Iron ore interests are held largely by the U.S. Steel Corporation and Bethlehem Steel Company, both of the U.S.A. The Corporation Venezolana de Guyana has the distinction of inventing Udy process of pig iron manufacture. The process is based on utilising non-coking coal, even lignite, and iron ore fines and adequate supply of electricity. Venezuela is probably the third country after the Brazil and Guyana in Latin America producing diamond. Its annual production is nearly half a million carat. Production of sulphur comes as a by-product of petroleum refining. Coal resources are small. Small deposits of phosphate rock with estimated reserves of 40 million tonnes containing 30% P_2O_5 are located near Lezardo in Falco. New lead-zinc deposits and nickel deposits have been proved at Bailadores and Loma-de-Hierro respectively. Nickel deposits are being developed by a French Company, Societe-Le-Nickel.

Guyana has vast resources of bauxite and is a major source of bauxite to the European countries, U.S.A. and Canada and also to the U.S.S.R. and China. Bauxite mining is State owned since 1971. Formerly, Alkan Limited of Canada was the principal bauxite mine owner and the producer of alumina in Guyana which was nationalised in 1971 and taken over by the State owned Guyana Bauxite Company. Production of manganese ore, alluvial diamond and gold is also reported.

Surinam is the second largest producer of bauxite in the world. Bauxite mining is the mainstay of Surinam mining activities. It is the major supplier of bauxite to the U.S.A., Canada and the Netherlands. Surinam is also a major exporter of alumina and aluminium metal. Most of the mining activities are concentrated in Bakhuys mountain areas containing some 200 million tonnes of proven bauxite. This country meets the entire requirement of alumina and bauxite of Delfzijl smelter in the Netherlands. There are four important companies engaged in mining of bauxite, namely, Aluminium Company of America, Aluminium Company of Canada, the Billiton Company (a Dutch Company) and Suriname Aluminium Company.

Colombia has fastly come up as an important mineral producing country. It has well established workable deposits of five important minerals, namely, copper ore, nickel ore, asbestos, emerald, mercury and coal. Two major copper porphyries have been discovered in north-west of Medellin containing 625 million tonnes of ore with 0.7% Cu. Nickel ore is found at Cerro Matoso; asbestos in Antoquia State; emerald at Guateque and coal at Cerrejon in Gaojiva State. Asbestos property is reported to be quite good and is expected to yield 30,000 tonnes of chrysotile asbestos per annum. Coal deposits have been found in four major basins which contain coking coal as well. The reserves of coal are placed at 100 million tonnes. Mercury is produced at Aranzazu in the Department of Caldas. It is also a small producer of platinum, gold, silver and sulphur (from sour gas). Colombia is self-sufficient in oil production, producing on an average ten million tonnes per annum. A number of new fields have been brought into production, the latest one is in Villavicencio.

Peru is one of the most heavily mineralised countries in Latin America. It is the largest producer of bismuth in the world and is a leading producer of copper, lead, zinc, silver, iron ore, tungsten and barytes. Bismuth is recovered during the treatment of multi-metal complex at La Oroya Refinery, formerly operated by Cerro de Pasco Corporation. Silver and cadmium are also recovered as by-products. The operation of Cerro de Pasco Corporation has been nationalised by the Peruvian Government w.e.f. 1st January, 1974. Peru has emerged as a major source of copper by developing Cuajone mine owned by Southern Peru Copper Ltd. Cuajone is regarded as one of the world's biggest porphyry deposits. Cuajone has a proven reserve of over 468 million tonnes with 1% Cu and 0.02% to 0.025% molybdenum. The same company holds Toquepala copper ore mines. Toquepala is the third largest open pit in the world on the basis of tonnage mined—the first and second being Utah copper mine, U.S.A. and Chuquicamata mine, Chile. This open pit is situated at the altitude of 3000 metres above the sea level. The Southern Peru Copper Corporation is held jointly by four companies: American Smelting and Refining Company 51.50; Cerro de Pasco Corporation 22.25%; Phelps Dodge Corporation 16.0% and Newmont Mining Corporation 10.25%. Peru is one of the important suppliers of iron slurry containing 68.52% to 70.05% Fe to Japan by Marconaflo system. The slurry is shipped from San Juan port. Japan has considerable interest in development of several non-ferrous metal deposits in Peru. A group of several Japanese companies have set up a copper electrolytic plant near Ilo utilising blister copper from Ilo smelter of Southern Peru Copper Corporation. Peru has good resources of potash mercury and their production are in exportable surplus. Tin is produced in a small quantity, from San Rafael mine in Puno held by British interest. Oil production is of the order of three million tonnes a year. Production comes both from offshore and inshore areas. Major inshore fields are Brea-Parinas, Lobitos, and Mirador and the offshore fields are Humboldt, Litoraly and Procidencia. The Lake Titica region is the highest oil field in the world, nearly 3900 metres above level.

Brazil is an important contributor to the world supplies of piezoelectric quartz crystals, beryl, sheet mica, manganese ore, monazite, columbite, zircon, bismuth, iron ore, lithium minerals and tungsten ore. It has enough reserves of nickel ore (garnierite) which is exported in the form of ferro-nickel. Vast reserves of bauxite have been discovered in the Amazon basin and in the Trombetas river valley, north of Santaram in the State of Para. The reserves in the latter area are placed over 165 million tonnes of high grade bauxite. In the Amazon basin area, Alcan Aluminium Ltd. in association with six other companies chiefly from Norway, Holland, U.K. and the U.S.A. have developed bauxite mining capability for 3.3 million tonnes per year. Japanese have also shown a great interest in the mining and production of aluminium metal in Brazil. Six Japanese companies have joined together and set up six lakh tonnes a year capacity of smelter in the State of Para. Mitsui Mining and Smelting Company of Japan is the major partner in this venture. Brazil forms an important source of iron and manganese ores to the U.S.A. Cia Vale do Rio Doce (CVRD) is the largest iron ore miner in Brazil. It has laid 528 km long narrow gauge railroad from Itabira to part Tubarao. It ships 17 types and grades of iron ore products including pellets. The bulk of the output comes from extensive hematite deposits in the Quadrilatero Ferrifero of Minas Gerias. The reserves are placed at 33,000 million tonnes which are among the largest in the world. Brazil exports about 42 to 44 million tonnes of iron ore annually through the ports of Tubarao and Vitoria. Reserves of manganese ore are estimated at 100 to 150 million tonnes; about two-third occurs in deposits in the Urucum district of Mata Grosso and most of the remainder in deposits in the Territory of Amapa and the States of Minas Gerias and Bahia. The bulk of production which started in 1956, has come from the Serra do Navio deposits of Amapa worked by Industria Comercio de Minerio, S.A. (ICOMI) under a fifty-year concessions. In this, Bethlehem Steel Co., of U.S.A. has 49% share and it is the best developed large-scale mining operation in Brazil. In addition to modern mining, crushing, washing and auxiliary facilities, the Company operates 193 km standard gauge railroad from mine to Port of Santana where there is a good arrangement for docking and loading facilities for ocean-going ships. Nearly a million tonnes of manganese ore of 45 to 48% Mn is mined for shipment.

It is the first country outside African continent where production of petalite (a lithium ore) is produced. The deposits occur near Aracuai in the Jequitinhonha Basin in Minas Gerias. Diamond, mostly of industrial variety, is recovered from the beds of Jequitinhonha river; all production is exported to the U.S.A. Brazil is in short supply of crude petroleum. Nearly 50% of the requirements are imported from Russia and Rumania.

Bolivia ranks second in the production of tin, antimony and bismuth in the world. Its production of lead, zinc, mercury and cadmium are also substantial. The tin ore is being mined by American and Irish Companies from the beds containing only 0.87% tin ore. Bismuth is recovered predominantly from the Tasha copper-bismuth ore body. Bolivia has revised its mining code and investment law to encourage foreign capital. The Decree of Royalties on mineral production was also reduced considerably. As a landlocked country, it has many handicaps in transports to the ports, and also all the mines are situated more than 3600 metres above sea level. Despite this, mining is the mainstay of this country and 90% of the foreign earnings come through the export of minerals.

It is a substantial exporter of oil. The chief producing centres are in the foothills belt along eastern slope of the Andes in east-central Bolivia. The petroliferous formations are Petaca (Tertiary) and the Cojones (Cretaceous) and are productive between 990 metres and 1143 metres. The Taiquati (Permo-Carboniferous) formation is productive from about 1800 metres. Oil is worked both by the government agency, Yacimientos Petroliferous Fiscales Bolivians (YPFB), and private firms. The major exporter is Bolivians Gulf Oil Co. The crude is transported for Siciasica in Bolivia to Africa in the northern end of Chile on the Pacific Coast, and Santa Cruz in the southern end of Argentina on the Atlantic Coast, through a pipeline for export purposes.

Chile is the third largest producer of copper ore in the world after the U.S.A. and the U.S.S.R. All the copper mines of Chile have been nationalised in 1971 under the scheme of gradual Chilean-isation of mining industry. It has a number of big, medium size and small copper mines of which five mines are of world-fame. These are Chuquicamata, Exotica, El Salvador, Andina and El Teniente. Chuquicamata open-pit is the largest high grade copper mine in the world, mining 1.88 to 1.80% copper in the ore on an average. This pit alone accounts for nearly 50% of the total copper metal production of Chile. All Chilean copper ore contains recoverable quantity of molybdenum which make this country the second largest producer of this strategic metal in the world after the U.S.A. El Teniente mines ore contains 1.15% Cu and 0.044% Mo. Chile is progressing fast in the development of iron ore mining industry as well and its

Fig. 6.   Location map of copper and iron ores mining centres in Chile.

annual export equals to 12 million tonnes. The iron ore is exported chiefly to the U.S.A. and Japan. Chile is nearly self-sufficient in oil. In this country, the government agency Empresa Nacional del Petroleo (ENAP) operates all fields. Major producing fields are Straits of Majellan, Tierra del Fuego Daniel, Calafata and Posesian. The last one is also a major gas producer.

Argentina is the leading producer of beryl in Latin America, and third in the world after the U.S.S.R. and Brazil, and contributes about 13% of the world supply. Production of high quality beryl containing 10-12% BeO comes from Pampean Ranges, mined [primarily in the Provinces of Catamarca La Rioja, Cordoba, San Luis, and occasionally Sara Juan. Columbite is obtained as co-products of beryl mining. There are numerous deposits of wolframite and scheelite occurring as dissemination in granite and aplite. WO_3 content is 60%. Bismuth is found associated with tungsten which is recovered. Both find export to the U.S.A., the U.K. and West Germany. The reserves of boron minerals are extensive, five major deposits are in Jujuy and six in Salta. Principal minerals are ulexite and tincal. Production of lead and zinc are also in surplus quantity. They are mined and refined mainly for export. This is the only country in South America which imports iron ore for domestic requirements. Resources of manganese ore are small, scattered and of low grade. Small production of sulphur comes from its Azufre de Norte mine in the Province of Salta controlled by military establishment. In 1970, a large deposit of fluorite has been located south of Sierra Grande by Kaiser Aluminium and Chemical Corporation. In oil this country is moving from the direction of self-sufficiency to deficiency because of fast rising consumption.

CENTRAL AMERICA

In Central America, Mexico is the only mineral-rich country. It possesses almost all minerals including coal, coking coal and petroleum. Only the resources of asbestos, chromite, magnesite and mica are negligible. With production of over 1.2 million kilogrammes of silver annually, this country is the largest producer in the world followed closely by Peru and U.S.A. It is also leading producer of fluorspar in the world, second in bismuth, strontium and sulphur and third in arsenic. Mexico is the leading world supplier of acid grade fluorspar. The major production comes from the lead-zinc mines worked by San Francisco Mines of Mexico Ltd. in the State of Chihuahua. The ore contains 12.16% fluorite, besides copper 0.61%, lead 5.5%, zinc 8.2%, gold 0.42 gm and silver 1.86 gm per tonne. One of the longest open pit has been developed at La Caridad comparable to copper deposit of Bougainville containing appreciable quantity of molybdenum about 37%. The production of elemental sulphur by Frasch process comes mainly from salt domes (more suitably to be called sulphur domes) in the Gulf of Mexico. The major field is the Golden Lane which has been prolific producer for half a century. The best prospective source of oil is the Poza Rica field, discovered in 1930. The search for new oil reservoirs is expected to be centred in the Gulf of Mexico. The petro-chemical industry of Mexico is well advanced coupled with its richness in natural gas and sulphur. It has joint venture projects in the works with British and American companies to produce large number of chemicals. Mexican exports of minerals and ores mostly go to the U.S.A.

In general, Central American countries, except Mexico, are poor in mineral resources. Salt, limestone and cement are the main industries based on indigenous resources. *Honduras* and *Nicarugua* produce small quantities of copper, lead and zinc with recovery of silver and gold as by-products. None of the Central American countries, except Mexico, produce coal and coke and petroleum. Venezuela is the principal supplier of petroleum to these countries. Belize has no mineral of any significance except limestone used for road metal. *Costa Rica* possesses lateritic type of bauxite deposit, small deposits of manganese ore and limestone. *El Salvador* having 90% of the area covered by volcanic rocks, has pumice stone, the only mineral worth name. There is a petroleum refinery and petroleum-based fertiliser plant at Acajulta. Cement production is the main activity in *Gautemala* as raw materials like limestone, clay and gypsum are available. *Panama* has made a good name in making discovery of porphyry copper deposit in an area called Cerro-Colorado. This discovery has been made by Javeline Copper Corporation, a Canadian Company. The ore is associated with rocks of Tertiary age. The country rock is andesite with quartz-porphyry intrusions. Drill-proven reserves are expected to be in excess of 3000 million tonnes containing on an average 0.81% Cu with minor amount of gold, silver and molybdenum.

NORTH AMERICA

The U.S.A. is a richly endowed country in the world having varieties of minerals in abundance. It is a leading producer and consumer of mineral ores, and metals so much so that despite being a great producer, it is also a bulk purchaser of ores and minerals in the international market for supplementing the domestic supplies. To maintain uninterrupted and steady supply of raw materials, the U.S.A. has acquired mining rights and entered into long-term purchase agreements with a large number of countries in the Latin America, Africa, Caribbean, West Asia, Far East Asia and Australia. The foreign buying of the U.S.A. is so great that even the slightest change in the procurement programme due to release of material from stockpile or for some other reason may upset the mining activities of many countries. The mining, metallurgy, fabrication and petrochemical industries have reached the highest order of automation and perfection. Fuller advantages of the improved techniques of ore dressing, chemical principles and metallurgy are taken for the recovery of all possible useful elements and minerals as by- or co-products. Recovery of gallium, germanium, cadmium, thallium, bismuth and iridium as by-products of lead and zinc operation and selenium, tellurium, cobalt, molybdenum, gold, silver and rhenium from copper product residues are now usual ore dressing and metallurgical processes in the U.S.A.

In some of the cases, the by-products have become the leading source of world supply. Treatment of low grade iron ore, recovery of iron from the taconite deposits and preparation of pellets thereof; recovery of manganese from rock containing as low as 10 to 15% Mn; recovery of ilmenite from Florida dune containing only 2% ilmenite as against 80% found in Kerala beach coast of India; extensive use and re-use of scrap material and intensive research programme and invention of substitutes are added features of the U.S.A. mineral industry. Each big firm earmarks 5 to 10% of the profit towards research whether it be for mining or for improving the quality of products manufactured.

The U.S.A. is world's leading producer of phosphate, barytes, borax, petroleum, sulphur, coal including anthracite, uranium, ore, rare earths, aluminium, titanium, copper, molybdenum, lead, zinc, iron and steel and second in iron ore, fluorspar and gold. The U.S.A. alone produces 80% of the world's requirement of borax and nearly 40% of phosphate. But this country lacks in supply of many minerals like sheet mica, manganese ore, chromite, antimony ore, bauxite, crystalline, graphite, nickel, tungsten, sillimanite, and is also a bulk purchaser of iron ore.

The production of phosphate comes chiefly from Florida, Tennessee, North Carolina and Idaho. The production of phosphate in the U.S.A. is nearly 43 million tonnes annually, about 40% of the world total and is followed by Morocco with annual production of over 15 million tonnes, about third of the U.S.A. In northern Florida, north of Bone Valley district, another huge reserves of phosphate have been discovered by Florida Occidental Petroleum Corporation. At the present the Bone Valley is the largest producing centre of phosphate in the U.S.A. In the North Carolina, Texas Gulf Sulpher Company has established reserves at Lee Creek equivalent to 1500 million tonnes of saleable phosphate rock containing 31% P_2O_5. The deposits from this area are utilised by the Company mainly for phosphoric and manufacture by Wet Process.

The world's largest production of borax and other boron minerals like kernite, ulexite and colemanite comes from Kramer and Searles Lakes of California. The lakes have solid salt beds of complex composition, and boron minerals are recovered as by-products of potash. Large deposits of ulexite occur in Clark County, Nevada.

The Climax mine in Colorado of Climax Molybdenum Company is the largest source of molybdenum in the world. Amax Exploration Ltd., holding company of Climax of Molybdenum have made discovery of 130,000,000 tonnes of molybdenum ore at Keystone mine (lead-zinc mine) contains 0.30 to 0.4% MoS_2. Another mine Henderson, situated 120 km north-east of Climax at Urad, Clear-Creek Country, belonging to the same Company has proved equally promising. The ore occurs as molybdenite (MoS_2). The reserves of the ore in Climax property are estimated at 415 million tonnes with cut-off-grade at 0.2% MoS_2. In Henderson property, 236 million tonnes of reserves (proved and Probable) have been established with 0.45% MoS_2. Each area produces over 50 million lbs of molybdenum a year. Pyrite, tungsten (second largest mine production in the U.S.A.), and tin (largest in the U.S.A.), fluorite, topaz and rare earths are recovered as by-products. The Questa open-pit mine in North Central New Mexico Molybdenum Corporation of America is another largest source of molybdenite in the world and the third largest in the U.S.A. Its production can be comparable to that of

Canada's Endako Mines Ltd. The ore is shattered albite and granite filled with a network of veinlets, stringers and veins consisting largely of molybdenite with quartz, rhodochrosite pyrite and fluorite as associated minerals. More than 20 million tonnes of ore assaying 0.297% MoS_2 are already proved. Questa contributes 12% of the U.S.A. molybdenum production. The U.S. Borax Company has improved its reserves at the Quarts Hill molybdenum porphyry to more than 250,000,000 tonnes assaying 0.18 to 0.25% MoS_2. New figure will make this property as fourth largest reserves in the world. Good amount of molybdenum is also recovered as a by-product during the processing of uranium-bearing lignite ash in South Dakota.

Again, rare earths mine in the Mountain Pass, Nevada is the world's largest source of rare earth ores worked by Molybdenum Corporation of America. This Company produces 10 million pounds of rare earths oxides, europium, praseodymium, neodymium, cerium, samarium, gadolinium and lanthanum per year. The ore body is a massive vein of basic magnetic rocks, 60 metres thick and dipping 35° to the west, the principal minerals being calcite, barytes, bastnasite, strontianite and silica. The solvent extraction method is used for processing rare earths. More than 10,000 pounds of rare earths ore is required to produce one pound of europium oxide. The colour-television has brought a boom in yttrium and europium oxides. The europium phosphors have solved a problem that had long baffled colour-television tube manufacturers. The new red phosphors containing yttrium and europium oxides give a truer red and improved brightness of the colour picture by up to 40%. Colour television depends on three phosphors—red, blue and green. Red phosphors used previously were not satisfactory because the red was too dull and the green and blue had to be deliberately deadened to stay in balance with the weaker red resulting in a dull colour picture which now can be avoided with the use of europium oxide.

In resources of copper ore, this country is the largest producer in the world followed by Chile, Zambia and Canada. Copper deposits are large and widespread. The Bingham copper mine in Utah worked by Kennecott Copper Corporation is the largest opencast copper mine in the world with 100,000 tonnes of daily copper ore production. The tenor is 0.9% Cu only. The copper mineralisation is associated with intrusion of monzonite-porphyries. The primary minerals consist of pyrite, chalcopyrite, and bornite with minor proportions of sphalerite, galena and molybdenite. In the cap rocks, the primary minerals have leached out and voids filled by limonite. Below the capping, the sulphides are replaced by covellite and chalcocite. Replacement deposits of copper, zinc and silver-lead are zonally arranged. The molybdenite is unaffected and won in considerable quantity making the Kennecott Copper Corporation the largest producer of molybdenum in the world. The Butte Copper mine in Montana worked by Anaconda is the largest underground workings for copper in the world. The mining area covers hardly more than 3×7 km. It is 1.2 km deep and underground working exceeds 1600 km. It is also an important source of gold, silver, zinc, lead and some manganese beside the chief metal copper. The primary minerals are chalcocite, chalcopyrite, enargite, bornite, tennantite, tetrahedrite, and covellite. Pyrite and quartz are dominant. Sphalerite is locally dominant. Galena is rare. Butte is called the richest hill on earth and is comparable to Rand gold deposits of Republic of South Africa and Sudbury nickel deposit of Canada, in metallic wealth. New Mexico, Arizona, Colorado, Idaho and Nevada are the other important provinces where copper ore is worked on a large scale. In New Mexico, new deposit of copper ore near Bayard has been proved to have 16,200,000 tonnes reserves containing 2.15% Cu. In Arizona, a number of copper prospects are worked, the important being Ithica peak and Christmas mine. Molybdenum and silver are produced as co-products. In silver production, the U.S.A. is the third largest producer after Mexico and Peru. The metal is mainly obtained as a co-product of lead, zinc, copper or gold refining. Idaho produces good amount of silver from silver-bearing deposits. Bismuth is recovered as a by-product from the lead-silver ores of Montana, Utah, Nevada and Arizona.

The output of iron ore and concentrates comes chiefly from Minnesota, Missouri, Michigan, California and Alaska. The trend is towards the use of low grade ore, like magnetic taconite and magnetite after pelletisation. The taconite deposits of Minnesota and of Lake superior region are greatly being worked. Minnesota alone produces on an average 60% of the total United States iron ore. The invention of grate kiln process has brought a great boom in production of pellets not only in United States but all over the world. The United States produces annually about 50 million tonnes of pellets out of the total iron ore and concentrates production of about 90 million tonnes. Kaiser Steel Corporation has set up the largest pelletisation plant in the world at its Eagle Mountain mine,

Riverside Country, having production capacity of 2.2 million tonnes. This Corporation has also signed a contract for export of 800,000 tonnes a year of pellets of Mitsubishi International Corporation of Japan. Very recently, a huge deposit of low grade iron ore has been proved at Chenik, west of Anchorage, in Alaska containing almost 1000 million tonnes of iron. It can be regarded among top seven or eight deposits in the world. Ilmenite is recovered as co-product of iron ore mining in Essen County, New York. Although the U.S.A. is the second largest producer of iron ore in the world, it is also a bulk importer of iron ore, over 30 million tonnes annually for the production of iron and steel which is of the order of 125 million tonnes annually. The imports are made from Liberia, Sierra Leone and Gabon in Africa; Peru, Chile, Brazil in Latin America and from Canada. Good use of scraps is also made in the production of steel. Arkankas is a great source of vanadium oxide from the deposit at Wilson Springs in Garland County. Arkankas is the only source of bauxite in the U.S.A. producing nearly 1.9 million tonnes annually. An additional requirement of bauxite to the tune of 11 million tonnes annually is mostly imported from Jamaica and Surinam where the U.S.A. has mining interests. The U.S.A. with an annual production of over 4.5 million tonnes of aluminium metal is the largest producer of aluminium in the world, followed by Canada with 0.8 million tonnes only, both countries base their production on the import of bauxite and alumina. Domestic production of antimony ore being small, most of the supply is met from Bolivia, Mexico, Peru, Turkey and Republic of South Africa. Nickel ore is worked near Riddle, Douglas County in Oregon, by Hanna Nickel Smelting Company. It operates only a ferro-nickel smelter. The production is small. There is only one nickel refinery in the U.S.A. at Braithwaite, Louisiana owned by Amak. It treats copper-nickel matte obtained from Botswana. It meets most of the requirements of nickel from Canada. The U.S.A. is the largest consumer of nickel in the world and consumes about 136,000 tonnes annually. The chief sources of mercury are Mt. Jackson, Sonoma County; Dawson mine, King County; Chileno Valley mine, Marin County; Knoxville mine, Nepa County; Polar Star and Cambri mines in San Luis Obispo County in California and Red Devil mine in Alaska.

The U.S.A. is the third largest producer of chrysotile asbestos in the world (after U.S.S.R. and Canada) producing over 100,000 tonnes annually from California, Vermont and Arizona. At the same time this country imports nearly half of the production of Canada for its domestic consumption. The imports from Republic of South Africa, Rhodesia, Swaziland and Zambia are also substantial. In gypsum and diatomite, this country again leads in world production. Gypsum is produced mainly in California, Iowa, Michigan, Texas and New York, and accounts for one-fourth of the world production. The world's most important diatomite deposit is situated near Lampoc in California, the major producer is Johns Manville Corporation.

The country is also an important producer of uranium. The Lucky Mc Division of Utah is one of the biggest producers and suppliers of uranium oxide. Large production also comes from Wyoming, North Dakota, and in New Mexico from McKinley and Valenia which are large producing centres. The Mitsubishi Mining and Smelting Co. of Japan has entered into an agreement with Rio Algam Corporation of the U.S. affiliated with Rio Algam Mines Ltd. of Canada on joint exploration programme of a uranium mine in Shirley Basin, Wyoming. Up to 50% of the uranium concentrates will be made available for shipment to Japan. Wyoming is also famous for trona (Na_2CO_3) deposit.

The U.S.A. is the biggest producer and exporter of bituminous coal and anthracite in the world. The production of coal averages 450 million tonnes and the export 53 million tonnes annually. Canada, Japan and the European Coal and Steel Community are the principal importers. Production of coal chiefly comes from Virginia, Tennessee and Ohio. Dakota and Vermount are famous for peat and lignite.

Texas and Louisiana are the chief sources of sulphur from sulphur domes. Considerable quantities of sulphur is recovered from the 'sour gas' during the refining of natural oil and gas in Texas, Louisiana, Oklahoma, Arkansas, California and also from Tennessee and Ohio. Domestic production of crude meets 70% of the internal consumption. The remaining quantities are imported from West Asia, Venezuela and Canada. The reserves of the crude oil at the end of 1964 were estimated at 41,800 million barrels just sufficient to last for 10 years with the current rate of production which is of the order of 11 million barrels a day. The U.S.A. possesses only 10% of the total world reserves. However, the exploratory activities of this country are so great that every year they go on proving sufficient quantities of oil reserves to compensate what they have consumed in the previous year. Great efforts are being made to recover vast reserves of oil from oil-shale deposit, 150 metres thick, located in the Green River formation of Colorado, Utah and Wyoming. The U.S. Geological Survey has estimated this deposit to contain equivalent to 4,000,000 million barrels of oil. If all of this oil were commercially recoverable, it would sustain more than 360 years' production for the entire world based on the oil production rate of 1965. The oil-shale is impregnated with a waxy organic compound Kerogen which is composed of about 80% of carbon, along with hydrogen, oxygen and nitrogen. The mixture varies a little from shale to shale. Many oil firms of the U.S.A. including the U.S. Bureau of Mines are engaged in working out the process for recovery of oil which is beset with not easy problems, like that of elimination of elements like sulphur and nitrogen and disposal of hot shale also. Production of oil from low grade coal is also being tried vigorously on pilot plant scale. The necessity of these efforts has arisen mainly due to strategic reason of impending danger of cutting of supply from the West Asia in case of war or due to other political changes. At present there is only one project on stream producing 50 to 60 barrels of gasoline a day from coal in pilot plant of Consolidation Coal Company based on Company's own Consol Process at Cresap in West Virginia. The office of Coal Research, Washington, is greatly engrossed in coal to petrol research programme.

The results of a survey of 329 salt domes in Texas, Louisiana, Mississippi, Alabama and offshore tide land of Gulf area have indicated enough salt to last for thousands of years, as well as more than 70 million tonnes of elemental sulphur, and formations associated with domes are estimated to hold more than five million barrels of recoverable petroleum.

Canada with an area more than double that of India and the population of only about 20 million or 1/25th of India, is another country of promise possessing vast quantities and varieties of mineral resources for the growing world demand. Most of the area, especially of northern Canada, is snow-covered, has severe climate, yet it is no deterrent to mining activities. Most of the minerals mined are exported in the form of concentrates or refined products chiefly to the U.S.A. The mining and metallurgical industry is highly advanced and practically complete automation has been introduced everywhere, for example, the production of petroleum in Alberta is remotely controlled from centralised unit at Edmonton city. Similarly, sophisticated automation has been introduced in the iron ore mining between the Labrador-Quebec region. The abundance of cheap hydro-electricity has enabled this country to develop large aluminium industry based on imported bauxite and alumina mainly from Jamaica, Surinam and other Caribbean countries.

The country leads the world production in nickel, zinc, platinum group of metals, chrysotile asbestos, potash and nepheline syenite. The output of nickel contributes 90% of the total world production (excluding U.S.S.R.). It ranks second in place in the production of uranium, cobalt, ilmenite, gold, cadmium, sulphur and gypsum, and stands high in the production of iron ore, copper, lead, silver, molybdenum, pyrite, fluorspar, magnesium from sea water and petroleum. A wide range of minerals are recovered as by-products. These include antimony, bismuth, cobalt, platinum and its group metals, selenium and tellurium. Primary ores of columbium and tungsten are found in British Columbia.

This country has, however, practically nil to negligible resource of manganese ore, chromite, bauxite, phosphate, magnesite, kyanite, sillimanite, mica, mercury and tin. Mica, mainly of phlogopite variety, was mined in Gatineau Valley, Quebec. It was mostly powdery in nature.

Sudbury in Ontario is the largest source of nickel in the world. Over 70% of the total world supply of nickel comes from this area. The ore minerals are primarily nickel-copper sulphides associated with intrusions of norite dyke, covering an area of 2 to 3 km in width and 60 km in length. The ore contains 1.45% Ni and 0.08% Cu. There are two leading producers, International Nickel Company (IN CO) and Falconbridge Nickel Mines Ltd. INCO holds 16 mines, while the Falconbridge has three mines. The former produces 230,000 tonnes of nickel a year and the latter 40,000 tonnes a year. The reserves in INCO area are estimated at 297 million tonnes and that in Falconbridge at 46 million tonnes. Both firms recover several other elements during the refining of nickel. The elements produced as co-products are copper, cobalt, arsenic, bismuth, gold, silver, selenium, tellurium and platinum and its group metals like palladium, rhodium iridium, osmium, ruthenium, besides the major element nickel. The primary ore minerals found in the deposit are pyrrhotite, chalcopyrite, pentlandite, minor cubanite; bornite, niccolite, maucherite, sperrylite, gersdorffite, violarite, complex arsenides, nickel-bismuth sulphides parkerite, tellurides of gold, silver and platinum metals. Canada derives platinum entirely from treatment of nickel ore containing 0.025 troy oz. per tonne. Nickel is also mined in Manitoba and British Columbia by other companies. The Lynne Lake mine in northern Manitoba feeds the refinery at Fort Saskatchewan.

There are a number of copper-lead-zinc deposits of large dimension in British Columbia, Ontario and Quebec. The Sullivan lead-zinc mine in British Columbia is one of the largest lead-zinc mines in the world. Bismuth, cadmium and antimony are obtained as by-products. Canada's entire production of antimony comes from antimonial lead resulting from lead refining. The Consolidated Mining and Smelting Company of Canada Ltd. (COMINCO) is the only producer of antimonial lead. The Granisle Copper Ltd.'s pit in British Columbia with 0.53% Cu is probably the second largest mine working the lowest grade of copper; Aitik pit of Sweden being the first having average grade of 0.5% Cu. A fantastically low stripping ratio of 0.23 to 1 has made possible for economically working the ore. Several copper-lead-zinc mines have been newly developed in North-west Territories and British Columbia. Texas Gulf Sulphur has established substantial reserves of copper-zinc-silver ores body near Timmias, Ontario. Most of copper concentrates produced in British Columbia are exported to Japan. British Columbia is also famous for Endako molybdenum deposit occurring in north central part; and is regarded as the second largest deposit in the world with production capacity of 23 million lbs of molybdenum a year. It is worked by Endako Mines Ltd. The ore is found as fracture and fissure fillings with quartz in an intensively sheared and fractured granite mass. The grade of the ore varies from 0.12% MoS_2 to 0.266% MoS_2. Indicated reserves are over 50 million tonnes with 0.25% MoS_2 (0.33 per cent molybdenite is equal to 0.25 per cent molybdenum). The grade is quite comparable to Climax Molybdenum Company, Colarado, U.S.A., where the recovered grade is 0.20% MoS_2.

Quebec, Ontario and British Columbia account for nearly total output of iron ore which is of the order of 42 million tonnes a year. About 75% of the total output is exported to the U.S.A., Japan and some of the countries of Europe. Three pelletisation plants are operating in Quebec-Labrador area since 1965 with total installed capacity of seven million tonnes. Steep rock mine in Ontario is historically famous for iron ore production in Canada.

Saskatchewan is famous for potash, El Dorado uranium mine and

Flin Flong lead-zinc mine. Potash deposit is the largest in the world. Several companies hold lease for potash in Saskatchewan and Manitoba, the important ones are International Minerals and Chemicals (Canada) Ltd., and Kalium Chemicals Ltd. Production is over five million tonnes. Potash mining is an unique feature in this country. The potash is an overlain by quicksand containing water under pressure. To overcome the water during shaft sinking, a technique of freezing the side-walls into ice has been adopted. This makes the working conditions chilly but it has proved effective in controlling the water. Walton Lake in Yukon territory is the only place where wolfram is worked. The grade is 2.53% WO_3 in the ore. Quebec is internationally famous for four mineral deposits: (i) lithium ore, (ii) hematite-ilmenite rock type deposit, (iii) asbestos, and (iv) copper-lead-zinc ores.

Spodumene (lithium ore) deposit found in Lacorne Township, Quebec, is regarded as one of the largest in the world similar to lepidolite deposit of Bikitia in Rhodesia. The ore occurs in parallel pegmetite dykes. The reserves are estimated at 20 million tonnes averaging 1.15% Li_2O. Lac Tio area of Allard region is famous for massive ilmenite-hematite deposit. It is the largest ilmenite deposit in the world containing 100 million tonnes of ilmenite. Ilmenite and hematite occur in ratio of 2:1, with anorthosite as principal gangue with minor pyroxene, biotite and pyrite. The ore is crushed and smelted to produce pig iron and the slag containing titanium dioxide is sold to pigment manufacturers throughout the world.

Jeffery in Quebec, owned by Johns Manville Co., is the largest mine in the world producing chrysotile asbestos. The mineral is associated with serpentine. Mineralisation is quite uniform. On an average, 6% asbestos fibre is recovered per tonne of rock mined. There is a well-equipped milling and fibrisation plant owned by the company which exports asbestos after grading according to universally adopted Quebec Test Grading Standard. Large tonnage of chrysotile is also mined in Newfoundland, British Columbia, Ontario and Yukon Territory. This country ranks high in production of sulphur. Sour gas of Alberta is the main source of supply of elemental sulphur. The resources of coal and lignite are in abundance occurring in British Columbia, Alberta, New Brunswick and Saskatchewan. The latter province has huge reserves of lignite. Again, Alberta is the main centre of oil and gas. Production of oil is more than a million barrels a day and is in surplus which finds export to the U.S.A., U.K. and Europe. This country possesses vast reserves of tar-sand in Alberta which are reckoned to contain more oil than what exists in West Asia. Great efforts are being made to obtain oil from sand-tar and they may ultimately affect the monopolist position of West Asia and adjoining African countries.

Greenland is the largest island on the earth. It is under Denmark's possession. It has a unique distinction of being the only country possessing natural cryolite. This mineral was being mined at Ivigtut (Southern Greenland) for over 100 years, and the mine was closed down during 1962 due to depletion of the reserves. Presently, cryolite is being obtained from mine's stocks, dumps and material reclaimed from the harbour where the rock obtained from the mine was used for construction purposes in early stages. Presently, over 50,000 tonnes of cryolite besides over 25,000 tonnes of concentrates are exported out annually. The reserve from the reclaimed material is likely to last for another 15 years with the above rate of production. Appreciable quantity of coal is mined on Disko island, South Greenland, which is mainly exported to Denmark. A great deal of geological work conducted by a team of geologists from Denmark and few companies of Canada and the U.S.A. has revealed the possibilities of finding rich deposits of uranium, oil and gas. Greenland has suddenly become a refreshing ground for mineral exploration. In 45-page Report No. 56-A Survey of Economic Geology of Greenland brought out by the Department of Geological Survey of Greenland illustrates many examples of finding several useful metallic mineral deposits in this country. Greenland at one time of cryolite fame has turned out to be a good source of lead-zinc ores, chromium, molybdenum, uranium and rare earths. A good deposit of lead-zinc ores have been proved in Black Angel area in West Greenland. The reserves of proved, indicated categories are placed at four million tonnes with 20% combined lead and zinc metals. It is worked by Vestgron Mines Limited, Greenex A/S, a subsidiary of Cominco Ltd. Large deposits of molybdenite have been found in Werner Bjerge in the Central East Greenland. The reserves are placed at 117 million tonnes with grade of 0.25% MoS_2.

Iceland has come into prominence very recently for the discovery of rich deposits of diatomite found on the bed of Lake Myvatyn which averages less than six metres in depth. The lake is located in northern Iceland, some 48 km from the nearest port at Husavik on the Arctic Ocean. It is regarded to be the largest deposit outside

America. It has been developed by a new corporation, Kisilidjan, h.f., in which Johns Manvile Corporation of U.S.A. is holding a minor share. It is Iceland's first industry based on mineral resource.

CARIBBEAN ISLANDS

Caribbean islands are the largest source of bauxite of the world. *Jamaica* is the leading world producer with annual production of 12 million tonnes closely followed by Australia. *Dominican Republic* and *Haiti* are the other two bauxite producers in the Caribbean Sea area. All the bauxite from these island countries is chiefly shipped to the U.S.A. and Canada. In Jamaica vast stretch of land is covered by high grade Bauxite. Several important companies hold the lease. Kaiser Bauxite Company, a subsidiary of Kaiser Aluminium and Chemical Corporation of the U.S.A. has built Port Rhoades on the north coast to handle six million tonnes of export of bauxite. Other important companies engaged in bauxite and production of alumina for export purposes are Reynolds Jamaica Mines Ltd., Alcoa Minerals of Jamaica, Inc. and Alcan Jamaica Ltd. In Haiti, besides bauxite, the prospects of finding copper deposits are there. Extension of copper lode of Meme mine containing 1.50 to 1.75% Cu has been found in Terre Neuve district. Dominican Republic has a good resource of lateritic nickel. It was prospected by Falconbridge of Canada. Proved reserves are placed at 70 million tonnes with 1.58% nickel. Production of ferro-nickel has started during 1972. *Puerto Rico* has a rich source of copper ore in Utuado-Lares region. It is worked by American investors. It is also an important producer of cement in the Caribbean sea islands.

Cuba is famous for lateritic nickel deposits of Nicaro and Moa Bay both in Oriente Province at the extreme eastern end of the country. The ore being fine grained weathered ferruginous serpentine containing 1.2 to 1.6% Ni and 0.15% Co. The production of manganese ore, chromite, gypsum, pyrite, limestone is small. Cuba on an average produces 40,000 tonnes of nickel per annum.

INDIAN OCEAN ISLANDS

In the Indian Ocean, *Christmas Island* (British), located 315 km south of Java, is important for guano phosphate. The shore-line of this island consists of steep cliffs rising straight up to an island plateau some 300 metres above the sea level. This plateau is covered by dense tropical vegetation beneath which are millions of tonnes of guano phosphate reserves estimated over 200 million tonnes. Mining of phosphate is the sole activity, accounting some 700–800 thousand tonnes a year. It is worked by the British Phosphate Commissioners jointly administered by Australia and New Zealand.

PACIFIC OCEAN ISLANDS

There are numerous clusters of small islands in the Pacific Ocean. Most of these islands are under the possession of the U.S.A., the U.K. and French. These islands are important from military point of view but some of them are also a great source of guano phosphate. At present, Makatea and Nauru islands are the most important sources of phosphate. They are worked by the British Phosphate Commissioners. Between the World War I and II, phosphate deposits on a number of islands of Caroline, Ryuku and Marshall groups were worked by Japan. Mining activities were abandoned due to exhaustion of the ore.

Of all the Pacific islands, *British Solomon Island, Fiji* and *New Caledonia* are very important from metal resources point of view. British Solomon islands offer extensive source of nickeliferous laterite, copper ore, chromite and bauxite. Nickeliferous laterite deposits occur on San Jorge and Guadal-canal islands. Fiji offers a good source of gold and copper. Gold is worked by Emperor Gold Mining Co. Ltd., at Vatukoula on Viti Levu island. A Japanese firm, Banno Mining Company has negotiated to work 30,000 tonnes of 5% Cu ore annually by open pit on an enriched sulphide copper deposit in rhyolitic tuff and breccia at Udu Point. New Caledonia has emerged as a good source of nickel ore containing about 3% nickel and valuable cobalt. There is a nickel boom in New Caledonia. The entire island consists of nickeliferous laterite. The concentrates of nickel ore are exported to Japan, Canada, France and the U.S.A. It ranks second in nickel production (in concentrates) after Canada. Iron ore is also produced on a small scale. Iron content in the ore is not high and averages 55% Fe only.

CHINA

A rapid development of mineral industry is the key to the progress of China. This country possesses most of the useful minerals. The supply position of diamond, mica, bauxite, chromite and asbestos is, however, critical. The shortage of crude oil, once felt and supplemented by the supply from the U.S.S.R., is no longer there. China is one of the leading producers of tin, tungsten and antimony in the world. It has exportable surplus of bismuth, mercury, fluorite, manganese ore and molybdenum. Import of aluminium ingots is substantial.

Ku-Chiu in Yunan and Eu-ho-Chung in Kewangsi are the active tin mining centres. A number of tin ore mining centres are located bordering N. Vietnam, Laos and Burma. Output of manganese ore mainly comes from south of Yangtse river. Production of mercury comes chiefly from Tung Jen Kweichow areas in south-west China.

The famous wolframite mine of Kiangsi, located in south Kiangsi, is under production for over half a century. The richest deposits are situated in the Yiling Lohsiao mountain ranges which extend through a large part of South Kiangsi and continue into Hanan and Kwangtung, where there are also wolframite mines. Average annual output is estimated at 20,000 tonnes containing 65% WO_3. Molybdenum is obtained as a by-product during the refining of tungsten. Iron ore deposits are in the Yangtse Valley between Kiangsu and Hupei. The reserves are estimated at 11,000 million tonnes. The iron ore mines of large dimensions are situated in Anshan in southern Manchuria, Steel Plants of Anshan town are situated bordering North Korea. Other large-scale steel plants are situated in Paotow, Wuhan and Shanghai. China produces nearly 25 million tonnes of steel per year.

Shihmien is the largest asbestos mine in China. The reserves of phosphate minerals are in abundance, mostly coming from Hupeh and Kweichow provinces. The requirement of sulphur and sulphuric acid is entirely met from pyrite. Shortage of copper is met to some extent by imports from Chile. Cuba is reported to have supplied nickel to this country. Bauxite requirements are met by imports from Guyana.

It has large reserves of coal situated in the provinces of Shansi, Shensi, Kansu and Hupei. Coal mining is limited to the northern part of the country in Hupei and South Manchuria. Coal output is second largest in the world, but most of the output is of inferior type used for domestic fuel only. The coal reserves are estimated at 1,200,000 million tonnes. Domestic production of oil, which is now estimated of the order of 50 million tonnes, come from Karamai, Taching, Shengli and Po-Hai Bay oil field in Sinkiang. Gas fields are located in Szechuan Province. Chinese have fastly developed the oil field of Manchuria. In fact, Manchuria is the key centre for the production of oil, coal and steel and major supplies of all coal and oil come from there. Oil wells are at Tachinv. In the province of Sinkiang, large deposits of petroleum have been discovered. A big oil refinery is situated near Yumen on the western end of the great Chinese Wall in the province of Kansu, Sinkiang province is reported to produce considerable quantities of fissionable minerals.

U.S.S.R.

U.S.S.R. occupies about 1/6th of the land surface of the earth's crust and is practically self-sufficient in all mineral resources. Russia presently has about half of the world coal, iron, potassium and manganese reserves; 1/3rd of oil and natural gas; and a substantial part of non-ferrous and precious metals. It is, however, slightly deficient in resources of lead, tin, aluminium, antimony, tungsten, cadmium, mica, fluorspar, barytes and high grade talc. The U.S.S.R. has emerged as one of the biggest producer of iron ore, mining about 250 million tonnes of iron ore annually. It also exports about 40 million tonnes of iron ore per year mostly to Comecon countries. Reserves of high grade iron ore are placed at 50,000 million tonnes and that of ferruginous quartzite at 10,000 million tonnes. The main mining centres are the Central and South Western part of the European part of the U.S.S.R. in Siberia, Kazakhstan and Trans-caucasia. More than 80% of the iron ore production in this country comes from open pit and remaining by underground mines. The U.S.S.R. is the largest producer of platinum group metals estimated to produce 50% of the world production followed by the Republic of South Africa (Rustenburg platinum mine). The reserves in the Norilsek and Kola Peninsula are estimated to contain 10 million ounce of platinum group of metals. The major production comes from Norilsek as a by-product of copper-nickel mines. The ore averages 0.75% Cu, 0.5% Ni and up to 11 gm/t platinum group metals—mainly palladium and platinum. It is also the second largest producer of gold in the world after the Republic of South Africa. Cobalt is mined at Khovu-Aksyn in the Tuva Republic. Cobalt extraction plant is also located at the mine site. Most of the cobalt production, however, comes from nickel mines in the far north of Norilsk and from the Petsamo nickel mine in the Kola Peninsula. There is a large nickel smelting plant attached to Rechengenikel mill in Murmansk region, bordering Finland. Russia possesses the largest apatite deposits in the world occurring in the Kola Peninsula.

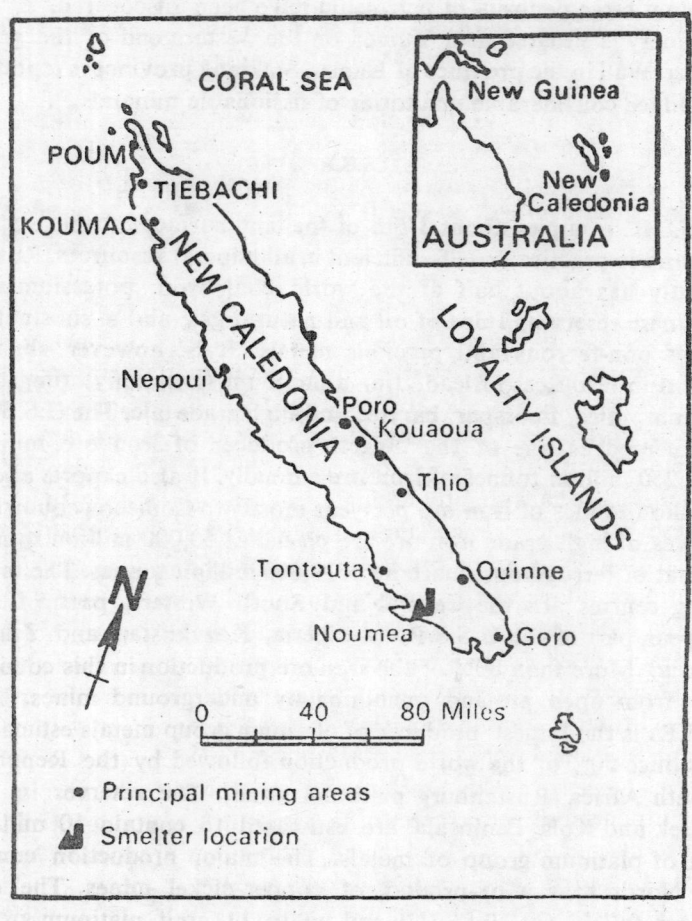

Fig. 7.

Apatite occurs in nepheline-syentite rock and is separated by flotation. About 12 million tonnes of apatite concentrates are produced from this area annually. Apatite occurs as a single sheet like body complicated by a number of pinches and swells which extends from Kukisvmtchor mountain in the north to the Koastiva mountain in the east. Apatite concentrate grades 36.5% P_2O_5. The tailing from the apatite concentration from the useful source for alumina which is recovered for the production of aluminium metal. Bauxite short Russia is pioneer in producing alumina from nepheline-syenite for the extraction of aluminium metal. The biggest plant is Achinsk aluminium complex on the Chalin river in Central Sibera. Nepheline-syenite is mined in Kuznatsk-Altan about 300 km from Achinsk for this plant.

Since 1977, the Soviet Union has become the leading producer of mercury. It has developed deposits in the Caucasus region and in the Chukchi peninsula. Chukchi peninsula is directly opposite the seaward peninsula of Alaska.

It is second largest producer of gold in the world afwi the Republic of South Africa. It has the largest reserves of manganese ore in the world, but the deposits are distributed geographically apart. Most of the known reserves are situated in Caucasus and southern Ukraine and in smaller proportion in Urals and in central and eastern Siberia. It is also the largest producer of manganese ore in the world, producing three times more than the next producer to Republic of South Africa and India, running parallel in production since 1965, the former slightly taking a lead over the latter. Russian manganese ore mines are worked in a giant scale at the Shevchenkovsky and Bodgonovsky by open pits near Ordzhonilsidge. The two pits mentioned above work a two-metre thick flat bed covering 125 sq km area. The stripping ratio is 25–30 to 1. Shovel-truck, shovel-electric train, dragline and bucket wheel strippers are used. It is richly endowed with vast reserves of other useful minerals like iron ore, coal, chromite, apatite, potash and petroleum.

Russia is the largest producer of asbestos, second being Canada. It possesses however mostly amosite and short chrysotile fibre deposits. Output mainly comes from the Bazhenovo region of Urals, from Kustanai deposits in Kazakhstan. Three new areas have been developed in Tadrikistan and Buriatskaya.

This country is striving hard for achieving self-sufficiency in copper. Copper ore is mined mainly in Ural, Uzbekstan, Armenia and

Kazakhstan. A number of copper ore deposits are being developed and mines' production expanded for increasing supply. Kounrad is the largest open pit copper-molybdenum mine located on the northern side of lake Balkhash. It contains sulphide and oxidised copper ores, the principal minerals being chalcocite and to a lesser extent covellite, chalcopyrite associated with pyrite. In Ural the mines are of medium size. The new big Gai mine and flotation mill is situated north of Orsk. In Uzbekstan, copper is produced from mixed lead, zinc and copper sulphides and from the recently developed Kalmakyr porphyry. In Armenia, there are three smelters near the border which produce copper and molybdenum. The ore is obtained from the three mines situated nearby. Cobalt is mined and extracted at Khovu-Aksyn in the Tuva Republic. Most of the cobalt production however comes from the nickel mines in the north of Norilsk and from Petsame nickel mine, on the Kola peninsula. The deposits of pyrite in Ural are expected to yield appreciable quantities of copper, zinc, tellurium and selenium. Till now, Bulgaria has been supplying increasing quantities of copper ore to Russia. Domestic supply position of lead and zinc is meagre. They are mostly imported from North Korea, China, Persia, Bulgaria and Yugoslavia. Bauxite deposit is also scarce. The major supplies of bauxite are obtained from Greece and some quantity also imported from Hungary. Two of the biggest chromite deposits in the world are located in this country at South Kimpersai district near Akhtyubinsk in Kazakhstan and in Sarany areas in Southern Ural. This country possesses 11 times more reserves of iron ore when compared to the U.S.A. The deposits are mostly concentrated in Ukraine, Central Russia, Kazakhstan and Ural. The iron content averages 40-50% only. Russia has made a great stride in iron and steel production. From production of 51 million tonnes of steel in 1961, it reached 130 million tonnes in 1978. Mineable reserves of coal are estimated at 7765,300 million tonnes down to 1800 metres depth. Coal reserves are the largest known in the world. The reserves may be distributed as brown coal 37%; high grade blendable and coking coal 31% and anthracite 32%. Major coal basins are Kuzbass and Donbass.

The U.S.S.R. is the second largest producer of crude petroleum after U.S.A. Proved oil reserves at the end of 1964 are estimated at 4.5 billion tonnes (32.70 billion bbl). The total sedimentary prospective area of oil and gas comprises nine million sq km of which

¾th area is classified as favourable for oil and gas. New fields have been located in Siberia. There are more than 500 oil fields under production. Nearly 130 wells are producing oil from depth below 4200 metres. The chief producing centres of oil and gas are Azerbaidzhan, Ukraine, Kazakhstan, Uzbekistan, Turkmania, and RSFSR. In Azerbaidzhan, half the output comes from 400 offshore wells in the Caspian Sea. The average annual output is over 225 million tonnes which far exceeds the domestic requirements. Nearly half the production is exported out to several East European countries, India and Japan.

TABLE 2

World Reserves of Some Important Mineral Commodities

('000 tonnes)

Asbestos (*in terms of fibre content*)

Canada	41,658
Central Economy Countries	31,498
Other Market Economy Countries	14,225
South Africa Republic	6,096
United States	4,064
	97,541

Barytes

India	64,000
U.S.A.	60,000
Canada	3,600
France	3,600
Germany, West	6,400
Greece	3,600
Ireland	5,500
Italy	4,600
Mexico	3,600
Thailand	4,500

TABLE 2 (Contd.)

Morocco	5,500
Peru	3,600
Other Market Economy Countries	50,000
Yugoslavia	2,700
Other Central Economy Countries	30,000
	251,200

Bauxite

United States	40,000
Australia	4,500,000
Central Economy Countries	950,000
France	40,000
Greece	750,000
Guinea	8,200,000
Guyana	1,000,000
Jamaica	2,000,000
India	2,200,000
Other Market Economy Countries	6,700,000
Surinam	490,000
World Total	26,870,000

Chromite

Philippines	2,032
Rhodesia	609,630
South Africa Republic	2,032,100
India	17,270
Turkey	7,112

TABLE 2 *(Contd.)*

Other Market Economy Countries	33,530
Central Economy Countries	27,433
	2,729,107

Copper Ores *(in terms of fibre content)*

Australia	8,128
Canada	34,546
Chile	94,492
Papua New Guinea	10,160
Peru	35,562
Philippines	19,305
Poland	14,225
South Africa Republic	3,048
India	5,840
United States	94,492
U.S.S.R.	40,642
Zaire	28,449
Zambia	32,514
Other Market Economy Countries	83,316
Other Central Economy Countries	12,193
World Total	516,912

Iron Ore

Australia	17.500.000
Brazil	26.800.000

TABLE 2 (*Contd.*)

Canada	36,000,000
China, Peoples Republic	6,000,000
France	4,000,000
India	10,000,000
Libya	1,400,000
Sweden	3,300,000
Venezuela	3,200,000
Other Market Economy Countries	20,300,000
U.S.S.R.	108,800,000
United States	17,000,000
Others	14,000,000
	268,300,000

Lead Ore (in terms of metal content)

Australia	19,112
Canada	13,107
Mexico	4,572
Peru	3,556
India	3,120
Yugoslavia	3,048
Other Latin America	4,572
Other Market Economy Countries	30,887
Central Economy Countries	30,481
United States	28,855
World Total	141,310

TABLE 2 (*Contd.*)

Zinc Ore (in terms of metal content)

Australia	13,209
Canada	37,594
Mexico	3,556
Peru	9,652
United States	30,481
India	3,810
Other Market Economy Countries	59,947
Central Economy Countries	23,369
World Total	181,618

4

Future Sources of Minerals Supply

IN THE preceding chapter, the resources of individual countries in respect of important minerals have been described. The description given therein, and particularly Table 2, provides a clear picture of the resource position of different minerals by individual countries. An idea of the change in the level of production of important minerals and metals, as also the composition of the top three producers of each mineral or metal, can be had from Tables 5 and 6 wherein a comparative position of countries leading in world production during 1958 and 1982 has been given. A comparison of Tables 5 and 6 indicates that production of most of the minerals and metals in 1982 has either doubled or trebled as compared to 1958. As regards the composition of the top three producers of the world, it is observed that whereas there was no change excepting change in the *inter-se* position of the top three producers in case of apatite/phosphorite, asbestos, coal, aluminium and gold, the top three producers were completely displaced by new countries in the case of diamond, ilmenite, and magnesite. In case of other minerals and metals one or two countries were replaced by new countries in the top three positions.

In the case of bauxite, iron ore, nickel and copper as a result of several new discoveries a distinct change in the relative world supply position has been possible. This change has resulted largely due to massive efforts undertaken by the various multinational mineral exploration and exploiting companies including government agencies.

Some authorities apprehend that the known reserves of minerals may not last long and most of them will exhaust well within 100 to

TABLE 3

Life of the Major Mineral Deposits Based on the Study of the Club of Rome

Commodity	World* reserves	Total life based on static rate of consumption (years)
Aluminium	1.17×10^9 tonnes	100
Chromium	7.75×10^8 tonnes	420
Coal	5×10^{12} tonnes	2300
Cobalt	4.8×10^9 lbs	110
Copper	308×10^6 tonnes	36
Gold	353×10^6 troy oz.	11
Iron	1×10^{11} tonnes	240
Lead	91×10^6 tonnes	26
Manganese	8×10^8 tonnes	97
Mercury	3.34×10^6 flasks	13
Molybdenum	10.8×10^9 lbs	79
Natural Gas	1.14×10^{15} cu ft.	38
Nickel	147×10^9 lbs	150
Petroleum	455×10^9 bbls	31
Platinum Group	429×10^6 troy oz.	130
Silver	5.5×10^9 troy oz.	16
Tin	4.3×10^6 lg tonnes	17
Tungsten	2.9×10^9 lbs	40
Zinc	123×10^6 tonnes	23

*These reserve figures have since been greatly improved

200 years. This prediction has got some validity in respect of expendable minerals like petroleum, and non-expendable metals like tungsten, tin, lead, zinc and mercury. In the book "Limits to Growth" by the Club of Rome a great apprehension has been shown about the life of many minerals. They have calculated the life of various mineral deposits by dividing the known reserves by the total consumption at a static rate and came to the conclusions shown in Table 3.

The prediction of the Club of Rome is true with the existing knowledge of the reserves but the rate at which new reserves are being established coupled with fastly changing concept of mineral resource linked with technology and price, the analysis of the Club of Rome that the minerals supplies are running out may be erroneous. A better understanding of the widespread physical availability of mineral reserves and of potential substitutability of most minerals supplies have contributed in removing such fears. The innovative development of substitute material is a constant process where price and research interact to moderate the use of scarcer minerals. With the rise in prices, resources that were identified as sub-economic at a current price will gradually become exploitable reserves. More so our knowledge about the mineral resource is still scanty. A beginning in thorough search of minerals in a scientific way has started only from a little earlier than the middle of the present century. Hardly a year passes now without significant new discoveries. New possibilities of opening mines beyond the land in the sea-bed has given a new dimension in the availability of new mineral resource and reserves.

Resources beyond land

At present ilmenite, rutile, tin ore, diamond, sulphur, coal, petroleum, sand and gravel, lime shell and iron sands are being mined from the surface of the sea-bed. In our own country, ilmenite, rutile and associated zircon, monazite and sillimanite are obtained from the beach sands of Kerala and Tamil Nadu coast. Mining of coral limestone in Tamil Nadu for industrial use like manufacture of calcium carbide, and of calcareous sand from the sea-bed from Dwarka area in Gujarat, for cement, is going on for a long time. Malaysia, Indonesia, Thailand and Burma recover tin ore from dredging of the sea-bed. Now-a-days considerable quantity of diamond is recovered from the concessions granted on the sea-bed in South-West Africa.

About 50 % of the world production of elemental sulphur is obtained from the domes occurring off the Mexico coast by Frasch method. Coal is recovered from under the sea-bed in England. Production of magnetite sand from sea-bed is reported from Japan since 1971. In Philippines iron sand is being mined off Luzon coast for export to Japan. A number of petroliferous structures occurring off the coast are being profitably mined. Similar favourable structures exist off Indian coast on the West known as Bombay High and Aliabat structures. Lying on the ocean floor (see Table 4) are enough minerals to meet the demand of the modern world for centuries. A great number of valuable minerals have been located in abundance on the ocean floor. A partial list would include phosphate, diamond, minerals of cobalt, copper, platinum, silver, gold, tin, nickel, manganese, iron, barium, chromium, iodine and rare earth's metals.

The most spectacular discovery of the decade is finding of 'ferromanganese' commonly known as 'manganese nodules' littered all over the oceanic floors generally below 4,000 metres depth. Scanning of the oceanic floors have shown concentration of manganese nodules at many places viz., (i) Kelvin Seamount, (ii) Blake Plateau, (iii) Red Clay Province, and (iv) Mid-Atlantic ridge in the North Atlantic region; Madagascar basin in Indian Ocean and Crozet Basin in South Indian Ocean but none of the concentration at these places have been found of economic importance because of low metal values. Most prolific concentration of manganese nodules with high metal values have been noted in North Pacific between Clipperton and Clarion Fracture zones situated between 139°W to 141°W and 4°N to 16°N. Nodules of this zone contain 1.3 to 1.6% Ni; 1.16% Cu; 0.23% Co; 25% Mn besides several other metals like iron, vanadium, molybdenum, silver, lead and platinum group of metals. Precise information about the reserves is not available but it is estimated that it may run several billion tonnes. It is estimated that 1 million tonnes of sea-bed manganese nodules may provide about 13,000 tonnes of nickel, 1000 tonnes of copper, 2500 tonnes of cobalt and 275,000 tonnes of manganese besides other metals. It appears nickel and copper are the mainstay of nodule mining.

Nodules from other areas have low metal values as said before. Average analysis of manganese nodules from different areas is given in Table 4.

Iron content is quite variable and it varies from 7 to 11%.

Mine site parameters have been stated to be (i) 30,000 sq km of

TABLE 4

Average Analysis of Manganese Nodules (Percentage in Dry Wet)

	Mn	Ni	Cu	Co
N. Pacific (silicious ooze)	24.5	1.28	1.16	.23
N. Pacific (red clays)	18.2	.76	47	.25
S. Pacific	14.6	.41	.13	.78
S. Pacific	15.1	.51	.23	.74
N. Atlantic	14.2	.38	.15	.34
S. Atlantic	18.0	.48	.15	.31
Indian Ocean	14.7	.50	.19	.28

an area, (ii) should have 10 dry kg of nodules per sq metre, and (iii) containing 1.3 to 1.4% nickel; 1.1 to 1.2% copper and 0.2 to 0.25% cobalt besides other metals. Alternatively (i) 10 dry kg of nodule per sq metre; (ii) combined copper/nickel content of 2.25% to 2.4% nickel with cut-off grade not less than 2% combined nickel, copper and cobalt; and (iii) enough material to mine at the rate of 3 million tonnes for 25 years.

The discoveries of the oceanic bed resource have greatly enhanced the availability of many valuable metals. It has been roughly estimated that the resources available in manganese nodules are several times more than the land resources in respect of metals content in these nodules—manganese 4000 times, nickel 1500 times, copper 150 times and cobalt 5000 times.

Recent Discoveries on the Land Surface

Iron and steel, aluminium, manganese, chromium, lead and zinc, cement, coal and sulphur are the major metals, minerals, or commodity that are required in bulk for daily use. Other minerals and metals though available comparatively lesser in quantity are also required in lesser quantities. Uptil now, there has been no report regarding the total exhaustion of mineral except for one viz., cryolite. Rather new reserves have been added year after year by the diligent efforts of the geologists. In the case of tin tungsten, whose reserves are regarded fastly depleting, have also improved their position by significant increase in output from China which has hitherto remained an iron curtain. China is regarded to possess potentially rich deposits of tungsten and tin. Moreover, recycling and substitution of these two metals have already begun by aluminium and molybdenum respectively, thus, enhancing their lifecycle. Very recently, large reserves of cinnar (mercury ore) have been found in the Russian territory opposite Alaska of the U.S.A. Moreover mercury is not an indispensable mineral.

Many high grade iron ore deposits averaging 65% Fe have been discovered in readily accessible areas of Western Australia, Brazil, Chile, Peru, Liberia, Sierra, Leone and Gabon. A new discovery has been made in Yukon, Canada, with possible reserves of 20,000 million tonnes with 46% Fe. Bolivia has estimated reserves of 50,000 million tonnes of all grades of iron ore. Russia claims reserves of 50,000 million tonnes of iron ore much of which is located in Western Siberia. In case of manganese ore, Brazil, Chile, Guyana, Bolivia, in Latin America; Mexico in Central America; Cuba in Caribbean sea; Gabon, Zaire, Congo, Morocco in Africa, Australia and Philippines in Asia have emerged as important producers in recent years. In nickel supply new sources have been established in New Caledonia and British Solomen island; the Dominican Republic and Guatemala in Caribbean sea and Nigeria, Botswana and Republic of South Africa in Africa. The ore is found to occur as nikeliferous laterite resulting from the weathering of ultrabasics, serpentine, dunite, etc. Typical weathered laterites contain approximately 1.0 to 1.5% Ni, 2% Cr_2O_3, 40% Fe and a very small percentage of cobalt.

Discovery of vast and extensive deposits of bauxite of the order of over 3000 million tonnes with 50% Al_2O_3 in Weipa area in the northern Queensland, Australia and nearly 2000 million tonnes of bauxite of the same grade on the east-coast covering Orissa and A.P. in India has vastly changed the supply position of bauxite to the world market for aluminium production. Bauxite deposit more or less of the same magnitude are reported from Guinea, Africa. New sources of chrysolite asbestos have been established in the New Zealand and Australia. The estimated reserves in New Zealand

(South Island) are of the order of 100 million tonnes in terms of fibres. This estimate was announced by Kennecott Copper Corporation. If the estimate be correct, this will ease to a great extent the supply position of chrysotile asbestos in the world market.

In the case of non-ferrous metals especially the copper, several major discoveries of porphyry copper* deposits on different parts of the globe have brightened the prospects of continued supply of this vital metal for considerable long time. During the last one decade porphyry copper deposits have been discovered and mined all the world over. The copper belts of Arizona, New Mexico and California in the U.S.A. and the northern Mexico are the well known world's great copper porphyry provinces. Another major copper porphyry region is in the Kazakhstan and Uzbekstan in the U.S.S.R. Yet another porphyry belt sweeps north westward from Sarcheshmeh, Central Iran through Turkey, Yugoslavia, Bulgaria and Rumania. In addition, there are three other major areas where known porphyries are clustered, they are: British Columbia and the southern Yukon in Canada, Andes mountain of South America stretching from central Chile through Peru into southern Equador and lapping over into western Argentina; and the great Pacific Fire belt from Luzon, Philippines arcing south east to Guadalcanal in Soloman islands. The newest discoveries of porphyry copper deposits are that of Bougainville in Papua and New Guinea, Cerro-Colorado deposit in Panama and two porphyry deposits of Columbia. Porphyry-coppers have become powerful source of valuable metals like molybdenum, gold, silver, cobalt, selenium, nickel and several other rare and precious metals in addition to copper. It is observed that all porphyry coppers by and large contain molybdenum as an associate metal unlike cobalt with non-porphyry deposit as is found in Zambia and Zaire.

The known mineable reserves of cobalt in the world, in addition to seabed resources, today are estimated between 3 and 5 million tonnes. Besides this, there are further 4 to 5 million tonnes classified as potential reserves associated with nickeliferous deposit, which are not economically recoverable. Such deposits are found in Cuba,

*By porphyry copper, it is generally meant large deposit, particularly with respect to its horizontal dimensions, containing uniformly disseminated copper minerals throughout the mass with low average copper content. Geologically, the porphyry coppers are igneous complexes intruding the wide variety of rocks.

TABLE 5

World Production of Important Minerals In Major Producing Countries during 1958

(In tonnes unless otherwise specified)

Mineral	Total world production	Largest producer			Second largest producer			Third largest producer		
		Country	Production	%	Country	Production	%	Country	Production	%
1	2	3	4	5	6	7	8	9	10	11
Apatite/phosphorite	34,870,000	U.S.A.	14,879,000	43	Morocco	6,238,000	17	U.S.S.R.	6,200,000	16
Asbestos	2,020,000	Canada	925,331	46	U.S.S.R.	550,000	27	S. Africa	175,366	9
Barytes	2,500,000	U.S.A.	486,287	19	Germany, West	409,105	16	Canada	201,329	8
Bauxite	20,700,000	Jamaica	5,722,000	28	Surinam	2,941,000	14	U.S.S.R	2,710,000	13
Chromite	4,050,000	U.S.S.R.	880,000	22	S. Africa	696,057	17	S. Rhodesia	618,841	15
Coal & Lignite (million tonnes)	2,400	U.S.S.R.	488	23	U.S.A.	383	15	China	270	11
Diamond (million carats)	22	Belgian Congo	16	72	Ghana (gold coast)	2	9	Sierra Leone	1	6

TABLE 5 (Contd.)

1	2	3	4	5	6	7	8	9	10	11
Felspar	1,025,000	U.S.A.	469,738	46	Germany, West	187,504	18	France	68,900	7
Fluorspar	1,760,000	U.S.A.	313,513	18	Mexico	244,982	14	U.S.S.R.	180,000	10
Gypsum	34,660,000	U.S.A.	9,600,000	26	U.K.	4,470,000	12	Canada	3,975,000	11
Ilmenite	1,710,900	U.S.A.	563,338	33	India	346,080	20	Norway	233,585	14
Rutile	102,750	Australia	92,900	90	U.S.A.	7,400	7	Senegal	1,157	1
Iron ore	397,036,000	U.S.S.R.	87,400,000	22	U.S.A.	67,947,000	17	France	58,516,000	15
Magnesite	5,900,000	Australia	1,346,133	23	U.S.A.	492,982	8	Yugoslavia	246,032	4
Manganese ore	13,049,000	U.S.S.R.	5,915,000	45	India	1,377,602	11	S. Africa	934,097	7
Mica (blocks and splittings)	11,000	India	8,800	—	Brazil	—	—	U.S.A.	—	—
Petroleum (crude) (million tonnes)	905	U.S.A.	335	37	Venezuela	137	15	U.S.S.R.	111	12
Sulphur	6,500,000	U.S.A.	4,645,577	71	Mexico	1,236,929	19	Japan	177,175	3
Tungsten	63,500	China	22,000	35	U.S.S.R.	8,300	13	North Korea	4,400	7

Ferrous

	Country		%	Country		%	Country		%
Crude steel	U.S.A.	269,000,000	28	U.S.S.R.	54,000,000	20	Germany Fed. Rep.	22,425,000	8
Cobalt	Belgian Congo	14,600	49	U.S.A.	2,012	14	Rhodesia	1,774	12
Molybdenum (pounds)	U.S.A.	56,500,000	73	U.S.S.R.	9,300,000	16	Chile	2,972,000	5
Nickel	Canada	245,000	57	U.S.S.R.	55,000	22	Cuba	19,782	8

Non-ferrous

	Country		%	Country		%	Country		%
Aluminium	U.S.A.	3,890,000	40	Canada	1,565,557	17	U.S.S.R.	605,000	15
Copper	U.S.A.	3,740,000	26	Chile	979,329	14	U.S.S.R.	470,000	13
Lead	Australia	2,520,000	15	U.S.S.R.	366,252	13	U.S.A.	267,377	11
Tin	Malaya	152,400	25	Indonesia	38,458	15	China	23,000	15
Zinc	Canada	3,350,000	13	U.S.A.	24,116	12	U.S.S.R.	400,000	12

Precious metal

	Country		%	Country		%	Country		%
Gold (in Troy ounces)	S.Africa	40,400,000	44	U.S.S.R.	17,665,739	25	Canada	4,537,007	11
Silver (in Troy ounces)	Mexico	236,800,000	21	U.S.A.	47,589,528	16	U.S.S.R.	25,000,000	11

TABLE 6

World Production of Important Minerals in Major Producing Countries during 1982
(Production in thousand tonnes unless otherwise indicated)

Mineral	Total world production	Largest producer Country	Production	%	Second largest producer Country	Production	%	Third largest producer Country	Production	%
1	2	3	4	5	6	7	8	9	10	11
Apatite/Phosphorite	122,633	U.S.A.	37,414	30	U.S.S.R.	26,100	21	Morocco	17,754	14
Asbestos	4,400	U.S.S.R.	2,180	54	Canada	834	21	S. Africa	212	5
Barytes	7,400	U.S.A.	1,647	22	China	900	12	Moroocco	540	7
Bauxite	77,200	Australia	23,625	31	Guinea	11,827	15	Jamaica	8,158	11
Chromite	7,700	U.S.S.R.	2,450	32	S. Africa	2,164	28	Albania	875	11
Coal & Lignite (million tonnes)	3,987	U.S.A.	756	19	U.S.S.R.	718	18	China	666	17
Diamond (million carats)	39	U.S.S.R.	11	27	S. Africa	9	23	Botswana	8	19
Felspar	3,800	Italy	783	21	U.S.A.	558	15	Japan	461	12
Fluorspar	4,500	Mongolia	660	15	Mexico	631	14	U.S.S.R.	540	12
Gypsum	63,500	U.S.A.	9,560	15	Belgium	6,102	10	Canada	5,987	9
Ilmenite	5,300	Canada	1,737	33	Australia	1,149	22	S. Africa	647	12
Rutile	341	Australia	221	65	Sierra Leone	48	14	S. Africa	47	14
Iron Ore	778,000	U.S.S.R.	244,411	31	Brazil	93,159	12	Australia	87,694	11

Magnesite	11,200	U.S.S.R.	2,150	19	China	2,000	18	Korea DPR	1,850	16
Manganese Ore	24,300	U.S.S.R.	9,821	40	S. Africa	5,216	21	Brazil	2,341	10
Mica (all grades)	179	U.S.A.	96	54	U.S.S.R.	48	27	India	13	7
Petroleum (crude) (million tonnes)	2,700	U.S.S.R.	613	23	U.S.A.	449	17	Saudi Arabia	324	12
Sulphur	50,776	U.S.A.	9,787	19	U.S.S.R.	9,640	19	Canada	6,281	12
Tungsten (tonne) (mine production-metal content)	46,000	China	12,500	27	U.S.S.R.	9,000	20	Australia	2,618	6
Ferrous										
Steel (Crude)	631,000	U.S.S.R.	147,165	23	Japan	99,548	16	U.S.A.	67,655	11
Cobalt (tonne) (Refined)	18,700	Zaire	5,573	30	U.S.S.R.	4,300	23	Zambia	2,446	13
Nickel (tonne) (Refined)	615,000	U.S.S.R.	190,000	31	Japan	87,300	14	Canada	61,544	10
Non-ferrous										
Aluminium	13,900	U.S.A.	3,274	28	U.S.S.R.	2,400	17	Canada	1,065	8
Copper (Refined)	9,600	U.S.A.	1,694	18	U.S.S.R.	1,490	15	Japan	5,107	11
Lead (Refined)	5,200	U.S.A.	1,032	20	U.S.S.R.	800	15	FRG	350	8
Tin (tonne) (Refined)	210,000	Malaysia	62,836	30	Indonesia	29,755	14	Thailand	25,479	12
Zinc (Refined)	5,900	U.S.S.R.	1,050	18	Japan	662	11	Canada	511	9

TABLE 6 (*Contd.*)

1	2	3	4	5	6	7	8	9	10	11
Precious Metal										
Gold Mine Production in kg. (metal content)	1,300,000	S. Africa	662,526	51	U.S.S.R.	266,000	20	Canada	64,735	5
Silver Mine Production in '000 kg. (metal content)	11,727	Peru	1,692	15	U.S.S.R.	1,595	14	Mexico	1,550	13

SOURCE: 1) World Mineral Statistics, 1979–83, British Geological Survey.
2) USBM 1983, Vol. 1.

Indonesia and also in Zaire and in India. Presently, cobalt is obtained chiefly as a byproduct of copper and nickel production. Zaire and Zambia are the principal sources of cobalt to the world market where it is obtained as a byproduct to copper production. It is estimated that at present less than 25% of the mineable reserves are being exploited. Another 25% of these reserves are under study and over 50% of the mineable reserves are receiving no attention whatsoever. Therefore, future supply of this strategic metal is quite secure. While on the land, an identified cobalt resources are placed at 10 million tonnes, it is 500 million tonnes on the sea-bed.

Large deposits of lead-zinc ores have been reported in the Cornwall Islands in Arctic sea and in Black Agel area in Greenland. Both are quite good deposits having combined metal content of 20%. One of the world's biggest and largest flourine-tin deposit is reported from Seward Peninsula, Alaska with 20% fluorite content.

In the energy minerals like coal and petroleum, there is abundant supply in the case of former. The world's coal reserves are estimated at 4000 million tonnes as against the current production of about 2500 million tonnes per annum. With this rate, the estimated reserves are likely to last for nearly two million years. There is no world shortage of petroleum as well. But in this case most of the resources are concentrated in the Middle East countries which accounts for 60% of the world trade in this commodity. The world reserves of oil and gas are estimated at 640,000 million barrel and 1.9×10^{15} cu. ft. respectively. The ratio of world offshore to onshore 'reserves' is placed nearly of 1 : 5 but that of 'resources' is estimated to be equal. This shows very clearly the great potential for undersea oil and gas. Known reserves are sufficient to last for 60 years with the current rate of consumption. Especially after the Arab-Israel wars, fivefold increase in the posted price of the crude by OPEC and certain embargo put forth by them on export, several advanced as well as developing countries in the world intensified search for petroleum to meet the challenge of energy crisis. As a result a number of oil producing countries reactivated the producing wells and increased the production considerably. Two most tangible discoveries which can be regarded to be of great magnitude and significance were reported by China and the U.K. The offshore drilling programme carried out by the Chinese have located vast reserves of petroleum in offshore Po Hai Bay, from where production has already started. Chinese annual production is reported to be over 50 million tonnes.

They have entered into the world market for sale of petroleum. One of the most spectacular discovery has been made by the U.K. and also by Norway in their respective continental shelves of the North Sea. It is estimated that the U.K. alone can meet a minimum of 15% requirements of petroleum and its products of entire Europe from their newly discovered oil field in the North Sea.

Technological Developments

The distinction between high grade and low grade is now vanishing. The only distinction remains these days is whether the ore is amenable to beneficiation or not. The considerable improvement in the mineral beneficiation and metallurgical techniques has added greatly to our known reserves and increased the life of a deposit. Only less than two decades ago, a copper deposit containing less than 2% Cu (on an average) was considered not workable, but now copper ore having cut-off grade 0.5% Cu is being worked profitably. In India, iron ore deposit containing less than 58% Fe is not mined, but at the same time France, West Germany and England are mining low grade colitic iron ore containing only 30% Fe and using it after improving the grade mainly adopting dry high-intensity magnetic separation. France is mining annually nearly 50 million tonnes of such low grade ore from Lorraine area for her steel mills. Now nearly 200 million tonnes of iron ore pellets are being produced utilising low grade siliceous iron ore called taconite containing only 30% Fe and iron ore fines which were considered a waste only a few years before.

The events witnessed in seventies provide a picture of unprecedented growth in the mineral production and discoveries, mining and utilisation of lean ores which were hitherto discarded as uneconomic. The present decade, it appears, will continue to witness over supply position of iron ore, manganese ore, chromite, bauxite, titanium, nickel, molybdenum and sulphur. The position regarding copper, lead-zinc are considered quite satisfactory for some time to come. The worldwide shortage of mercury, tin, tungsten and to some extent of chrysotile is felt. Restricted supply position of fossil fuel is likely to continue unless some alternative source has been found by individual countries having inadequate supply position in this vital commodity. Surprisingly, no new discovery of uranium ores except that of Australia, Niger and South Africa have been reported in recent years. It appears there is world shortage of this vital fissionable material. It might be that because of the strategic reason, most of the discoveries are not being reported by the respective governments.

TABLE 7

Recoverable Minerals from the Marine Environment

Mineral	Geographical location	Water depth in metre
Sand and gravel	Atlantic and Pacific coasts—U.S.A.	30
Glass and foundry sand	Atlantic and Pacific coasts—U.S.A.	60
Magnetite	Australia, India, Japan, Pacific coasts—U.S.A.	30 to 120
Glauconite	Pacific coasts—U.S.A.	9 to 1800
Rutile	Australia, Atlantic coast—U.S.A.	30
Zircon	Australia	30
Cassiterite (Tin ore)	Malaysia, Indonesia, Thailand, Alaska, Great Britain	120
Silver	Pacific and Alaskan coasts—U.S.A.	120
Gold	Pacific and Alaskan coasts—U.S.A.	120
Platinum	Pacific and Alaskan coasts—U.S.A., Greenland	120
Diamonds	S.W. Africa	60
Manganese ore	Atlantic and Pacific Oceans, Mediterranean Sea	1200 to 5400
Phosphorite	Atlantic and Pacific coasts—U.S.A., Australia, Africa, Indian Ocean, India off Andhra Pradesh coast	30 to 1200
Coal	Canada, Great Britain, Japan	120
Ilmenite, Monazite	S. India, Ceylon, Australia	0 to 60
Shell	Gulf and Pacific coasts—U.S.A., Iceland	30
Sulphur	Gulf coast—U.S.A.	30
Calcareous sea sand	North West Coast, India	9

Judging from the world-wide efforts in exploration, discoveries made and new prospects established, the future sources of mineral raw material, besides obtainable from ocean bed, can be listed as follows:

Bolivia
Peru
Australia   } Antimony
China

Canada
New Zealand
Rhodesia
Republic of South Africa   } Asbestos
Swaziland
U.S.S.R.

Guyana
Surinam
Caribbean Sea Islands
Greece
India
Indonesia   } Bauxite
Malaysia
Australia
Guinea

Brazil
India
Afghanistan
Argentina   } Beryl
Republic of South Africa

Peru
Bolivia   } Bismuth

Turkey
U.S.A.   } Borax

India
Philippines
Turkey
Rhodesia
Republic of South Africa   } Chromite
Albania
U.S.S.R.

Congo
New Caledonia
Zaire
Zambia
Morocco
Canada
} Cobalt

Chile
U.S.A.
U.S.S.R.
Peru
Zambia
Zaire
Rhodesia
Uganda
Mexico
Canada
Australia
Fiji
Bougainvalle Island
Malaysia
} Copper

Congo
Angola
Tanzania
S.W. Africa
Sierra Leone
Ghana
Venezuela
} Diamond

Ceylon
Malagasy
} Flake graphite

Mexico
U.S.A.
Thailand
U.K.
Republic of South Africa
China
} Fluorspar

Republic of South Africa
Canada
Rhodesia
} Gold

India
Sri Lanka
Malaysia } Ilmenite
Australia
U.S.A.

India
Australia
Liberia
Brazil
Chile
Peru
Sierra-Leone
Gabon } Iron ore
Bolivia
Brazil
Canada
Sweden
Mauritania
Philippines
Sudan

Brazil
Canada
Rhodesia } Lithium ores
U.S.A.
Mexico

U.S.A.
Australia
Peru
Morocco
Algeria } Lead and zinc
Canada
Tunisia
Congo
Iran

India
Rep. of South Africa
Gabon
Brazil } Manganese ore
Equador
Ghana
U.S S.R.

Italy Spain	} Mercury
India Brazil Republic of South Africa Tanzania Malagasy U.S.S.R.	} Mica
Chile Canada U.S.A. U.S.S.R. Guyana Rumania	} Molybdenum
Canada Cuba Republic of South Africa New Caledonia British Solomon Islands Niger	} Nickel
Iran Iraq Saudi Arabia Kuwait Abu Dhabhi Qatar Bahrain Islands Colombia Libya Canada Rumania U.K. U.S.S.R. Venezuela Malaysia Indonesia	} Petroleum
Republic of South Africa U.S.S.R. Canada	} Platinum and its group of metals

U.S.A.
Morocco
Jordan
Egypt
Tunisia   } Rock phosphate
Algeria
Spanish Sahara
Pacific Ocean Islands

Australia
Sierra Leone   } Rutile

Mexico
U.S.A.
Canada   } Elemental Sulphur
France
Poland

Malaysia
Indonesia
Thailand
Burma
Laos
Niger
Bolivia   } Tin
Zaire
Rwanda
Australia
China
U.S.S.R.

Portugal
Thailand
South Korea
Bolivia
U.S.S.R.
U.S.A.   } Tungsten
China
North Korea
Burma
Rwanda

U.S.A.
Republic of South Africa
Niger
Congo
Gabon
Namibia
U.S.S.R.   } Uranium ores
Canada
Yugoslavia
India
China
Australia

5

Mining Laws in
Various Lands

MINING LAWS are enacted to solve the technical, economic, political and legal problems involved in the development and exploration of the mineral deposits. The technical problems are mainly
physical connected with climatic, topographical and other factors
such as the treatment of complex ores, rock pressure, temperature
gradient, ground water condition, and consequent complications in
mining, processing, haulage, methods of deep mining and off-shore
mining. The other problems are interrelated with the stage of industrial development, attitude of government towards mineral development, nature of mining, mineral trade controls, international relations and legal aspects of title to mineral properties to be worked.
In most of the countries the mineral rights vest in the state which
leases out areas to its nationals as well as to foreigners. Only few
countries have no such restriction imposed on foreigners of acquiring
mineral rights, and developing, exploiting and disposing of minerals, as are not applied to nationals engaged in similar pursuits.
Mining being capital intensive enterprise, care is always taken to
pass on the titles to such individuals, firms, companies or corporations who are able to finance exploration and mining activities in
a systematic way conforming to the principles of conservation and
development.

Some countries follow the policy of opening mines wholly controlled by the Central or Provincial Governments or both and also by
joining hands with the private enterprise, besides the ingenuity of
individual entrepreneurs. The philosophy behind such an attempt
is to make the private and public sectors supplement each others'

effort in development. The State owned agencies are functioning in Finland, France, Norway, Sweden, the U.K., the Republic of South Africa, Egypt and many other countries. The LKAB, the largest iron ore mining company in Sweden is State-owned. The Mineral Exploration Corporation in Quebec, owned by the Quebec Government, is the first venture opened in Canada under State control. In the recent years there has been increasing involvement of the federal govt. of Canada according to the National Energy Policy in the exploration activities. Coal is worked in the U.K. by the State for more than two decades. The setting up of the American Wartime Strategic Mineral Development Programme body is another example of the State-owned organisation for mineral development. A number of such State-owned organisations exist in India.

Other considerations of significant importance are technical knowhow and finance. A country may have adequate mineral resources but not sufficient finance and advanced technological know-how in exploration, mechanisation, processing and ore handling methods. It is but natural for a country, in such circumstances, to seek help and depend upon foreign aid. Mutual cooperation and collaboration have proved quick results and, thus to accelerate pace of development a number of countries have modified their legislations to suit foreign investment. In most economies joint venture, by way of technical collaboration and/or financial participation, are preferred because of their distinct and mutual advantage to both the collaborators. Especially in the case of developing economies, technical and financial collaboration is a great boon as it helps to build expertise over a period. While the local partner furnishes his intimate knowledge of local conditions and requirements, the foreign collaborator blends his technical know-how with the market requirements. Such joint ventures in the equity ratio or in the ratio 51: 49 have become the common feature i.e., the aiding country invests to the context of 49 per cent in the share capital either by way of investment in terms of money or supply of experts, machinery, designs and other details required for the projects. The climate for foreign investment depends on the general background of economic and political conditions prevailing as well as political relation with the aiding country in addition to laws and regulations applying directly. Political relationship is an important factor governing such prospects of collaboration. Factors such as manpower, material resources, market conditions, transport facilities, restrictions on ownership, management and production, extent of taxation and royalty, stability or instability of exchange rates, each play a part. The attitude towards foreign investment in general may be said to be of restricted and conditional welcome. Most of the developing countries have felt the necessity of greater governmental control in the mining enterprise, as it is found in most of the South American, African countries.

The fear of the newly independent countries or under-developed and developing countries is real. During the colonial regime, the main purpose of the ruling nations for opening mines in the countries under their control was exploitation i.e., getting the mineral raw materials at far lesser cost than the then existing international prices. Such situation has persisted for a long time before the World War II. The situation thereafter changed considerably. After the World War II, the pressure of political sovereignty resulted in a greater control over the mining rights by the domestic government. Although multi-national private mineral companies operating in developing countries have changed, rather went on modifying their attitude and policies in relation with the host government, local economies and local people providing increasing opportunities of employment to local talents, still attitude of the host government has remained that of suspicion.

In the context of conflicting situation, at the same time, an anxiety of the developing countries to leap forward for increasing the standard of living of their people, they continue to make an effort to attract foreign capital for rapid development. Three types of code of investments are generally fostered by the developing countries for attracting foreign capital: (i) providing equity participation and expansion of the existing facilities, (ii) some preferred a barter system capital, in return for manufactured and semi-manufactured products, and (iii) outright payment for the import of technical know-how and machinery through the world financial institution.

Each country follows its own specific code suiting its political conditions. Most, however, require that the State should own at least major equity shares. Tax is another irritant which deter most of the investing companies. Therefore, most of the developing countries are evolving or have evolved a balanced taxation system by suitably restructuring the system for sharing returns equitably. As a matter of fact, tax structure has been the major concern and negotiating item by the investors with the host government. How to use foreign capital and technology and at the same time not to lose control over production and pricing remain the major objective in the policy decision of the developing countries.

Depending upon the above factors, the laws and regulations of one country vary greatly with another. In the countries having socialistic form of government, the framing of laws regulating grant of leases except for safety measures is seldom taken into account. It is because the State owns all the enterprise, and in such cases the elements of competitiveness and costs factor are lost in oblivion and not kept at the helm of considerations. It is true for all socialist countries whether it be Niger or like Tanzania in Africa.

Four stages are generally recognised in the development of mineral resources. The first stage may be described as a preliminary prospecting or reconnaissance. At this stage a vast track of land is rapidly examined to demarcate favourably area for further examination. The preliminary prospecting can be rightly described as an operation of narrowing down the area of interest. The second stage may be described as prospecting or exploration. During this stage, an effort is made to further narrow down the area of existence of the overbodies and to determine its physical and chemical characteristics, dimensions, grade, reserves etc. This stage help in selection of blocks for opening up mines. In the case of bigger areas, it becomes essential to further evaluate the deposits for investment decision. This may be designated as a feasibility studies. (See Ch. 12 also.) This operation may be rated as a third stage of mineral exploration. In the fourth stage which is a last stage in which the actual mining and exploitation takes place. The nature of operation, the period of validity and area limitations differ from country to country and for minerals to minerals depended upon individual country's socio-economic conditions.

In the following paragraphs the rules and regulations prevalent in selected countries are described to provide a clear and comprehensive idea of the mining conditions available in different parts of the world.

MINING LAWS IN IMPORTANT COUNTRIES

African Countries

Algeria

Foreign participation is welcome on conditional basis. Most of the mining rights are held by the French. Under the "Ministerial Decision" (Decree) of 16 July 1964, the Government has restricted transfer of currency from Algeria to France. Accordingly foreign oil companies must situate at least 50% of their financial activities in Algerian banks, and thus have no right to transfer abroad any sum that would exceed the equivalent of 50% of their gross turnover in Algeria. As for other companies, they must situate their entire activities in Algerian banks and have no right to transfer any fund abroad without the permission of the Government.

Ghana

Investment of foreign capital is welcomed. The mining laws are governed by the Mines and Minerals (Conservation and Development) Act, which was passed in March 1965, repealing the old Mines and Mineral Ordinance. The new act provides for regulation and supervision of mines and minerals by the new Ministry of Mines and Mineral Resources. It authorises the Ministry to inspect mines and mineral prospects, cell for returns on production and quality of minerals, fix fees for mining leases and prospecting licences, and to take cores and specimens from any mine or prospect. These provisions are similar to Indian laws.

Mozambique

The mining leases are granted to her own nationals or to British subjects only.

The Republic of South Africa

The laws in operation are the Reserve Mineral Act, 1926; Precious Stones Act, 1927; Base Minerals Amendment Act, 1942; National Oil Act, 1942; Mines and Works Act, 1956; and the Atomic Energy Act, 1948. The mines come under the jurisdiction of the Central Government and not the Provincial Government. The laws are administered by the Department of Mines. Foreign participation is freely allowed. In fact most of the important mines are worked by firms having world-wide mineral interests. The Government also owns several mines. The Alexander Bay Diamond Digging operations and Iron and Steel Corporation are wholly State owned.

Zambia

Zambia, for giving a great boost in mineral exploration and development, brought out new legislative changes by enacting Mines and Minerals Act of 1970. The new Act provides the foreign companies to participate in mineral exploration and mining. Prior to this Act, prospecting and mining rights were largely vested in the two mining companies viz. Anglo-American Corporation of South Africa Ltd. and Roan Selection Trust Ltd. These two companies prior to 1970 Act had held practically all the mineralised areas in Zambia on perpetual lease. Under the new Act, the areas hitherto not prospected or very little developed by the two companies have been acquired and earmarked for granting, prospecting and exploration licences and mining leases to national and foreign companies. For those two companies also the areas under special grants have been reduced and converted to 25 years lease with a renewable clause. Under the new Act no limit has been kept for area under prospecting licence and is valid for 4 years. Exploration licences are valid for 3 years, with certain renewable rights. Maximum area which one can acquire is 10 sq miles. Under the Act, the Government has kept an option to take up to 51% of the equity in any future mine by paying its share of the prospecting and exploration costs. New tax incentives have also been announced, as a result a large number of foreign companies have been attracted to mining development programmes of Zambia.

Zimbabwe

The Government owns all mineral rights and leases out its nationals, and also offers attractive terms to foreign firms. Most of the Rhodesian chromite and asbestos mines are held by the U.S. and British interests.

The Sudan

The mineral rights vest in the Government. Mining and quarrying are regulated by the Mines and Quarries Ordinance, January, 1959 and the Mines and Quarries Regulations framed thereunder. The Ordinance of 1959 repeals the Mining (Prospecting Licence) Ordinance of 1899. A Committee of Ministers are charged with the responsibility for development and use of mineral and quarried material resources. A Board constituted by the Committee regulates the provisions of the Ordinance. The Board on behalf of the Government issues four types of licences and leases, namely (i) prospecting licence, authorising the holder to prospect for minerals, (ii) mining licence, authorising the holder to enter upon lands for the purpose of getting minerals, (iii) mining lease, authorising the lessee to get minerals from the area comprised in the lease, and (iv) quarry licence, authorising the holder to get quarried from a quarry. Quarried material means rock, stone (including limestone), gravel, sand and clay.

Foreign participation is allowed provided an alien, or a company incorporated outside the Sudan, shall incorporate a company in the Sudan for the purpose of receiving the grant and exploiting the mining lease.

Asian Countries

Indonesia

According to the Constitution, mineral resources are national property controlled by the State and the State has the exclusive mining rights. On certain conditions stipulated by the Mining Laws the Government may grant a so-called "Kuasa Pertambangan", or authorisation to mine, to a State enterprise, a national private business, or an Indonesian citizen. As for foreign participation in mining, article 8 of Law No. 1, 1967, stipulates that foreign capital investment in the field of mining shall be based on a co-operation with the Government on the basis of a working contract, or in other form, in line with the existing regulations.

Thus, foreign companies interested in mining in Indonesia must enter into a "contract of work" with the Government or with the holder of a "Kuasa Pertambangan". The principle underlying this system is that the foreign party will be conducting all stages of operations, which may include the marketing of products, for and on behalf of the Government. Mining rights remain in the hand of the Government and the foreign participant retains the status of contractor. Tools, equipment, and installations are recognised as contractors property and all risks have to be borne by the contractor. A "contract of work" for mining has to be signed by the foreign contractor and the Minister of mines, the latter representing the Government. To promote foreign capital investment in general, Law No. 11, 1967, stipulates also mitigation on taxation and other levies. Exemption from import duties are granted to foreign investors for the import of machineries, working tools, and instruments needed to run the enterprise. The Government gives due consideration and priority to projects which are of great economic importance and additional mitigations and allowances are granted to foreign enterprises willing to invest capital in such projects.

Thailand

The Government of Thailand provides a special incentive to foreign companies for mining tin in deeper off-shore areas. If the depth of off-shore area is less than 60 metres, the foreign companies will have to part with at least 51 per cent of the share to the Thailand Government. If the depth of water increases more than 60 metres, 100 per cent foreign owned companies can be allowed to operate provided that no less than 60 per cent of the issued share are sold to the Thai nationals within 10 years of the commencement of the operation.

Iran

Iranian laws provide very lucrative terms to foreign investors. Any investor may take out of Iran all the net profit each year and may also withdraw capital investment at any time. The only provision is that should the Government of Iran not have funds available, the investor may withdraw permissible goods to the value of the net profit and/or his investment, whichever it is he wishes to take out.

Pakistan

The mining laws in Pakistan are controlled by dual system, those of the Central Government and those of the State Governments. The mining laws in operation are the Regulation of Mines and Minerals Act of 1958 and the Oil Fields (Federal Control) Act of 1948. The Act of 1958 is the superseded version of the Oil Field and Minerals Development (Federal Control) Act of 1948, leaving the latter to control the affairs of oil only. Under these two acts two sets of concession rules namely the Pakistan Mining Concession Rules, 1960 and the Pakistan Petroleum (Production) Rules, 1949 have been issued. Under the 1958 Act, the Central Government assumes jurisdiction over all minerals except those controlled by the State Governments, atomic energy minerals, and oil and gas. According to Pakistani laws foreigners can acquire lease in collaboration with Pakistani citizen with equity shares limited to a maximum of 49% only.

Australasia

Australia

The mineral right, including petroleum in the offshore areas, also vest with the provincial governments. The grant of petroleum concession is governed by the Petroleum (Submerged Lands) Act, 1967. This act is uniformly adopted by all the provincial governments and is so framed as to avoid any territorial sea disputes between the States. The territorial sea bed and the continental shelf has been divided equitably among the coastal States helping in the principles Geneva Convention on the Continental Shelf, 1958. A two-title system has been adopted for all offshore developments including in the territories of Papua and New Guinea. In the first stage exploration permit is granted authorising all phases of exploration including exploration drilling, and in the second stage a production licence is granted authorising exploitation. The maximum area of a permit is 400 blocks about 10,000 sq miles. The minimum area is 40 sq miles. There is no statutory limitations on the number of permits which may be granted to any one company. Permits are granted for 6 years period with a right of renewal for additional 6 years period. However, the right of renewal extends to half of the original permit. This surrender arrangement is designed to encourage companies to concentrate their exploration in the most promising area. Production licence is granted over much lesser blocks and if the company wants to have right over more blocks it has to pay higher rate of royalty compared to blocks over which it was permitted production rights earlier.

European Countries

United Kingdom

The working of coal is governed by the Coal Act, 1938. Under this Act, all coal bearing areas and mines of coal have been passed on to the Coal Commission on July 1, 1942 as well as any other minerals comprised in coal mining leases existing on January 1, 1939.

On the abolition of the office of the Coal Commission on April 1, 1947, the power of the Coal Commission was transferred to the National Coal Board. The purpose of the 1938 Act was to nationalise coal but it was actually effected under Coal Industry Nationalisation Act of 1946. The persons formerly working on coal, whether as owners of lessees of the minerals, became lessees of the Coal Commission (later National Coal Board) and granted new mining leases from time to time.

Apart from the exceptional circumstances mentioned above minerals are privately mined and the lease-holder may work himself or let out to others subject to compliance with the planning law, the avoidance of nuisance, and the observance of the rights of the neighbouring landowner. The procedure for acquisition of land for mining purposes is governed by the Mines and Quarries Act, 1954.

By ancient rules of the common lands all unworked gold and silver deposits are owned by the Crown.

Under the Petroleum Production Act, 1934, petroleum in its natural condition in strata also belongs to the Crown.

In May 1964, the U.K. ratified the Geneva Convention on the continental shelf. North Sea licenses are issued only to a citizen resident in the U.K. or Corporate bodies. For the grant of the license, the sea falling under the U.K. area covering about 256,000 sq km has been divided into 250 sq km blocks. A production license gives exclusive exploration and exploitation rights of the area covered.

The atomic minerals are dealt under the Atomic Energy Act, 1946.

The Town and County Planning Act of 1947, with subsequent legislations, set the pattern for land restoration after open pit mining of all minerals. It includes returning the land to its original contour and drainage pattern as far as practicable. Roads, fences, and drainage ditches are rebuilt, replaced or restored so that normal agricultural use of land can be resumed. Both the mine-owner and the Government share the cost of land restoration. A well planned planting and grazing programme is carried out immediately to assure strong root growth and consequent soil stabilisation of all restored area.

France

Mining law in France is still governed by the Law of April 21,

1810. The concession creates a new property right to the mine in the name of the concession holder, which is distinct from the ownership of the surrounding surface land. It has been made clear by the law that the surface owner cannot be the owner of the mine or the mineral deposits unless he holds the concession. The unique feature of French law is that the leases are granted for an indefinite period rather for perpetuity just like those which existed in Goa during the Portuguese regime. The concessionnaires are not required to pay royalties or special payments except the annuity sum fixed at the time of grant of concession. The annuity sum is irrespective of the profit or annual production.

The French law for oil and gas exploration has undergone several changes during 1955 and 1956 to suit the expanding oil activities in France. Foreign collaborations are allowed if they offer a substantial participation for French private capital. The exploration permit is granted with exclusive right to prospect and explore over large areas. The exploration permit holder can obtain several exploitation concessions over the boundaries of the same permit. The concessionnaires are required to pay certain charges separately per hectare of the area under lease to the Government, district administration and a royalty on the annual crude production in excess of 50,000 tonnes. Certain rebate in royalty tax is given if the crude is transported to the nearest refinery or port of loading which is situated over 500 km from the well. The production of natural gas is exempted from royalty up to 300 million cubic metres, after which a 5% royalty is levied.

Spain

By Decree No. 411 of December 10, 1964, this country has liberalised foreign investments in the mining sector. Foreign participation up to 50% is freely authorised, but beyond that the permission of the Government is necessary.

Latin American Countries

Brazil

According to the Constitution, all mineral property belongs to the Federal Union, except on the lands held as private property that were registered before July 20, 1936. Law stipulates that only the Brazilian citizens and companies organised in Brazil can undertake mining. But this provision has not been strictly adhered to. A number of foreign companies are working in Brazil. The Americans have large interests in developing manganese mining in Amapa territory.

The leases are granted on the condition that the lessee must commence work within a year from the date of the decree and must submit a plan for proper utilisation and financial capacity. The cases for grant of leases are processed by the Government through the National Department of Mines.

The Mining Code does not apply to petroleum and related minerals. The 1953 statute declared a Government monopoly over all petroleum activities (both oil and gas) from prospecting and exploration, through the refining stages to the final pipeline and ocean transportation of all petroleum products produced in the country. Existing petroleum concessions, pipeline and tanker facilities, and refineries in operation till October 1953 or authorised before June 30, 1952, are excluded from the statute, may not be renewed when they expire.

Chile

Mining is controlled by Decree Law No. 488 of 1932 and Regulations Decree 2228 of December 21, 1932. Exploration for petroleum and natural gas has been a State monopoly since the enactment of Law 4109, December 23, 1926. Except for petroleum and natural gas, the law recognises no difference between a Chilean and a foreigner with regard to the acquisition and enjoyment of the civil rights governed by article 57 of the Civil Code.

The Mining Code specifies the grant of leases in private and public lands. The State, however, reserves exclusive rights to mine petroleum and natural gas whether found in private or public land. The area applied for mineral concession should be in multiples of claims, i.e. it should not be less than 1 hectare and more than 5 hectares for metals and precious stones and for other minerals up to 50 hectares. The concession must be of 50 metres minimum width.

Guatemala

According to Law 342 which becomes effective from 3rd June, 1965, prospecting, exploration and exploration concessions are granted to Guatemalan nationals and foreigners alike.

Holders of exploration licence are obliged to invest a minimum annual amount in their concessions. The amount is fixed at the time of granting concession.

Mexico

The mining concessions are governed by the mining law which became effective from October 1, 1930 superseding all earlier laws. The grant of mining rights in the free land is given by the National Government. The grant of lease in the national reserve land is administered by the Director of Mines and Petroleum and Regional Mining Agents. Three types of concessions are granted : (1) prospecting, (2) exploitation, and (3) treatment plants.

Prospecting licence is granted up to maximum area of 9 hectares for a period of two years. Only one prospecting licence is granted at a time to one prospector, or corporation. According to a decree of 31st December, 1943, a prospector has to submit a proof regarding the existence of mineral deposit and obtain a prospecting concession.

Exploitation concessions are granted to three classes of minerals and concession may be granted to only one class: metallic minerals, non-metallic minerals, or coal and graphite; but when the holder of a concession discovers another class of minerals he may apply for inclusion of the same. Concession is granted for unlimited period but up to the vertical depth of 100 metres each side. Like other countries there is no extra lateral right to work up to the centre of the earth, except with special permission.

The participation of a foreign national in a mining venture is not encouraged although according to the constitution, a foreign national can be granted a lease except within 10 km inland from the land borders and 50 km inland from the sea coast. A joint business with the participation of foreign capital can be established provided the majority of the shares are held by Mexicans.

North America

Canada

The State Governments hold the legal title of all lands except Indian Reserves, National Parks, lands in the North West Territories and the Yukon which are controlled by the Central Government. Foreigners have the same privilege for the acquisition of mining rights as that of a citizen of Canada provided 50% of the shares are owned by persons who are Canadian citizen. Any person who has attained the age of 18 years and any company authorised to do business in Canada can acquire a lease.

The grant of rights to mine and to extract ore is given generally on 21 years lease with renewal clause. In Newfoundland it is given on 50 years lease and in Ontario, an absolute title to certain mineral deposits may be obtained. Mineral resources on vacant and certain other Dominion lands may be developed under a 21 years lease with renewal clause. The three controlling statutes for these lands are the Territorial Land Act, the Yukon Placer Mining Act, 1952 and the Yukon Quartz Mining Act, 1952. Certain Dominion legislation affect all mining. The Atomic Energy Control Board created under the Atomic Energy Control Act 1946 has wider powers over radioactive minerals throughout Canada. In Canada each of the 10 provincial governments has its own mineral laws applicable to the disposition of mineral rights, conservation, mineral taxation, royalties, operating safety rules, and other control on mineral industries. Alberta and Saskatchewan, which are the major producers of oil and natural gas, follow the following rules and procedures. They issue three types of exploration permits: (i) geological or geophysical exploration licence for a fixed period at a special fee, (ii) exploration permit allowing prospecting and exploratory drilling on contiguous, rectangular areas effective up to 3 years with further extension possible, and (iii) a drilling permit valid for one year and may be renewed twice. Two types of production leases are given: (1) a natural gas production lease, and (2) an oil and natural gas lease. Drilling and production are closely regulated under the Oil and Gas Resources Conservation Act and Regulation passed in 1957.

United States

The land in the U.S.A. may belong to Federal, State or individuals. The acquisition of mineral rights in the U.S. is governed by either Federal or State or District law, depending upon whether the mineral property sought is owned by the Federal Government, the States, or individuals. The important laws governing the grant of lease in the Federal lands are: (1) the Mineral Location Law of 1872 (revised from time to time) governing lode and placer claims of metalliferous minerals, including uranium; (2) the amended Mineral Leasing Act of 1920 covering all the deposits of coal, phosphate, sodium, potassium, mineral oil, oil shale, or gas; (3) the Outer Continental Shelf Lands Act of 1953 that permits the development of minerals in the outer continental shelf beyond territorial water; and (4) the Multiple Mineral Development Act of 1954 that allows joint use of the same tracts of public lands for mineral and non-mineral development. The grant of leases in the State-owned land is governed by the laws enacted by the State legislatures. Such laws are generally modelled after the Federal Laws. In respect of individually owned lands, the use, development or transfer of mineral rights is accomplished within the framework of the laws of each State relating to property titles, sales and conveyances, leases, licences, and contracts. These private rights are subject to general Federal State legislation pertaining to such matters as conservation, taxation, safety, health, or other matters within the constitutional powers of the respective sovereigns.

The administration of all Federal mineral land is the responsibility of the Bureau of Land Management under the Secretary of the Interior.

According to the Act of 1872, only citizens of the U.S.A. and those who have declared their intention to become such are eligible to acquire rights to public mineral lands. Domestic corporations are considered citizens. Although an alien cannot acquire title by patent or location valid against the Federal Government, his location, inheritance, of purchase of an unpatented claim is not subject to question.

In the U.S.A. the law governing the grant of legal title (by a patent) to work lode deposits is quite interesting and unlike the law prevailing in India or other countries. Firstly, the discovery is the prime requisite for the establishment of a valid location. Claims may be located on veins or lodes, where the mineral is in place.

The mineral deposit is considered to have been discovered if that outcrop of the lode or vein, as the case may be, falls in the area projected vertically, and this locates the claim of the applicant. Having thus located the claim, the locator is granted permission to mine the vein/lode in the dip side irrespective of the fact that the vein/lode falls in another man's property in its dip side.

The number of mining claims that may be located by an individual, corporation, or association is unlimited.

To keep the claim valid, the Act of 1872 provides that at least $ 100 worth of development work must be done each year.

To obtain title to land covered by location the locator must apply to the Bureau of Land Management for a patent with specified fees.

For obtaining a lease-hold interest in oil and gas on the public Dominion falls into two classifications, depending on whether the land sought lies within or without the known geological structure of a producing oil or gas field. Lands lying within the geological structure of a producing field are subject to lease only by competitive bidding, all other lands may be prospected by non-competitive wild cat leases, which are issued to the first applicant.

6

Marine Mineral Resources and the Law of the Sea

OCEANS ARE a vast storehouse of minerals. At present, oil and natural gas and some 13 other minerals are being produced from near shore marine sources. Sand and gravel, limeshells, and several placer minerals such as ilmenite-rutile sands, tin, zircon, monazite and magnetite are recovered through dredging; sulphur and salt by solution mining through bore holes; barytes by sub-sea quarrying and coal and iron ore by underground mining with entry from the adjacent coastal area. Gold, platinum and diamond have been produced in the past. Potash and phosphorite are likely to be produced in future. However, oil and natural gas is the single most important mineral resource of the continental margins. The deep seabed area has attracted considerable attention due to the manganese nodules which lie on the surface of the ocean floor and contain several metals—nickel, cobalt, copper, manganese, molybdenum vanadium and titanium. Other potentially important deep seabed resources are metalliferous deposits associated with hydrothermal systems along the oceanic ridges or ocean floor spreading systems. The most notable of these is the metal bearing hot brines and muds of the Atlantis II deep in the Red Sea.

The law of the sea forms part of the International Law. The traditional international law recognised special interest of the coastal States in near shore areas. A three mile territorial sea and the freedom of the high seas beyond this territorial sea were the main elements of the traditional law of the sea. The superjacent waters, and not the seabed and its resources, were the primary concern of the traditional international law.

In 1945 the attention of the world was drawn towards the importance of seabed mineral resources when President Truman of United States of America, through a proclamation, claimed sovereign rights over the minerals of the United States continental shelf. The International Law Commission deliberated on the law of the sea between 1949 and 1956 and prepared basic proposals which were considered by the First United Nations Conference on the Law of the Sea held at Geneva in 1958. This Conference adopted four Conventions on the (i) Territorial Sea and the Contiguous Zone, (ii) High Seas, (iii) Fishing and Conservation of Living Resources of High Seas, and (iv) Continental Shelf. This Conference, however, was unable to agree on a precise limit of the territorial sea and the exclusive fishery zone. The continental shelf was defined to include the seabed and the sub-soil of submarine area adjacent to the coast but outside the area of the territorial sea to a depth of 200 metres or, beyond that limit, to where the depth of the superjacent waters admits the exploitation of the natural resources of such areas. The Second U.N. Conference on the Law of the Sea was convened in 1960 to decide the non-resolved issues relating to the outer limits of the territorial sea and the exclusive fishery zone. This Conference, was, however, unsuccessful in its objective.

In the meantime, many countries of Africa and Asia who had not participated in the above two UN Conferences on the Law of the Sea became independent. These countries were eager to become economically independent through the exploitation of their living and non-living marine resources. In the mid sixties, the world also became aware of the immense riches contained in the manganese nodule resources of the deep sea area and the possibility of their economic exploitation. It was apparent that in the absence of a legal regime, the exploitation of these resources will only benefit the few highly industrialised countries of the world. The 1958 Geneva Convention on the Continental Shelf also favoured the technologically advanced countries. There was, thus, a widespread dissatisfaction with existing international law of the sea and a demand for evolving a fairer legal regime.

It was in this climate that in 1967 on a proposal made by the Government of Malta, the General Assembly of the United Nations appointed a Committee on the Peaceful Uses of the Seabed and the Ocean Floor Beyond the Limits of National Jurisdiction. In 1970 the UN General Assembly adopted a Declaration of Principles Governing Seabed and the Ocean Floor and the Sub-Soil Thereof Beyond the Limits of National Jurisdiction, and solemnly declared, *inter alia*, that the area of the seabed and the ocean floor and the sub-soil thereof, beyond the limits of national jurisdiction, as well as its resources, are the *common heritage of mankind*, the exploration and exploitation of which shall be carried out for the benefit of mankind as a whole, irrespective of the geographical location of States. By another resolution, the General Assembly decided to convene the Third United Nations Conference on the Law of the Sea (UNCLOS).

The Third UNCLOS held eleven sessions between 1973 and 1982. On the 30th April, 1982 the Third UNCLOS adopted by vote the United Nations Convention on the Law of the Sea. This Convention will come into force 12 months after sixty, countries have ratified or acceded to it. After it comes into force all activities relating to the oceans will be governed in accordance with the provisions of this Convention. From the point of view of development of marine mineral resources the legal regime of the exclusive economic zone, the continental shelf, and the international seabed area are relevant. The salient provisions of the Convention in this regard are summarised below.

Exclusive Economic Zone (EEZ)

For the EEZ the Convention lays down an outer limit of 200 nautical miles from the shore. In the EEZ, the coastal State has sovereign rights for the purpose of exploring and exploiting, conserving and managing the natural resources, living as well as nonliving, of the seabed and sub-soil and of the superjacent waters, as also for production of energy from the water, currents and winds. The Coastal State has exclusive right to construct and to authorise and regulate the construction operation and use of artificial islands, installations and structures for the above purpose. It has also jurisdiction with regard to marine scientific research and the protection and preservation of marine environment.

Continental Shelf

The Convention stipulates that the continental shelf of a coastal State comprises the seabed and subsoil of the sub-marine areas that extend beyond its territorial sea throughout the natural prolongation of its land territory to the outer edge of the continental margin. The continental margin comprises the submerged prolongation of the land mass of the coastal State, and consists of the seabed and subsoil of the shelf, the slope and the rise. It does not include the deep ocean floor with its oceanic ridges or the sub-soil thereof.

The Convention lays down a minimum limit of 200 nautical miles and a maximum limit of 350 nautical miles or 100 nautical miles from the 2,500 metre isobath, for the continental shelf. These limits are to be measured from the baseline. The coastal State is required to delineate the actual boundary of the continental shelf between these two limits, by using either of the following two criteria:

(i)   the thickness of sedimentary rocks on each point of the boundary line is at least one per cent of the shortest distance from such point to the foot of the continental slope; or

(ii) every point on the boundary line is not more than 60 nautical miles from the foot of the continental slope.

For the above purpose, the foot of the continental slope shall be the point of maximum change in the gradient at its base, in the absence of evidence to the contrary.

It is evident from the above that the Convention uses the term 'continental shelf' in a legal sense which includes, from the geomorphological point of view, not only the continental shelf proper but also the continental slope and the continental rise. In other words, the 'continental shelf' of the Convention encompasses the entire continental margin. The geomorphologists use these terms in the following sense:

Continental shelf: This term includes the nearly flat or gently sloping surface of the seafloor bordering the continent. The average dip is about 0.1°. The inner edge of the continental shelf is the shore line. The outer edge is defined as the topographic break where the slope of the surface increases and the seafloor drops off to greater depths. The depth of the edge of the shelf varies from region to region but for convenience it is taken to correspond to the 200 metre or 100 fathom isobath. In India, the depth of the shelf edge is minimum off the coast of Bombay and maximum off Calcutta. Likewise, the width of the continental shelf also varies widely from a few metres to 1200 km. In India, generally, the width of the continental shelf is narrower in the eastern coast as compared to the western coast.

Continental slope: It comprises of the relatively steep slope extend-

ing seaward from the shelf-edge. The dip of the slope is usually
2° to 6°, but locally it may be 20° or more. The foot of the slope
is commonly taken as the point where the gradient becomes less
than 1:40, at depths that range from 1500 to 5000 metres.

Continental rise: The gently sloping surface of the apron of sediments spread on the oceanfloor beyond the foot of the continental
slope is described as the continental rise. The gradient of the continental rise is as much as 1:100 near the base of the slope but it
diminishes gradually to about 1:1000 where it meets the abyssal
plain.

Continental margin: The combined continental shelf, continental slope and the continental rise is described as the continental margin.

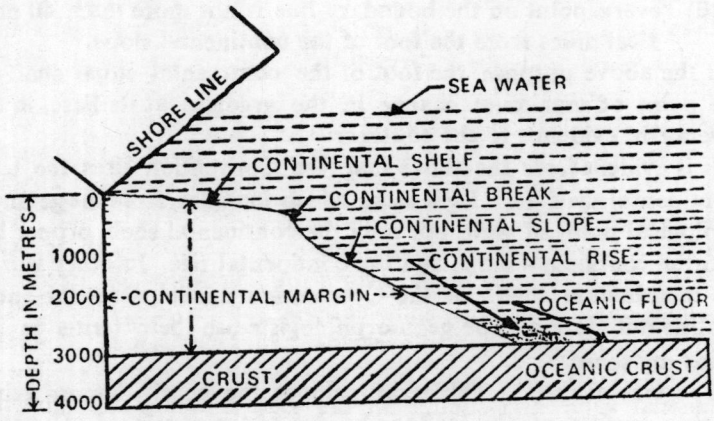

Fig. 8. Profile of continental shelf, slope and rise

The Convention recognises the coastal State's sovereign rights
over the continental shelf for the purpose of exploring and exploiting its natural resources. No one can undertake these activities
without the express consent of the coastal State. The coastal State
has exclusive right to authorise and regulate drilling on the continental shelf for all purposes.

International Seabed Area

The international seabed area means the seabed and oceanfloor
and subsoil thereof beyond the limits of national jurisdiction i.e.

area beyond the limits of territorial sea, the EEZ and the continental shelf. The international seabed area and its resources—all solid, liquid or gaseous mineral resources *in situ* in this area at or beneath the seabed, including polymetallic nodules—are the common heritage of mankind. All activities of exploration for, and exploitation of, the resources of the international seabed area must be carried out for the benefit of mankind as a whole. The Convention establishes an International Seabed Authority (ISA) to administer the resources of the international seabed area on behalf of the mankind as a whole. All States who become parties to the Convention will be members of the ISA. The Convention establishes, as principal organs of the ISA, an Assembly consisting of all the members of the ISA, a 36 member Council and a Secretariat. It also establishes Enterprise, as an organ of ISA throughout which the ISA shall directly carry out exploration and exploitation of the resources of the area as well as the transporting, processing and marketing of the minerals recovered from the international seabed area.

The exploration for and exploitation of the resources of the area can be undertaken by the Enterprise, States which are parties to the Convention, State enterprises or natural or juridical persons which possess the nationalities of and are sponsored by any State party to the Convention. Each application, other than those submitted by the Enterprise, shall cover a total area, which need not be a single continuous area, sufficiently large and of sufficient estimated commercial value to allow two mining operations. The applicant shall divide this area into two parts of equal estimated commercial value and submit all the data obtained by him with respect to both parts. The ISA will select one part to be reserved for exploitation by the Enterprise or in association with developing countries and grant the other part for exploration and exploitation by the applicant. The applicant's plan of work is required to be approved by the Council. The persons granted authority to operate in the international seabed area must abide by the rules and regulations of the ISA including payments of financial levies.

In order to prevent adverse effect on the economy of States producing minerals and metals from their land resources as a result of exploitation of the resources of international seabed area, the ISA will regulate the production from this area in accordance with a formula which, in principle, is based on sharing the growth in demand of nickel between the production from the land and the international seabed area. The Convention also provides for compensating the land-based producers in the event of adverse effects on their economy as a consequence of production from the international seabed area.

The UNCLOS established a Preparatory Commission which is functioning at the proposed seat of the ISA at Kingston, Jamaica. The Preparatory Commission has been entrusted with the responsibility of undertaking preparatory work for the establishment of the ISA and its various organs, the Enterprise and the formulation of draft rules and regulations.

The UNCLOS has accorded India the status of a pioneer investor, the only developing country to be given this status.

India has enacted "The Territorial Waters, Continental Shelf, Exclusive Economic Zone and other Maritime Zones Act, 1976" which lays down the limits and the jurisdiction of India over territorial waters, continental shelf, exclusive economic zone and contiguous zone. According to this legislation the limit of (reckoned from the base line) territorial waters extends to 12 nautical miles, "contiguous zone" to 24 nautical miles and the "exclusive economic zone" to 200 nautical miles. This Act also lays down that the continental shelf of India comprises the sea bed and sub-soil of the submarine areas that extend beyond the limit of its territorial waters throughout the natural prolongation of its land territory to the outer edge of the continental margin or to a distance of two hundred miles from the base line where the outer edge of the continental margin does not extend up to that distance.

7

Mines and Minerals Legislation of India

LAWS GOVERNING the mines and minerals in India are placed under two classifications, one for the regulation and development and other for the safety and welfare of miners. The objectives of the former are effected through two set of Acts, namely, the Oil Fields (Regulation and Development) Act, 1948 and the Mines and Minerals (Regulation and Development) Act, 1957, and that of latter through the Mines Act, 1952, and the rules framed thereunder. In addition to these two, other acts have been enacted for achieving some specific purpose as would be found from discussion later.

In the Indian Federal system, the mineral rights vest in the State Govts. under whose territorial jurisdiction the minerals lie and they are authorised to grant concessions and lease out areas to private individuals, companies, corporations, etc. But for the leases in the off-shore areas i.e. in the territorial sea, an exclusive economic zone and the continental shelf, the right of granting concessions vests in the Union government. Under the constitution, although all mineral properties belong to the State Government, the power of framing rules for regulating the grant of mineral concessions and development of minerals lie with the Union government. Similarly the power to make laws for development of oil fields and mineral oil resources vests in the Union government. Also the power to make laws for safety and welfare of the workers in mines and oil fields vests in the Union government.

Any citizen of India can acquire mineral concessions. A foreign national can also hold a concession, if approved by the Central Government, provided he holds minority shares in a public company as defined in the Companies Act, 1956, and only if majority of the directors of the company are citizens of India. In case of petroleum the rules have been suitably modified to encourage foreign participation.

CATEGORIES OF MINERALS FOR GRANT OF CONCESSIONS

Minerals have been placed under two categories for grant of concessions, one under mineral oil and gas and the remaining others under the second which are further subdivided into 'minor' and 'major' including coal and atomic minerals. The grant of concessions for mineral oil and gas falling under the first category is governed by the Oil Fields (Regulation and Development) Act, 1948 and that of the second category by the M.M. (R and D) Act, 1957, but with a procedural difference in respect of 'minor' and 'major' minerals. Under Section 15 of the Act, exclusive power has been vested to the State Governments for regulating grant of mineral concessions in respect of minor minerals. Thus each State Government has framed its own minor mineral concession rules. On the other hand, concessions for oil and natural gas and major minerals are regulated by the Petroleum and Natural Gas Rules, 1959 and the Mineral Concession Rules, 1960 respectively framed under the above two acts and are applicable uniformly all over India.

In consonance with the international trend, the Government of India enacted the Territorial Waters, Continental Shelf, Exclusive Economic Zones and other Marine Zones Act in 1976. This Act provides sovereign rights for the purposes of conservation, exploration, exploitation and management of all the resources within the exclusive economic zones and the continental shelf of the country covering about 2 million sq km.

MINOR MINERALS

Minor minerals, for the purpose of law, are those minerals as defined in Section 3(e) of the Mines and Minerals (Regulation and Development) Act 1957. A certain group of minerals of local importance or minerals which can be developed on small or cottage scale for the benefit of local people, have been specifically declared as minor minerals. The followings have been listed as minor minerals —building stones, marble, quartzite and sandstone for the purpose of making building, road metal and household utensils, stone used for household utensils, boulder, shingle, gravel, chalcedony pebbles used for ball mill purposes only; limeshell, kankar, and limestone used in kilns for manufacture of lime used as building material; murrum, brick earth, fuller's earth, bentonite, road metal, rehmatti, slate and shale used for building material, ordinary clay, ordinary sand used for purposes other than refractory, ceramics, metallurgical, optical and stowing in coal mines, manufacture of silvecrete (cement), sodium silicate, pottery and glass, and saltpetre. The Central Government may add or delete any mineral from the list of 'minor minerals' by notification in official gazette.

MAJOR MINERALS

Minerals other than those not declared as 'minor minerals', are major minerals. Under the 'major minerals', atomic minerals are those which have been declared as 'prescribed substance' under the Atomic Energy Act, 1962. The minerals falling under the prescribed substance are: minerals/ores of uranium, and its compounds including tailings containing uranium; thorium and its compounds and minerals containing thorium including monazite; zirconium, its compounds and minerals including zircon; beryllium and its minerals including beryl, lithium and its minerals including lepidolite, deuterium and its compounds, neptunium and its compounds, columbite and tantalite, ilmenite, and rutile.

PROCEDURE FOR OBTAINING MINERAL CONCESSIONS

Any person intending to go in for mining business has to obtain first the Certificate of Approval (CA). The CA is obtained by depositing Rs. 500 as a fee in the manner prescribed under the M.C.R., 1960, with the respective State Governments in whose State the concession is desired to be obtained. The CA issued by one State is valid for the particular State only and for acquiring mineral concession in other State, it is required to be obtained from that State Government as well. It is valid for only one year and can be renewed by paying Rs. 250 every year. Once a concession is acquired it is not necessary to renew it for the entire tenure of the lease unless a new concession is sought for.

A mineral concession confers right to prospect or mine within acquired properties. The concessions can be acquired in terms of prospecting licence or mining lease so far major minerals are concerned. The procedure for obtaining lease for atomic mineral is the same as applicable to other major minerals except that in the case of former, approval of the Department of Atomic Energy, Government of India, has also to be obtained and the intending lessee is required to be guided by the Atomic Energy Act, 1962, in matter of production, disposal and in maintaining secrecy about it. In case of petroleum and natural gas, the term 'exploration' licence instead of prospecting licence has been used. For obtaining a concession, an application is required to be filed in a prescribed form with requisite fees as provided in the rules to local government through district authorities. Full rights are enjoyed by the State Governments in grant of concessions, except for specified minerals given in the first schedule of M.M. (R and D) Act, 1957, for which prior approval of the Central Government is necessary. The specified minerals are: apatite and phosphatic rocks, chrome ore, coal and lignite, copper ore, gold, gypsum, iron ore, lead ore, zinc ore, vanadium ore, manganese ore, molybdenum ore, nickel ore, platinum and other precious metals and their ores, precious stones, rutile, silver, sulphur and its ores, tin and tungsten ores and the prescribed minerals under the Atomic Energy Act. In the case of petroleum and natural gas exploration licence or mining lease is granted only with the approval of the Central Government, on such terms and conditions as may be mutually agreed before between the Central Government and the applicant.

A lease or permit for mining or quarrying for minor minerals is obtained as per the procedures laid down in the Minor Mineral Concession Rules framed by the respective State Governments. For obtaining concession for minor minerals, a Certificate of Approval is not necessary. Only the three State Governments, namely, Punjab, Haryana and Jammu & Kashmir, stipulate that a separate Certificate of Approval is necessary for obtaining lease for minor minerals. Fee is Rs. 50 which can be renewed every year by paying Rs. 25. Only the Government of J & K stipulates that if a person is holding a Certificate of Approval for major mineral he need not obtain CA for minor mineral i.e. he can apply for concession for minor mineral.

Restriction on Area for Concessions

There is restriction on the maximum area one can acquire. Maximum area for which a prospecting licence or mining lease can be granted or obtained in any one State in respect of any one mineral or a group of associated minerals is as under:

(a) a total area covering not more than 25 sq km comprising one or more areas under prospecting licence; or

(b) a total area covering not more than 10 sq km comprising one or more areas under mining leases.

The provisions of maximum of area are, however, relaxable in the interest of mineral development by the Central Government. The maximum limit of area holds good only for one particular or a group of associated minerals. Associated minerals are those which occur usually together and cannot be mined independently, as listed in the Mineral Concession Rules, 1960. Law provides that any person already holder of a licence or lease can hold prospecting licence or mining lease in respect of other minerals as well to the maximum permissible limits in each case. For petroleum and natural gas, the area applied for licence should ordinarily be 3000 sq miles and in case of lease 100 sq miles. Maximum area of lease assigned to one party is 1000 sq miles.

Period of Concessions

The prospecting licence is obtainable for one year in case of mica and two years for any other mineral. It is renewable for one or more periods, each not exceeding the period for which the licence was originally granted. The period for which a mining lease is granted does not exceed thirty years in the case of coal, iron ore or bauxite and 20 years in the case of any other minerals. The lease can be renewed for equal period after the expiry of original lease. However, with the approval of the Central Government the lease can be extended beyond the renewed period. In case of petroleum and natural gas tenure of licence is four years which may be extended for further two periods of one year each; and for lease it is twenty years unless a shorter period is asked for and may be renewed for one term only not exceeding the original term. A tenure of 20 years or 30 years with renewable clause for the same period has been assigned taking into account the normal life of a deposit within which it is likely to be mined out.

TERMINATION, SURRENDER AND DETERMINATION OF MINING LEASE

Under the MMRD Act the Contral Government is empowered to prematurely terminate a mining lease and grant a fresh lease over the area to a Government Company, if it is of the opinion that it is expedient in the interest of regulation of mines and mineral development to do so. A lease could also be terminated by the State Government for breach of terms and conditions of lease or non-payment of rents and royalties, after following the procedure laid down in MCR, 1960. A lessee is also empowered to surrender the lease after giving 12 months' notice. Part surrender of the lease is also permitted in specified circumstances.

Royalty and Dead Rent

Royalty is the payment to the owner of mineral rights for the privilege granted by him for mining and producing mineral. The owner may be Government or private individual. When mineral property is leased out by the owner, he does so in consideration of receiving some share, the royalty in the value of mineral extracted. Royalty may be defined as share to a lessor in the value of mineral extracted by a lessee. After a lease has been obtained, a lessee has to pay royalty at the rate stipulated on the quantity removed or consumed. Amount of royalty is to be deposited every six month or after expiry of a year as directed by the State Government. The rate of royalty has been fixed on tonnage basis for all known recoverable minerals (see Table 8). In case of petroleum and casing head condensate, the rate of royalty is payable at the rate of Rs. 42 per metric tonne of crude oil produced. In the case of natural gas the rate of royalty is levied at the rate of 10% of this value. The new rates have been introduced w.e.f. 8th September, 1976. A provision has been made in the Petroleum and Natural Gas Rules, 1959, that the State Government may obtain royalty in terms of money or petroleum or natural gas. A 'Dead Rent' is the minimum royalty which a lessee has to pay to the owner of the property. Dead rent is payable only from the second year of the lease, if the property remain unexploited. No dead rent is charged in the first year of the lease, although royalty becomes payable in the very first year if the deposit has been opened up for mining and some quantities have been produced. A lessee is required to pay either dead rent or royalty, whichever is more, and not both. In the dead rent schedule, increasing rates have been prescribed depending upon the period (see Table 9) to prevent lessee indirectly from keeping the area unexploited. For petroleum, dead rent is payable at the rate of Rs. 5 per acre or part thereof for the first 50 square miles; and Rs. 10 per acre or part thereof for area exceeding 50 square miles.

Surface and Water Rents

Such rents are levied in addition to royalty or dead rent on the area used for actual mining operations out of the total lease which comes about Rs. 2.50 per hectare a year. A lessee may also be required to pay compensation for damages to the land and falling of forest trees.

MAXIMUM QUANTITIES REMOVABLE FREE OF ROYALTY

During the tenure of prospecting licence, a provision has been made for licensee to take out stipulated quantities of minerals free of royalty for the purpose other than commercial (see Table 10). Intention of such stipulation is that the quantity ordinarily required for preliminary tests, beneficiation or research may only be removed free of royalty. A licensee can also remove minerals in excess of the limits on payment of royalty but not exceeding twice such limits won during prospecting. More relaxation has been given in case of mica and limestone where maximum quantity which could be removed is 10 tonnes and 500 tonnes respectively. In the case of gold, silver and precious stones, a licensee may carry away any quantity on payment of royalty.

The different acts and rules in force, regulating the grant of concession, production, conservation, safety and welfare of miners are briefly detailed below:

THE OIL FIELDS (REGULATION & DEVELOPMENT) ACT, 1948
(Act No. 53 of 1948)

The act relates only to regulation and development of petroleum and natural gas. Only one set of rules, namely the Petroleum and Natural Gas Rules, 1959, have been framed under the act, which provide procedure for obtaining concession, rates of royalty and dead rent. A lessee is required to submit monthly production return.

TABLE 8

Rates of Royalty on Minerals

(As per Section 9 and Second Schedule of the Mines and Minerals
Regulation and Development Act, 1957)

| *Minerals* | *Rates of Royalty* |

---

1. Coal:

   (i) Group I Coals:

    (a) Coking Coal Steel Grade I
        T Steel Grade II Washery
        Grade I

    (b) Hand picked Coal produced in   Seven rupees only per tonne.
        Assam, Arunachal Pradesh,
        Meghalaya and Nagaland.

   (ii) Group II Coals:

    (a) Coking Coal Washery Grade II
        Coking Coal Washery Grade III

    (b) Semi-coking coal Grade I   Six rupees and fifty paise
        Semi-coking coal Grade II   only per tonne.

    (c) Non-coking coal Grade A
        Non-coking coal Grade B

    (d) Ungraded R.O.M. Coal produced
        in Assam, Arunachal Pradesh,
        Meghalaya and Nagaland.

   (iii) Group III Coals:

    (a) Coking Coal Washery Grade IV   Five rupees and fifty paise

    (b) Non-coking coal Grade C   only per tonne.

   (iv) Group IV Coals:
     Non-coking coal Grade D   Four rupees and thirty
     Non-coking coal Grade E   paise only per tonne.

TABLE 8 *(Contd.)*

Minerals	Rates of Royalty
(v) Group V Coals: Non-coking coal Grade F Non-coking coal Grade G	Two rupees and fifty paise only per tonne
(vi) Group VI Coals:* Coal produced in Andhra Pradesh (Singareni Collieries Company Ltd)	Five rupees only per tonne.
2. Mica:	
(a) Crude mica	Eight rupees per 100 kg.
(b) Trimmed mica of qualities other than heavy-stained, dense-stained or spotted second quality.	Sixteen rupees per 100 kg.
(c) Trimmed mica of heavy-stained dense-stained or spotted second quality.	Eight rupees and forty paise per 100 kg.
(d) Waste and scrap mica	Two rupees and eighty paise per 100 kg.
(e) Waste rounds	Three rupees and fifty paise per 100 kg.
3. (a) Gold	Two rupees per one gram of gold per tonne of ore and on pro rata basis.
(b) Silver	One hundred rupees per kg. of metal.
4. Iron:	
(i) Ore lumps:	
(a) with 65% Fe or more	Four rupees per tonne.
(b) with 62% Fe or more but less than 65% Fe	Three rupees per tonne.

*Explanation—For the purpose of this item the specification of each such grade of coal shall be as prescribed under clause 3 of the Colliery Control Order, 1945.

TABLE 8 (*Contd.*)

Minerals	Rates of Royalty
(c) with 60% Fe or more but less than 62% Fe	Two rupees per tonne.
(d) with less than 60% Fe	One rupee and fifty paise per tonne.
(ii) Ore Fines:	
(A) fines (including natural fines and fines produced incidental to mining and sizing of ore)	
(a) with 65% Fe or more	Two rupees and fifty paise per tonne.
(b) with 62% Fe or more but less than 65% Fe.	One rupee and fifty paise per tonne.
(c) with less than 62% Fe	One rupee per tonne.
(B) Concentrates prepared by beneficiation and/or concentration of low grade ore, containing 40% Fe. or less than 40%	Fifty paise per tonne.
5. (a) All precious and semi-precious stones (except diamond and agate).	Twenty per cent of the sale price at the pit's mouth.
(b) Diamond	Fifteen per cent of the sale price at the pit's mouth.
(c) Agate	Fifty rupees per tonne.
6. Manganese Ore:	
(a) Manganese dioxide (containing 78 per cent or more of MnO_2 and 4 per cent or below Fe).	Thirty rupees per tonne.
(b) 46% Mn and above	Twelve rupees per tonne.
(c) 35% Mn and above but below 46% Mn.	Seven rupees and fifty paise per tonne.
(d) Below 35% Mn but above 25% Mn.	Five rupees per tonne.
(e) 25% Mn or below	Two rupees per tonne.

TABLE 8 (*Contd.*)

Minerals	Rates of Royalty
7. Chromite—(both lumpy non-friable ore and concentrates)	
(a) Containing 48% Cr_2O_3 and above	Fifty rupees per tonne.
(b) Containing less than 48% Cr_2O_3 and more than 40% Cr_2O_3	Twenty-five rupees per tonne.
(c) Containing less than 40% Cr_2O_3	Fifteen rupees per tonne
8. Limestone: (including lime Kankar)	Four rupees and fifty paise per tonne.
9. Dolomite	Five rupees per tonne.
10. Graphite:	
(a) with 80% or more fixed carbon	Fifty rupees per tonne.
(b) with 40% or more fixed carbon but less than 80% fixed carbon	Thirty rupees per tonne.
(c) with less than 40% fixed carbon	Twelve rupees per tonne.
11. China clay: (including ball clay)	
(a) Crude	Four rupees per tonne.
(b) Washed	Eighteen rupees per tonne.
12. Kyanite	Thirty rupees per tonne.
13. Gypsum	Four rupees per tonne.
14. Limeshell (including calcareous sand and chalk)	Five rupees per tonne.
15. Fire clay (including plastic, pipe, lithographic and natural (pozzolantic) clay)	Four rupees per tonne.
16. Ilmenite	Six rupees per tonne.

TABLE 8 (*Contd.*)

Minerals	Rates of Royalty
17. Copper Ore	Four rupees per unit per cent of copper metal per tonne of ore and on pro rata basis.
18. Lead Ore	Three rupees per unit per cent of metal per tonne of ore and on pro rata basis.
19. Zinc Ore	Four rupees per unit per cent of zinc metal contained per tonne of ore and on pro rata basis.
20. Garnet (Abrasive)	Ten rupees per tonne.
21. Sillimanite	Forty rupees per tonne.
22. Barytes:	
(a) White (including snow-white)	Fifteen rupees per tonne.
(b) Off-colour	Eight rupees per tonne.
23. (a) Quartzite	Two rupees and fifty paise per tonne.
(b) Sand for stowing	Forty paise per tonne.
24. Quartz and silica sand	Two rupees and fifty paise per tonne.
25. Corundum	Sixty five rupees per tonne.
26. Bauxite	Eight rupees per tonne.
27. Ochre	Five rupees per tonne.

TABLE 8 *(Contd.)*

Minerals	Rates of Royalty
28. Talc, Steatite and Soap stone	Eight rupees per tonne.
29. Apatite and Rock Phosphate:	
(a) Ores with 25% or more P_2O_5	Fifteen rupees per tonne.
(b) Ores with less than 25% P_2O_5	Ten rupees per tonne.
30. Asbestos:	
(a) Chrysotile	Two hundred and fifty rupees per tonne.
(b) Amphibole	Fifteen rupees per tonne.
31. Cadmium	Sixteen rupees per unit per cent of cadmium metal per tonne of ore and on pro rata basis.
32. Calcite	Six rupees per tonne.
33. Diaspore	Twenty rupees per tonne.
34. Felspar	Four rupees per tonne.
35. Fluorspar:	
(a) containing 85% CaF_2 or more	Seventy-five rupees per tonne.
(b) containing 70% CaF_2 or more but less than 85% CaF_2	Fifty rupees per tonne.
(c) containing more than 30% CaF_2 but less more than 70% CaF_2	Thirty-five rupees per tonne.
(d) containing 30% CaF_2 or less	Fifteen rupees per tonne.
36. Magnesite	Six rupees per tonne.

TABLE 8 (*Contd.*)

Minerals	Rates of Royalty
37. Nickel Ore	Two rupees per unit per cent of nickel metal per tonne and on pro rata basis.
38. Pyrites	Twelve paise per unit per cent of sulphur per tonne of ore and on pro rata basis.
39. Pyrophyllite	Five rupees per tonne.
40. Rutile	Seventy rupees per tonne.
41. Vermiculite	Four rupees per tonne.
42. Tungsten Ore	Ten rupees per unit per cent of WO_3 per tonne and on pro rata basis.
43. Wollastonite	Ten rupees per tonne.
44. All other minerals not here-in-before specified	Ten per cent of sale price at the pit's head.

TABLE 9

Rates of Dead Rent
(As per Section 9A and Third Schedule of MM (R & D) Act, 1957)

Period of the mining lease	Rates of dead rent per hectare
1st year	Ni
2nd year to 5th year	Rs. 12.50
6th year to 10th year	Rs. 25.00
11th year onwards	Rs. 37.50

THE MINES AND MINERALS (REGULATION & DEVELOPMENT ACT, 1957.

The Act came into force on 1st June, 1958 superseding the MM (R & D) Act of 1948. It is an important act for regulating grant of concessions and conservation of minerals. In this Act the provisions of minimum and maximum area under prospecting licence and mining lease required to be taken by lessee have been prescribed. Schedules of specified minerals, royalty and dead rent are appended in this Act. One significant point in the act is that it also authorises the Central Government to undertake prospecting or mining in any area not already held under a concession. A new provision has been introduced by an amendment in 1972 enabling the Central Government to make premature termination of mining leases in the interest of regulation of mines and minerals development and to authorise the G.S.I. or such other authority to carry out investigation in the leased out areas for obtaining specific information. The following three rules have been framed under the Act:

TABLE 10

Maximum Quantities of Ores and Minerals Removable Free of Royalty

(Schedule III, Minerals Concession Rules, 1960)

[*See Rule* 14(1) (*ii*) (*a*)]

Class	Mineral	Quantity
1.	Auriferous rock and gravel containing no visible gold.	2 tonnes
2.	Metalliferous ores meant for extracting aluminium, iron and manganese.	10 tonnes
3.	Metalliferous ores meant for extracting antimony, arsenic, bismuth, chromium, copper, lead, nickel, tin, titanium, tungsten and zinc.	5 tonnes
4.	Metalliferous ores meant for extracting cadmium, cobalt, mercury, molybdenum, silver, thallium and vanadium.	2 tonnes

TABLE 10 (*Contd.*)

Class	Mineral	Quantity
5.	Compound ores containing the metals of class 4 in smaller quantities than those of class 3.	5 tonnes
6.	Concentrates of the ores enumerated in classes 3 to 5.	100 kg
7.	Minerals of the rare-earth's groups.	250 kg
8.	Gypsum, limestone, iron pyrites, shales, red oxide and yellow ochre.	5 tonnes
9.	Barytes, bitumen, borax, corundum, emery, grossularite, felspar, flourspar, and calcite.	500 kg
10.	Asbestos, graphite, mica and native sulphur.	50 kg
11.	Sillimanite, kyanite, magnesite, serpentine, steatite, vermiculite, fire-clay, kaolin and other refractory materials.	50 tonnes and any additional quantities subject to the approval of the State Government on payment of royalty
12.	Beryl	10 kg
13.	Coal and lignite	50 tonnes
14.	Uraniferous rock without visible uranium minerals.	2 tonnes
15.	Uraniferous rock with visible uranium minerals.	250 kg
16.	Uranium bearing minerals (including uraniferous columbite-tentalite).	5 kg
17.	Columbite-tentalite (non-uraniferous),	10 kg
18.	All other minerals not herein before specified.	Quantity to be fixed by the State Govt. according to circumstances of each case.

MINERAL CONCESSION RULES, 1960

The rules define procedure of obtaining mineral concessions in government as well as private lands, rights and obligations of a licencee and a lessee. The rules prescribe maximum quantity of ores and minerals recoverable free of royalty. A provision has been made under rule 54 that any person aggrieved by the orders of the State Government can approach the Central Government to redress grievances. The judgement of the Central Government in this respect is taken to be final.

THE MINING LEASES (MODIFICATION OF TERMS) RULES, 1956

The purpose of framing the rules was to bring all the leases granted prior to 1949 in conformity with the existing Mineral Concession Rules. A necessity of such provision was felt after the integration of the feudal States numbering over 600 and odd with Indian Union with effect from 1st April 1950. In those times leases were granted over large areas for indefinite periods extending up to 999 years to the tenants. In Goa, leases granted during Portuguese regime were on perpetual basis. Also the rate of royalty, dead rent, etc. was variant. Power to modify terms does not apply to natural gas, petroleum, coal or minor minerals.

MINERALS CONSERVATION AND DEVELOPMENT RULES, 1958

As the name implies these rules are framed for the conservation and systematic development of minerals. The rules do not apply to minor minerals, atomic minerals, coal, mineral oil and gas. Under the rules, mine owners are required to submit notices for opening, closing, reopening of mines and for stopping operations; and to submit monthly and annual production data in respect of each mine and the explosive consumption returns every quarter to the Indian Bureau of Mines in the prescribed proforma appended to the rules. Mine owners are required to appoint geologists and mining engineers under Rule 21 for efficient prospecting and mining.

THE COAL MINES (CONSERVATION & DEVELOPMENT) ACT, 1974

The above Act which came into force w.e.f. 26th August, 1974 repeals and replaces the earlier Act viz., The Coal Mines (Conservation, Safety and Development) Act, 1952. The Coal Board has been abolished and the functions earlier carried out by the Board under the repealed act are now being discharged by the two public sector undertakings viz., Bharat Coking Coal Ltd., and Coal India Ltd.

THE COAL BEARING AREAS (ACQUISITION AND DEVELOPMENT) ACT, 1957

It was enacted with the object of obtaining greater public control over the coal mining industry and its development. Necessity of such legislation was felt after the formation of National Coal Development Corporation (now merged with CIL) for mining coal under the public sector. The act vests power in the central government to acquire unworked coal bearing areas in both government and private lands and even take over areas under a licence or lease provided they are not being worked.

THE ATOMIC ENERGY ACT, 1962
(Act No. 33 of 1962)

It came into force from 14th September, 1962 repealing the Atomic Energy Act, 1948. The Act provides wide powers to the Central Government for regulating the control of radioactive substance. As per the provisions nobody can mine, sell or use the prescribed substance except with a licence from the Central Government. The government may compulsorily acquire the rights to work the prescribed minerals and also have rights to acquire plants, buildings, railway siding, etc , serving such plants. The mine owners are also required to maintain secrecy about the production and disposal. The information regarding atomic minerals is not disclosed to the public.

THE MINES ACT, 1952
(Act No. 35 of 1952)

It is most important piece of legislation regulating welfare, working conditions and safety of workers engaged in mining activities. It came into force on 1st July, 1952 repealing the Indian Mines Act, 1923. On functional basis, the Act can be divided into two parts, one related with technical directions and other safety and welfare. Power of the Central Government to make rules on technical directives and that of safety and welfare has been vested in sections 58 and 57 respectively. Under section 58, six rules have been framed, namely: (1) The Mines Rules, 1955, (2) The Mysore Gold Mines Rules, 1952. (3) The Mines Vocational Training Rules, 1966, (4) The Coal Mines Pithead Bath Rules, 1950, (5) The Coal Mines Rescue Rules, 1959, and (6) The Mines Creche Rules, 1963. The regulations framed under section 57 are three, namely: (i) The Coal Mines Regulations, 1957, (ii) The Metalliferous Mines Regulations, 1961, and (iii) The Oil Mines Regulation, 1983.

Provisions of the Act, rules and regulations are administered by the Director General, Mines Safety, who has been empowered along with other inspecting officers of the Directorate to inspect mines, and ensure safety, health and welfare of every person employed in a mine.

Under Section 24 there is provision to set up a court of enquiry in cases of accidents. Central Government is vested with wide powers to constitute Mining Boards and Committees to look into any matter connected with the mines or group of mines. Under Section 22 inspectors have been empowered to take immediate action when employment of persons is dangerous in any mine.

Stringent punishments of imprisonment up to two years or fine up to five thousand rupees or both have been provided for contravention of various provisions. Offenes under the Act cannot be tried in a court inferior to that of a resident magistrate or magistrate of the first class.

Minimum age of employment on surface has been raised up to 15 years and in underground up to 16 years. Women are not permitted to be employed underground. Women can, however, be employed above ground between 6 a.m. and 7 p.m. No person is allowed to work for more than 6 days in a week and up to 10 hours inclusive of overtime in a day. Overtime up to twice the ordinary rate of wages have been prescribed. Entitlement of annual leave is directly linked with number of attendance which are 190 for underground workers and 240 for other workers in each calendar year. One day of leave with wages have been prescribed for every 16 and 20 days of underground and other attendance respectively.

MINES RULES, 1955

These rules came into force on 2nd July, 1955. They mainly deal with employment and welfare of miners. These rules provide terms and powers of Mining Boards, Committees, representations of different strata of personnels in the Boards and committee; health and sanitary conditions, first aid and medical facilities, employment, leave with wages and overtime to the workpersons. Management are required to submit periodical notices, regarding time of relays and shifts of different categories of workers. Types and nature of different records to be maintained are prescribed in details. These rules have also rescinded the Mysore Gold Mines Rules 1952 except the rules contained in Chapters 4 and 5 which continue to apply to gold mines located in Karnataka.

COAL MINES REGULATION, 1957

These regulations came into force on 24th October 1957. These deal with the working of coal mines and safety of workers. Mine managements are required to submit monthly and annual returns, notices regarding occurrences of accidents and incidence of diseases, monthly returns relating to quantities of material stowed, notices regarding appointment or termination of services of managers, assistant manager, engineers, surveyors, etc. Qualification and number of officials including that of managers have been laid. Chapter 3 governs formation of Board of Mining examinations for purpose of holding different examination. Retiring age of 60 years for all officials have been prescribed unless otherwise found medically fit. Comprehensive technical provisions covering safety in mine in different aspects relating to maintenance of accurate plans and sections, means of access and egress, transport of men and material, safety in actual mining operations, precautions against fire, dust, gas and water, ventilation and lighting, safe use of explosive and machinery have been prescribed.

METALLIFEROUS MINES REGULATION, 1961

This regulation came into force on 11th March 1961. The provisions in this regulation are similar in nature to those in Coal Mines Regulation except that these are applicable to metalliferous mines.

Legislations affecting the labour working in the mines are also embodied in: (1) Employees Provident Funds and Family Pension Fund Act, 1952; (2) Payment of Gratuity Act, 1972; (3) Workmen's Compensation Act, 1923; (4) Employees State Insurance Act, 1948; (5) Coal Mines Provident Fund, Family Pension and Bonus Schemes Act, 1948; and (6) Industrial Disputes Act, 1947. The mine labourers are further governed by such other regulations of the country in vogue viz. the Minimum Wages Act. 1948 which have got applications on wage earners and industrial labourers. In addition to the legislation mentioned above, the Maternity Benefit Act, 1961 is also applicable to mines and for this purpose, Maternity Benefit (Mines) Rules, 1963 have been promulgated.

8

Mineral Taxation and Incentive Measures

THE METHOD of taxation applied to mineral industry and the principles involved in it form an important aspect of study to students of mineral economics for the development of mineral properties and trade. A mineral economist is not expected to be a practising legal expert on taxation but he is supposed to have professional knowledge of it, that will enable him to study the incidence of taxes on cost at each stage of development; for the laws governing taxation may, at some stage, be a factor limiting the success of a proposed venture and likewise constitute a problem to an operating venture.

Taxes can be of Central, State and Municipal. These add to the selling price over and above transport cost by road and rail, port charges, custom duty, shipping freight, etc. the burden of which falls on the consumer. In mining the incidence of taxes commences from the very beginning of acquiring lease to the ultimate selling point.

Incentive, on the other hand, is a measure largely coming under the ambit of income-tax structure providing special rebates and allowances to an investor on his earnings, for risking his capital. Although taxes and incentives are diametrically opposite in principles, the former aiming at collection of revenues and the latter in providing fiscal reliefs, yet both go hand in hand. Revenue is necessary for the Government, at the same time peculiarities inherent in mining warrant careful consideration in taxation which should be set forth in such a manner so as to serve both as a revenue earner and at the same time provide sufficient incentives for attracting required capital to this field. As a rule, it should be based on the principle by which the investor is allowed to recoup his capital in the shortest possible time, and the profits must cover abortive expenses in prospecting and supply of needed capital for the development of new mines. If the life of a mine is estimated to be thirty years, the capital invested must be allowed to be recouped within a period of less than fifteen years. The net return after paying the taxes, therefore, should be high enough not only to cover the normal profit but also to attract risk capital. The method of incentive, however, differ from country to country depending upon socioeconomic conditions. A local investor in a developing country may seek more liberal tax concessions than those prevalent in advanced countries for the simple reason that the industries in advanced countries have attained a high degree of stability in technical knowhow, capital gains and finance, capable of solving most of the problems without seeking much aid from the Government. Such conditions obviously lack in the developing countries.

INCIDENCE OF TAXES

The taxes levied in India can be grouped as under, depending upon the revenues collected by the State, Municipal and the Centre.

States	*Municipal*
(i) Dead rent and royalty.	(i) Octroi duty.
(ii) Surface rent.	
(iii) Sales tax.	
(iv) Road tax.	
(v) Education cess.	
(vi) Panchayat tax.	

Central

(i) Coal Mines Labour Welfare Fund Act, 1947.

(ii) Mica Mines Labour Welfare Fund Act, 1946.

(iii) Limestone and Dolomite Mines Labour Welfare Fund Act, 1972.

(iv) Iron Ore Mines and Manganese Ores Mines Labour Welfare Fund Act, 1976. (It came into force from 1st Sept., 1978).

The incidence of taxes accruing from the dead rent, royalty and surface rents has been dealt in Chapter 7 on pages 132, 133 and 141. Sales tax is of two types, the Central sales tax and the States sales tax. Both are administered by the State Government. The Central sales tax is levied on the inter-States sales, and the State sales tax on the intra-State sales. In the case of minerals, the provisions in the Central as well as States sales tax laws have been made in such a manner that minerals do not bear more than 2% of tax on the sale value. Minerals meant for export are exempted from sales tax.

Octroi duty is charged by municipality for all the incoming commodities in the municipal area. Each municipality has its own rate. Road tax is levied by the State Government of Maharashtra, as per rates prescribed by Maharashtra Tax on Goods (Carried by Road) Act, 1962. Panchayat Tax is also levied by the Government of Maharashtra. Similarly, the State Government of Orissa, under Orissa Mining Area Development Fund Act and Orissa Taxation Act, levies some nominal cess on minerals produced. The Government of Andhra Pradesh levies Education Cess calculated on royalty paid.

Under the Central Act, funds are collected for providing amenities to the labourers employed in iron ore, coal, limestone, dolomite and mica mines. The present rate of cess on iron ore is 50 paise per tonne excluding unmarketable fines and rejects. Cess on coal is levied on coal and coke despatched from collieries subject to minimum of 25 paise and maximum of 75 paise per tonne. Cess on mica (of all kinds) is levied presently at the rate of $3\frac{1}{2}$ ad valorem of the F.A.S. price w.e.f. 17.7.74. The cess on dolomite and limestone has been fixed at 20 paise per tonne on the quantities sold or consumed. The maximum limit of cess on dolomite and limestone as per the act is Re. 1. The cess on iron ore and manganese ore as per the Act will not exceed Re. 1 and Rs. 6 respectively. The cess on manganese ore is still to be levied.

In addition to cesses discussed above there is appreciable impact in the selling price of mineral commodities due to the central excise duty. Now about one per cent central excise duty is levied on practically all minerals produced in processed form, washed or beneficiated. On coal different rates of excise duty as mentioned in the table on next page are levied. The net proceeds from such excise duty on coal is utilised for conservation and other activities eligible for assistance under the Coal Mines (Conservation and Development) Act, 1974.

Description of coal and coke	Rate of excise duty/tonne
(i) On all coals including soft coke but excluding coking coal and hard coke.	Rs. 1.65 (Rupees one and paise sixty-five only).
(ii) On all coking coal.	Rs. 2.40 (Rupees two and paise forty only).
(iii) On hard coke.	Rs. 3.60 (Rupees three and paise sixty only).

INCENTIVE MEASURES

Various methods are adopted in providing incentives to the mining sector. These can be classified under three groups: (i) concessions in income-tax, (ii) facility of loans on easy terms and government's sharing the risk, and (iii) simplification of taxation laws.

Concessions Allowed in Income-Tax

Income beyond certain limit of net earning becomes liable for tax. The expenditure for taxation purposes is broadly classified under two heads, namely, (i) revenue, and (ii) capital. Revenue expenditure comprising salary, wages, office expenses, interest, etc. is allowed to be deducted from the gross income in a particular year. Any asset acquired for producing income forms the capital and the expenditure made on the same by way of purchase of machinery and plant and their installation, prospecting, exploration, shaft sinking, laying of pipe lines, road and all other pre-production expenditure like construction of building, staff quarters, etc. comes under capital expenditure. Such expenses are generally disallowable from the gross income, instead certain reliefs are granted. The scope of incentive lies in the capital expenditure. To invite capital to mining industry, various countries provide attractive rebate and allowances on the capital expenditure which can be placed under the following subheads: (*i*) depreciation allowance on machinery and plant, (*ii*) pre-production expenditure like prospecting, etc., (*iii*) development rebate on new machinery purchased, (*iv*) depletion allowance, (*v*) sinking fund, and (*vi*) tax holiday.

Under the Indian Income Tax Act, 1961 (I.T. Act), the mining industry is treated at par with any other industry. There is no such provision like depletion allowance (except in the case of mineral oil) and the liberalised tax holiday as prevalent in advanced countries, as it will be found from the discussion later in this chapter. For the first time a new provision has been introduced under Section 35 E of the Income Tax Act brought out by the Taxation Laws Amendment Act 1970 to amortise expenditure incurred on prospecting for minerals and development of mines. Prior to this, expenditure on prospecting and exploration was admissible only when some tangible results have been achieved by way of proving reserves. A new provision has also been introduced under section 35D for amortising preliminary expenses incurred for specific purposes. In case of business consisting of prospecting for or extraction or production of mineral oil in relation to which the Central Government has entered into an agreement with any individual or firm, there is a provision under Section 42 of the Income Tax Act to allow such expenditure, in addition to expenditure admissible under the Act, which normally would not have been allowable. The incentives allowed under the Income Tax Act, 1961 are given below.

	Incentives	*Section under which allowed*
1.	Depreciation allowance	Section 32
2.	Investment allowance	Section 32 A
3.	Rehabilitation allowance	Section 33 B
4.	Expenditure on scientific research	Section 35
5.	Amortisation of preliminary expenses	Section 35 D
6.	Deduction of expenditure in prospecting, etc.	Section 35 E
7.	Tax holiday	Section 80 J

Earlier in the Income Tax Act there was a provision of classifying some industry including mining of certain minerals, namely, ores of aluminium, copper, lead-zinc, iron and manganese, coal, lignite, dolomite, magnesite, and oil and gas, as 'priority' industry. The industry listed under the 'priority' industry was receiving special treatment for the purpose of taxation. However, a distinction between 'priority' and 'non-priority' industry has been abolished w.e.f. 1.4.74. In the Act, there was also a provision of giving 'development rebate'. It was a kind of relief allowed in first year only on the cost of installation of new machinery and plant in addition to depreciation allowance. Development rebate has been replaced by investment allowance brought into effect from 1.4.76. The concession enjoyed under development rebate was however withdrawn much earlier. No concession was allowed under the said provision on plant and machinery installed after 31.5.74.

Depreciation

It is defined as capital cost allowance. To commence mining and to keep a mine running, a long list of machinery and spares is required. These machinery are prone to wear and tear due to constant use, have limited years of life, and require replacement. Therefore, almost all countries provide in their income-tax rules, certain annual deductions on the capital expenditure computed on a diminishing balance basis as depreciation allowance from the gross income for the purpose of computing income tax. It is one of the most important measures of incentive. If an appropriate depreciation commensurate with the risk and hazard involved is given it may attract large capital to mining industry. The maximum rate of depreciation allowance varies from country to country. A number of countries for example, the U.S.A., Canada, Australia allow a full tax rebate on the preproduction expenditure incurred during the course of prospecting, exploratory mining, etc. and grant higher depreciation on the capital investment and in some cases like sinking of shaft, laying of pipe lines, installation of haulage allow 100% depreciation in the very first year of assessment. The main idea behind granting higher depreciation is to allow an investor to get back his capital as quickly as possible for further investment by postponing the tax liability.

Depreciation allowed, vide Rule 5 of Income Tax Rules, 1962 as in force from the 1st day of April, 1970, in respect of the plant and machinery generally used in mining industry is as given below.

Items on which depreciation is allowable	Rate: On written down value basis
A. Mines and quarries (No extra shift allowance allowed)	
(a) Surface and underground machinery (other than electrical machinery, boilers and portable underground machinery), headgear, moving parts and rails	15%
(b) Portable underground machinery and earth moving machinery used in open cast mining	30%
(c) Ropeway structures—ropeways, ropes and trestle sheaves and connected parts	30%
(d) Tubs, winding ropes, haulage, ropes and sand stowing pipes	100%
(e) Safety lamps	100%
(f) All other items like boilers and headgears (excluding moving parts), shafts and inclines and tramways on the surface	10%
B. Machinery and plant coming into contact with corrosive chemicals (NESA)	15%
C. Machine tools (NESA)	15%
(a) Automatic and Semi-automatic	
(b) Precision machine tools, e.g. grinding machines	
D. Mineral oil (NESA)	
(a) Field operations (distribution). kerside pumps including underground tanks and fittings	15%
(b) Field operations above ground—portable boilers, drilling tools, wellhead tanks, rigs, etc.	30%
(c) Fields operations below ground excluding items (a) and (b) above	100%

(d) Plant used in field operations (above ground)— 100%
Returnable packages

(e) Pipe lines above ground, boilers, prime movers 10%
process plant, storage tanks above ground

E. Ships (NESA)

(a) Ocean going ships

5% to be
calculated
on the
actual
cost

(b) Vessels ordinarily operating in Inland waters

   (i) Speed boats 20%

   (ii) Other vessels 10%

F. Aerial Photography Apparatus (NESA) 30%
Aeroplanes—Air craft (NESA) 30%
Aeroplanes—Aero-engines (NESA) 40%

G. Buildings	Factory	Non-factory
First class	5%	2.5%
Second class	10%	5%
Third class	15%	7.5%
Purely temporary	100%	100%
Other class	—	2.5 to 5%

H. Motor lorries, Motor tractors, Motor buses (NESA) 30%

I. Weighing machines (NESA) 10%

J. Furniture and fittings (general rate) 10%

K. Office machinery (NESA) 15%

L. Road making plant and machinery (NESA) 15%

W. Railway sidings (NESA) 10%

D. Locomotives, rolling stock, tramways and railways 10%
used by concerns, excluding railway concerns (NESA)

Rehabilitation Allowance

Under Section 33 B of the Income Tax Act 1961 introduced w.e.f. 1.4.67, where the business of an individual undertaking carried on in India is discontinued in any year by reason of extensive damage to or destruction of any building, machinery, plant or furniture owned by the tax payer and used for the purposes of the business as a result of natural calamities like flood, typhoon, hurricane, cyclone, earthquake, etc., or riots and civil disturbance, or accidental fire or explosion, or enemy action with or without a declaration of war and if within a period of three years from the end of the year in which the discontinuance takes place, the business is re-established or revived by the tax payer then in the year in which the rehabilitation takes place, the tax payer is entitled to a rehabilitation allowance which is equal to 60% of the loss worked out under Section 32 (1) (iii) for the year in which the destruction has taken place, allowance allowable in respect of that building, machinery, plant, etc., damaged or destroyed. Since the word "industrial-undertaking" in Section 33 B has been explained as including mining, this rehabilitation allowance is also admissible to mining business in the circumstances stated above.

Expenditure on Scientific Research

Under this provision capital expenditure on scientific research if incurred before 1.4.67, 1/5th of the expenditure is allowable as a deduction for each of the 5 years including the year in which the expenditure was first incurred. If, however, the capital expenditure is incurred after 31.3.67, the whole of such capital expenditure is allowable in the year in which it has been incurred. Further, where any capital expenditure is incurred for scientific research before the commencement of the business, the total expenditure for a period of 3 years immediately prior to the commencement of the business will also be considered as having been incurred in the year of commencement of the business and will qualify for deduction.

Amortisation of Preliminary Expenses

The Taxation Laws Amendment Act, 1970 has introduced for the first time a provision for amortisation of preliminary expenses incurred by ..dian Companies and business concerns to be formed or where expansion of existing business concerns takes place. These are contained in section 35D of Income Tax Act, 1961. Certain conditions are, of course, prescribed. The kinds of expenditure which can be amortised are expenses for preparation of feasibility report and project report, conducting market survey, engineering services related to the business, legal charges for drafting of agreements. expenses for drafting and printing of memorandum and articles of the company, fees for registration of the company, issue of shares, underwriting commission, brokerage and advertisement of prospectus and such other expenses as may be prescribed. The amortisation is achieved by deduction of 1/10th of expenditure for 10 years from the year of commencement of business of completion of the expansion of the existing undertaking, as the case may be.

Deduction of Expenditure in Prospecting etc.

The Taxation Laws Amendment Act, 1970 has introduced w.e.f. 1.1.1971 measures to amortise expenditure incurred for prospecting for minerals and development of mines by introducing a new section, called Section 35E. A list of minerals to which this provision attracts is given in the 7th schedule of the Act which covers all minerals, except those declared as 'minor minerals' under the Mines and Minerals (Regulations and Development) Act, 1957. As per this section, expenditure for acquisition of site or rights over the site, acquisition of deposits of minerals or any rights over the same etc. can be amortised by Indian companies and other assessees resident in India who are engaged in any operations relating to prospecting for or extraction of production of any mineral or group of minerals. The expenditure, however, should have been incurred after 31st March, 1970. The expenditure should relate to the year of commercial production and any one or more of the four years immediately preceding that year. The deduction allowable is 1/10th of the expenditure incurred and will be allowable to a maximum of 10 years from the year of commercial production.

Tax Holiday

Originally Section 84 of the I.T. Act provided that no income-tax is required to be paid in respect of the profits from an industrial undertaking to which the provisions of the Section applied to the extent of 6% per annum on the capital employed in the undertaking. In other words, a deduction from the total income up to 6% of the capital employed in the undertaking would be allowable subject to certain conditions. This section has been omitted w.e.f. 1.4.68 and from that date Section 80J has been introduced which provides for a straight deduction from the income of the undertaking at the rate of $7\frac{1}{2}\%$ of the capital employed in the undertaking and this deduction is admissible for a total period of 5 years, viz., the year in which the industrial undertaking began to produce or manufacture and the 4 years immediately succeeding that year. How the capital employed in the undertaking is to be computed and what are the conditions which make an industry eligible for this are given in the Rules. For tax holiday purpose the capital employed is generally taken to be paid up capital plus the borrowed capital if any.

Tax holiday is a concession allowed at the initial stage of any undertaking including mining. Under this provision, no income tax is charged for a stipulated period. Through the tax holiday industry is given opportunity of rapidly recovering the capital expenditure which was necessary to achieve production. In the Indian Income Tax Act, the provision of tax holiday is limited to the extent $7\frac{1}{2}\%$ profit only on capital invested within the stipulated period of five years. The tax becomes payable on the income exceeding $7\frac{1}{2}\%$ of the capital in any one of the five years from the year of first commencement of production. In Canada and the U.S.A. a complete tax holiday for a period of three years is allowed irrespective of profits from the year of commencement of sale in reasonable quantities. Contrary to this under the Indian Income Tax Act, the period of tax holiday begins from the very year of commencement of production, although the rate of production may not be compatible to the capacity. Many advanced countries have started offering tax holiday for a longer period. The Government of Colombia, for example, offers 10 years tax holiday on new mines.

DEPLETION ALLOWANCE

Depletion allowance is provided in recognition of the fact that mineral deposits are depleting or exhaustible assets. Through this allowance it is intended that a mine owner will make expenditure at least equivalent to the taxes saved in order to prolong its productive life or locate new deposits. The countries like the U.S.A., Canada and Australia allow some proportion as depletion allowance which may extend up to $33\frac{1}{3}\%$ of the total profit in addition to depreciation allowance.

SINKING FUND

The sinking fund or redemption is an amount to which yearly or periodic payments are made from the earnings and the amount is improved at a safe rate of compound interest so that in a given period of time allowed, redemption, the actual payment plus the accrued interest, will equal the capital invested. In the Republic of South Africa, a complete redemption of capital expenditure is allowed over the life span of the mine but no separate depreciation or depletion allowance is allowed.

FINANCIAL PARTICIPATION

It is one of measures introduced in recent times to encourage mineral development activities. A number of countries have laid down procedures for aiding development by physical and monetary participations.

In the United States, the Office of Minerals Exploration (OME) gives direct assistance both financially and physically to small domestic companies and prospectors. Minerals have been listed for such joint participation and OME contributes up to 50% of the costs. In some cases like scarce metals, silver and gold OME's contribution is extended up to 75%. The return of the investment is effected through royalty on production if ore is found; otherwise the government's share is written off. The Government of Australia provides financial assistance in respect of exploration for minerals, oil and gas only. On the other hand, the British Government offers financial assistance to the companies exploring for non-ferrous and non-metallic deposits in U.K. Under this scheme, the Government assists the companies up to 35% of the exploration and evaluation cost.

The Government of Rhodesia provides substantial financial assistance to cover mining risk. If the government is satisfied that the owner is unable to raise sufficient capital through private sources, it may grant loan to carry out work. Two or three types of government's loan are prevalent, one of which is on no-risk-loan basis. Under this scheme, if the operation is successful the miner will repay the loan at 12% interest and will sign a provision permitting government to sell the property if he does no development of the mine at a specified rate. If such project is unsuccessful, the miner will, have no formal risk. Other types of development incentives are also known where the cost of development is borne by 50-50 basis. The miner's liability is limited to 50% of the total sum if the project is unsuccessful. Presently, Rhodesia offers freely on 6% interest loan on full personal risk.

SIMPLIFICATION OF TAXATION LAWS

The idea of simplification in the procedure of collection and payment of taxes is often mooted with a view to remove the cumbersome burden of maintenance of accounts and payment of multiple taxes at different stages and thereby promote investment in mining. To achieve this objective, the Government of Brazil passed a bill in 1965 creating a single tax on minerals which is uniform at the rate of 10% on minerals other than coal and 8% on coal determined on the basis of f.o.b. value less 40% allowance towards transport cost. The tax is shared by the Federal, Municipality and the State, in the ratio of 10%, 20% and 70% respectively.

9

Tenor, Grade and Specification

BEFORE THE subject matter of the topic is dealt it is necessary to understand first the distinguishing features between *mineral* and *ore*. According to the conventional usage 'ore' is a natural material from which metal or its alloy is extracted at profit. It is described that all ores are minerals or aggregate of minerals (rock-ore as sometimes ore-dressing engineers call it) but reverse is not true; indicating clearly that the application of the word 'ore' is restricted to mean metallic mineral constituents only. This orthodox concept has, however, undergone a profound change since early fifties. The meaning of ore has been extended to connote non-metallic minerals also, although with some reservations. Pyrite is now often called an ore of sulphur and iron. Fluorite, barytes, asbestos, vermiculite ores, etc., are found to be commonly used by the various authorities in the geological literature. The ore in the case of non-metallic minerals generally connotes the crude, which requires processing for elimination of superficial gangue before making it marketable. As in case of asbestos, unless the rock containing fibres are mined, crushed and milled to separate out the fibres, it cannot be put to use or marketed. The rock containing fibres in this case is considered as an ore of asbestos.

The term 'ore' therefore is to be understood and appreciated in the context it has been used, and may be defined 'as natural aggregate of one or more minerals from which one or more metals and economic minerals can be recovered profitably.' It is necessary to emphasise here 'economic minerals' as a mineral may or may not have any economic significance and may form the gangue material only. The emphasis on profitability, although significant and important, is also undergoing a change in outlook. In case of strategic or atomic energy minerals, it may be necessary to work a deposit by subsidy.

TENOR

Originally the word tenor was used to mean metal content in an ore. Its scope has been enlarged like an ore to cover non-metallic deposits also. It may thus be understood to mean as the incidence of mineral or metal content in an ore. The tenor in the case of non-metallic minerals conveys the percentage of mineral content in the ore (rock ore) mined; whereas in the case of metallic minerals it signifies metal content in an ore mined. As for example, when we call, say in the case of asbestos mine that the tenor is 6%, it means that the incidence of asbestos fibre in the ore is 6%. In the case of metallic deposits, say copper ore mine, when it is referred to as 2%, it means that the ore contains 2% copper metal. The tenor may reach up to 100% in the case of native deposits. The tenor may vary from deposit to deposit. The lower limit of tenor up to which a deposit can be worked at profit is dependent upon the various technoeconomic factors like location, size, price, advancement of mining and extractive technology and the presence of other recoverable substances.

The tenor has a great economic significance to a mine-life or a deposit. It signifies much more than the mere incidence of a mineral or metal content in a mineral property and it can be rightly described as an economic indicator. Once again taking an example of an asbestos deposit, it may not be sufficient to say only that the tenor is 6% for economic evaluation. It is necessary to indicate the percentage recovery of longer fibres in overall recovery. In the case of chrysotile asbestos the tenor is generally found varying from 3 to 6% as is known from the various deposits, being worked throughout the world. It is found that unless in property the proportion of longer fibre is more than 50% of grade 4 or at least 5 of the Quebec Standard, it is difficult to establish the mine which can well guarantee the reasonable return on investment under varying market conditions. For this reason the data on tenor in the case of asbestos ore bodies should reveal not just the percentage recovery but percentage recovery of each grade as well. Similarly, in the

case of mica mines if the crude mica mined does not give a minimum of 10% block mica (on an average generally 4, 5 and 5.5 are
produced) after processing, in that event the same mine is likely
to be rendered uneconomic. In mica also the tenor of the crude
varies from 3–6% usually.

GRADE

The grade, broadly signifies: (i) the commercial classification and
takes into account the chemical and physical properties as well,
and (ii) an expression of the quality of deposit and may be expressed in terms of metal contents in case of metallic deposits like
iron ore, manganese ore, base metal minerals, etc.; in terms of
penny weight, ounce, troy ounce, gramme, etc., per unit weight in
case of precious metals; in terms of percentage of oxides in the case
of many metallic and non-metallic minerals e.g., Cr_2O_3 in chrome
ore, WO_3 in tungsten ore and P_2O_5 in apatite and rock phosphate,
In the case of coal, limestone and such other deposits more than
one criterion are adopted for grading i.e. in the case of coal, the
carbon content, ash content, volatile matter, calorific value, moisture and the caking quality, all these together determine a grade
of coal. In some cases, impurities also determine a grade, for example, the presence of sulphur and phosphorus in iron ore, manganese ore, and coal; the combined iron oxide and alumina content
in rock-phosphate for the manufacture of superphosphate. Again,
in some cases, the strength of the material and colour are taken
into account for grading purpose. Spinning and non-spinning types
in case of asbestos; snow white, white and off colour in case of
barytes; and friable, compact or massive and crystalline in case of
limestone are some of the examples. China clay in trade is graded
as super textile, paper, rubber, paint grades, etc., depending upon
mainly the brightness and grit content. Rubber manufacturers
specify absolutely copper free china clay as even small trace of
copper results in early decay of rubber. Yet, in other case the physical properties and the size are the only two factors considered for
grading, the chemical composition does not come into the picture
at all; as for example in the case of mica and asbestos. Grading of
minerals on size (fines, lumps, etc.) is not uncommon. Examples
of prevalent grades for commercial classifications of various ores
and minerals are given in Table 11. Classification and gradation of

coal as per Colliery Control Order, 1945, and continued in force by Section 16 of Essential Commodities Act, 1955, is given in Table 12.

SPECIFICATION

The specification is intimately related with the grade. It pinpoints the tolerance limits of all constituents present in it. Individual consumers may prescribe different specifications for the same grade of mineral which is dependent upon two factors: (i) the technique of manufacturing process adopted by individual units, and (ii) the grade of other raw material required to be used (as a whole), if necessary, to obtain the end product. A little clarification by way of example will make this point more clear. When we talk about flux grade limestone, immediately we think of two grades, namely, the blast furnace grade and the steel melting shop grade with special relation to its suitability in the two furnaces depending upon the total insolubles present in it. It should not only satisfy the chemical constituents but also should be compact and non-fritting. Further, for the same blast furnace or steel melting shop grade the individual steel units may specify a different set of specifications describing the minimum of lime content required and tolerance limits of other constituents present as impurities, besides the size of the material. This specification again may be depending upon the chemical constituents of other raw material used. In a blast furnace if the coke which is to be used contains very low ash, then the permissible limit of alumina in limestone, dolomite and even in iron ore can be greatly raised. It has been found that, if the alumina percentage in the slag of blast furnace during pig iron making process is either below 15% or higher than 22%, the slag becomes viscous. To reduce the viscosity, it is necessary, that the alumina percentage should be kept between 15 and 22%; the optimum being 18%. The Indian coke contains high ash usually above 18% ultimately introducing more percentage of alumina. In such cases, it is necessary that the alumina percentage in other raw material for iron making is prescribed carefully. In other case, take the example of Australian coke which contains invariably less than 5% ash, therefore, while using coke of such type, the specification of other raw material used in iron and steel industry will vary from those when coke containing high ash is used. The presence of deleterious elements like phosphorus and sulphur in one particular mineral supply may call for stringent stipulation of specification in other raw material so that the total may not exceed the maximum permissible limit.

Specification again highlights the special characteristics of a mineral which may render it suitable or unsuitable for a particular industry. As in case of bauxite for aluminium extraction, the presence of reactive silica more than 2% makes the ore unsuitable. Presence of free silica and iron is not regarded deleterious in aluminium industry. The refractory manufacturers, on the other hand, give more stress on low silica and iron contents. The presence of SiO_2 and Fe_2O_3, more than 3% each, render the bauxite unsuitable for the manufacture of high class refractory bricks. The cement manufacturers prefer to have limestone containing less than 3% MgO; up to 5% is tolerated. Magnesia forms periclase during the manufacture of cement, which continues to expand slowly on hydration resulting formation of cracks even after normal period of setting.

Again for purification of sugar cane juice, the manufacturers prescribe high purity of lime containing not less than 95% CaO, 1% MgO maximum and low in silica and iron oxide. The combined insolubles $(SiO_2 + Al_2O_3 + Fe_2O_3)$ should not be more than 4%. The juice is purified by double carbonation and sulphitation process. Sulphur dioxide gas combines with magnesium oxide and form magnesium sulphate which is laxative and may give adverse effect on the quality of sugar refined. Silica precipitates as gelatinous mass and coats the sugar crystals. Iron oxide goes into the solution and colour the product.

Further study of few typical examples of specification provided by the industry would be of interest (see Table 13). Let us take the case of standard grade ferro-manganese. Suitability of manganese ore for the purpose is mainly dependent on Mn : Fe ratio which should be minimum 7 : 1 or more because, it maintains high production capacity in the furnace. No standard grade ferro-manganese can be manufactured if Mn : Fe ratio falls below the stipulated ratio. Besides, it should have very low phosphorus content as it is most unwanted impurity. The entire quantity of phosphorus, present in ore, goes in the alloy which deteriorates the quality. When ferro-manganese containing high quantity of phosphorus is used in steel making, it causes cold shortness i.e. the steel becomes brittle when cold. The

TABLE 11
Gradation of Principal Minerals and Ores

	Metal content	Commercial classification	Other desirable characteristics
Manganese ore	MnO_2, 78% min	Battery grade	Iron soluble in HCl should not be more than 4%
	46–48% Mn	First grade	
	44–46% Mn	Second grade	
	40–44% Mn	Medium grade	
	Below 35% Mn	Low grade	
Chromite	48% Cr_2O_3 min	Metallurgical grade	Cr: Fe = 2.8:1 min
	38–48% Cr_2O_3	Refractory grade	Al_2O_3, 12 to 24% combined Cr_2O_3 + Al_2O_3 should be above 60%
	48–50% Cr_2O_2	Chemical grade	Cr: Fe = 1.6:1
Bauxite	Al_2O_3, 50% min	Metal grade	Total silica less than 5%
	Al_2O_3, 58% min	Chemical grade	Fe_2O_3 content less than 3%
	Al_2O_3, 55% min	Refractory grade	SiO_2 and Fe_2O_3 should be less than 3% each
Fluorite	CaF_2, 70–80% min	Metallurgical grade	Available CaF_2 is calculated by deducting 4% from the total for each per cent of silica
	CaF_2, 91% min (available)	Acid grade	
	CaF_2, 95–98%	Ceramic grade	
Dolomite	CaO, 28–33% min MgO, 18–20% min	Blast furnace grade	Total insoluble 7% max.
	CaO, 29% min MgO, 20% min	Steel melting shop grade	Total insoluble 4% max.
	consistent chemical composition	Glass grade	Fe_2O_3 less than 0.2%
Limestone	CaO, 45%	Cement grade	MgO, 3% max
	CaO, 46–48% min	Blast furnace grade	Total insoluble 11.3% max
	CaO, 48% and above preferred	Steel melting shop grade	Total insoluble less than 4%
Gypsum	$CaSO_4$ $2H_2O$ +85%	Fertiliser grade	
	$CaSO_4$ $2H_2O$ +70%	Cement grade	

TABLE 12

Classification of Coal
(effective from 14th Feb. 1981)

Indian coals are mainly bituminous. The exceptions are—lignite of Tamil Nadu, Rajasthan and J & K. The bituminous coal is broadly divided into two categories—coking and non-coking. The bituminous coal which can turn into hard lumpy mass on devolatisation in an inert atmosphere is called coking coal. The hard coke is used in blast furnaces, in ferrous and non-ferrous foundries and in chemical industries. The non-coking coal after distillation makes excellent soft coke for domestic fuel. Depending upon the quality, coking coal is further sub-divided into prime-coking, medium-coking and semi-coking and placed under 8 different grades based on the ash content in the case of coking coal and ash plus moisture in semi-coking coal.

Non-coking coal which till July, 1975 was classified on the basis of ash and moisture contents is now classified into 7 different grades on the basis of useful heat value. The new grading system based on the heat value is considered most scientific both from the point of view of pricing and in providing of combustion parameters. Wide price differentiation between grades are expected to help conservation of quality coal and encourage quality control as well as designing of burning equipment for lower grades which are abundant whereas the higher grades are limited.

TABLE 12

Revised Classification and Gradation of Coal
(effective from 14.2.81)

Sl. No.	Class	Grade	Grade/specification
1	2	3	4
1.	Non-coking coal produced in all States other than Assam, Andhra Pradesh,	A	Useful heat value exceeding 6200 kilocalories per kilogram.
	Meghalaya, Arunachal Pradesh and Nagaland	B	Useful heat value exceeding 5600 kilocalories per kilogram but not exceeding 6200 kilocalories per kilogram.

TABLE 12 (*Contd.*)

1	2	3	4
		C	Useful heat value exceeding 4940 kilocalories per kilogram but not exceeding 5600 kilocalories per kilogram.
		D	Useful heat value exceeding 4200 kilocalories per kilogram but not exceeding 4940 kilocalories per kilogram.
		E	Useful heat value exceeding 3360 kilocalories per kilogram but not exceeding 4200 kilocalories per kilogram.
		F	Useful heat value exceeding 2400 kilocalories per kilogram but not exceeding 3360 kilocalories per kilogram.
		G	Useful heat value exceeding 1300 kilocalories per kilogram but not exceeding 2400 kilocalories per kilogram.
2.	Non-coking coal produced in the States of Andhra Pradesh, Assam, Arunachal Pradesh, Meghalaya and Nagaland.		Not graded
3.	Coking Coal	Steel Grade I	Ash content not exceeding 15 per cent.
		Steel Grade II	Ash content exceeding 15 per cent but not exceeding 18 per cent.
		Washery Grade I	Ash content exceeding 18 per cent but not exceeding 21 per cent.

	Washery Grade II	Ash content 21 per cent but not exceeding 24 per cent.
	Washery Grade III	Ash content exceeding 24 per cent but not exceeding 28 per cent.
	Washery Grade IV	Ash content exceeding 28 per cent but not exceeding 35 per cent.
4. Semi-coking and weakly coking coals	Semi-coking I	Ash plus moisture content not exceeding 19 per cent.
	Semi-coking II	Ash plus moisture content exceeding 19 per cent but not exceeding 24 per cent.
5. Hard Coke	By-product Premium	Ash content not exceeding 25 per cent.
	By-product ordinary	Ash content exceeding 25 per cent but not exceeding 30 per cent.
	Beehive Premium	Ash content not exceeding 27 per cent.
	Beehive Superior	Ash content exceeding 27 per cent but not exceeding 31 per cent.
	Beehive Ordinary	Ash content exceeding 31 per cent but not exceeding 36 per cent.

Notes—1. Coking Coals are such coals as have been declared as coking coals by the erstwhile Coal Board under the Coal Mines (Conservation, Safety and Development) Act, 1952 or such coals as have been declared or may be declared as coking coal by the Central Government under the Colliery Control Order, 1945 or the Coal Mines (Conservation and Development) Act, 1974 and the rules or regulation made thereunder.

2. 'Semi-coking Coals' and 'weakly coking coals' are such coals as were declared as 'blendable coals' by the erstwhile Coal Board under the Coal Mines Conservation (Safety and Development) Act, 1952 or as may be declared as 'Semi-coking' or ''Weakly coking' coals by the Central Government.

3. Coals other than Coking or Semi-coking or weakly coking coals are non-coking coals.

4. 'Useful Heat Value' is defined by the following formula:
HU$=8900-138\times$(A plus M).
Where HU$=$Useful Heat value in kilocalories per kilogram.
A$=$Ash content in percentage.
M$=$Moisture content in percentage.

TABLE 12 (Contd.)

In the case of coal having moisture less than 2% and volatile content less than 19% the useful heat value shall be the value arrived at as above reduced by 150 kilo calories per kilogram for each 1% reduction in volatile content below 19% fractions prorata.

Both moisture and ash shall be determined after equilibrating at 60 per cent relative humidity and 40°C temperature as per relevant clauses of Indian Standard Specification No. IS: 1350-1959.

5. Ash percentage of Coking Coal and Hard Coke shall be determined after air-drying as per IS: 1350-1959. If the moisture so determined is more than 2 per cent, the determination shall be after equilibrating at 60 per cent relative humidity and 40°C temperature as per IS: 1350-1959.

6. The above classification shall not apply to coals other than Bituminous or Sub-Bituminous coals as specified under Indian Standard Specification No. IS: 770-1964.

presence of sulphur and alkalies in the ore does not generally matter much; however, the presence of considerable amount of alkalies causes great difficulties in handling ferro-manganese gas. The presence of sulphur in small quantity, as usually found in manganese ore, is immaterial because the high manganese content in the alloy eliminates it by a process of segregation.

Size is an important consideration, as excessive amount of fines leads to heavy dust losses. Since any ore is seldom of ideal composition most ferro-manganese manufacturers obtain from more than one source and blend them to attain the stipulated contents as,

Mn/Fe	7 : 1 and above
Mn	48% min
Fe	6% max
SiO_2	4% max
P	0.12% max

with limiting the size not to exceed 15 cm in diameter and the fines below 20 mesh not to exceed 5%.

The manufacturers of dry battery cells on the other hand require pyrolusite having not only high content of manganese dioxide (MnO_2) usually above 78% but it should also have gamma and delta type crystal structure. The manganese dioxide having alpha or beta type crystal structure is reported to impede flow of current and is not considered very much suitable for the manufacture of dry cells. Also

TABLE 13 (a)

Specification of Minerals Given by Important Industries in India Limestone

Specification / Mineral constituents in percentage	Blast furnace grade					Steel melting shop grade				
	TISCO	IISCO	Bhilai	Rourkela	Durgapur	TISCO	IISCO	Bhilai	Rourkela	Durgapur
CaO	44 min	47 min	41 min	46 min		48 to 51	50 min	48 min	46 min	52
MgO	5 max		4 to 5						4	4 max
SiO_2	9 max		12 max			5 max				
Fe_2O_3	1 max		4 max			1 max				1 max
Al_2O_2	2 max					2 max				
Total insolubles		10 max		not exceeding 14, preferably less than 11.5	12 to 14		6 max	6 max	4.5 max	3 Penalty for above 3 to 5% above 5% rejection
Size	−125+50 mm	−80+25 mm	−80+12 mm	−80+25 mm	−130+50 mm	−125+50 mm	−130+50 mm	−84+12 mm		−50+30 mm

Limestone (Contd.)

Specification Mineral constituents in percentage	Flux grade (Ferro-manganese plant)		Mysore Iron and Steel Co. Ltd.
	Khandelwal Ferro-Alloys Ltd.	Universal Ferro and Allied Chemicals Ltd.	
CaO	45–48	—	39–52
MgO	5–8	—	2–9
SiO$_2$	4–6	5 max	3–8
Fe$_2$O$_3$	—	—	
Al$_2$O$_3$	—	—	4
Insolubles	—	—	
Size	−25 mm +12 mm	−25 mm +6 mm	−50 mm +12 mm

TABLE 13 (b)

Manganese Ore

Specification Mineral constituents in percentage	Battery grade		Chemical grade			
	Union Carbide of India Ltd., Calcutta	Estrella Batteries Ltd., Bombay	Atic Industries Ltd.	Universal Chemical and Industries Private Ltd., Bombay	The Swadeshi Chemicals Private Ltd., Bombay	India Dyestuff Industries Ltd., Bombay
MnO_2	84	80 min	80-90	80 min	84-85	90 min
Total iron content (including magnetic iron)	less than 2.5	—	—	3 max (including silica)	3 max	0.2 max
$Fe_2O_3 + Al_2O_3$		20 max	—	—	—	—
Cu	less than 0.3					
Acid soluble	—	10 max	—	—	—	—
Crystal structure	gamma type	gamma type				
Si_2			Use in the manufacture of vat dyes	Use in the manufacture of potassium permanganate	Use in the manufacture of potassium permanganate	Use in the manufacture of vat dyes and intermediates

Manganese Ore (Contd.)

Specification	Ferro-manganese grade						
Mineral constituents in percentage	Ferro-Alloys Corpn. Ltd., Garividi	Khandelwal Ferro-Alloys Ltd., Kanhan	Universal Ferro and Allied Chemicals Ltd., Tumsar	Visvesvaraya Iron and Steel Ltd., Bhadravati	Tata Iron and Steel Co. Ltd., Joda	Jeypore Sugar Co. Ltd., Rayagada	Dandeli Ferro-Alloys Pvt Ltd., Dandeli
Mn	45-48	47.5-48.5	(i) 49.25 (ii) 46-48	42-47	45-46	46-48	44/46 to 46/48
Fe	8.0	8.0	N.A.	6-12	9-10	N.A.	10-12
SiO_2	8.0	6.7	N.A.	6-12	5-8 ($SiO_2 + Al_2O_3$)	9.0	8 max
P	0.16	0.17	N.A.	0.10	0.10-0.12	0.15	1 max

TABLE 13 (c)

Bauxite

Specification	Metallurgical grade				Chemical grade (for the manufacture of sulphate of alumina (ferric), ammonium alum, etc.)			
Mineral constituents in percentage	Indian Aluminium Co. Limited, Muri	Aluminium Corporation of India, Jaykaynagar	Hindustan Aluminium Corporation, Renukut	Madras Aluminium Co. Ltd., Mettur	Dharamsi Morarji Chemical Co. Ltd., Durg and Ambernath	Mettur Chemicals and Industries Corporation	Cawnpore Chemicals Works, Kanpur	Bengal Chemical and Pharmaceuticals Works Limited
Al_2O_3	48–52	53 min	48–50	38–40	60	55 min	55 min	58–60
SiO_2	1–5	1.8*	3–4		2 max	4 max	—	—
Fe_2O_3	9–11		11–3		below 2	4 max	2 max	2–3
TiO_2	9–11		8–10		4 to 5	10 max	10 max	—
LOI	24–26		Balance		—	30	—	—

*Reactive silica

TABLE 13 (d)

Fluorspar

Specification mineral constituents in percentage	Arc welding electrodes industry	Acid grade		Chemical industry
	Advani-Orlikon Pvt. Ltd., Bombay	Indian Oxygen Ltd., Khardan (W.B.) & Ambattur (Madras)	National Flourine Corporation, Thana	Navin Fluorhe Industries, Bhestan (Gujarat)
CaF_2	98 min	96 min	98 min	97.0 min
SiO_2	0.5 max	—	0.6 max	1.0 max
Fe_2O_3	Traces	—	—	—
R_2O_3	1.0 max	—	—	—
BaO	0.5 max	—	—	—
CaO	—	—	—	—
$CaCO_3$	—	—	1 max	1.0 max acceptable up to 1.5 with penalty
S	0.025 max	—	—	S and Sulphate:0.15 max
P	0.02 max	—	—	P_2O_5: 0.3 max
Lead	—	—	0.15 max	0.25 organic matter
Size	Powder	—	250 mesh	No particle more than 0.5 mm and max retained on 0.2 mm screen 10%
Bulk density	—	—	—	1800 kg/m^3

Fluorspar (Contd.)

Specification mineral constituents in percentage	Aluminium industry—acid grade		Glass industry—ceramic grade		
	Indian Aluminium Co., Alupuram and Hirakud	Aluminium Corp. of India, Jaykaynagar	Krishna Silica and Glass Works Ltd., Calcutta and Thana	Alembic Glass Industries Ltd., Baroda	Sarabhai Glass (P) Ltd., Baroda
CaF_2	97	96.5	49	70-75	65-80
SiO_2	1	—	—	23-30	—
Fe_2O_3	1.5 max	—	—	—	0.2 max
Size	—	—	—	30-100 mesh	Lumps

Fluorspar (Contd.)

Specification mineral constituents in percentage	Steel industry				Iron and steel casting		
	I.I.S. Co., Burnpur	V.I.S.L. Bhadravati	H.S.L. Bhilai	H.S.L. Durgapur	Electro-Steel Casting Ltd., Sukchar Calcutta	The Singh Eng. Works, Kanpur	Telco, Jamshedpur
CaF_2	85 min	85	—	90–95	90	90	80–90
SiO_2	10 max	—	5 max	2–3 max	5–8	—	6–10
Size	$\frac{3}{8}''$ to 4″	1″ to $1\frac{1}{2}''$	5–25 mm	$\frac{1}{4}''$ to $\frac{1}{2}''$	—	—	—

pyrolusite must have low iron content not exceeding 4% and should possess a certain degree of porosity and be of moderate hardness. It should be free from the compound of metals which are electronegative to zinc (container) such as copper, nickel, cobalt, arsenic, lead and antimony.

In the manufacture of manganese chemicals like hydroquinone, potassium permanganate, sulphate, chloride, etc. which are used in welding rod, glass dyes, paint and varnishes, fertilizer, pharmaceutical and photographic industries, ore containing not less than 75% MnO_2 with not more than 1.5% Fe; 1% Al_2O_3; 6% SiO_2 and 0.02% Cu is suitable. It is clear that the presence of impurities and other deleterious elements beyond tolerance limits greatly affects the end product and thus has a bearing on a particular industry. The consumers, therefore, look for the minerals of right specifications. For further details about specifications and the effect of impurities on the manufacturing process, reader is advised to refer to literatures given under bibliography.

10

Marketing

MINERALS ARE mined either for domestic consumption or for export. A number of consumers may have their own captive mines and in such case the question of sale does not arise unless there is a surplus production. In other case minerals are raised for sale in domestic market or for export. In between the producers and exporters there may be intermediatories, the middleman, whose main business is to buy minerals and ores from different sources and market them after suitable blending, upgrading or pulverising according to market specification. A resourceful mine owner may be an exporter also. Thus in the market three classes of people, mine owner, intermediatory and exporter exist.

In the international trade the role of the shipping agency and the intermediatory is inescapable. The shipping agency handles all incoming and outgoing commodity at the port and takes care of the payment of port dues. The international trade normally commences by mutual negotiations between the buyer and the seller on the agreed price. However, in the international scene more often powerful groupings viz., cartels and monopolistic tendencies are encountered. Monopoly may be understood as an exclusive possession of the trade in some commodity or the exclusive privilege, conferred by the State, of selling commodity or trading with a particular place and country. The cartel on the other hand is a group of organisations joining together with a view to regulate production, price and marketing. In the international scene presently a number of groupings exist as for example, International Tin Council (ITC), Organisation of Petroleum Exporting Countries (OPEC), Intergovernmental Council of Copper Exporting Countries (CIPEC), International Bauxite Association (IBP), Association of Iron Ore Exporting Country (AIOEC), Diamond

Trading Company (DTC), and the lead and tungsten group, etc. Truly, all the above cartels or groupings except OPEC, DTC and ITC are not effective. These are more a forum for discussions rather than catalyst for action. ITC can rightly be called "cooperation" rather than a cartel. Only OPEC and DTC controlled by a powerful company De-Beers Limited, London, are the most effective cartels so far pricing, marketing of petroleum and diamond are concerned. DTC controls 90% of the world trade in diamond and OPEC about 60% in the crude petroleum. The OPEC consists of 13 members (one member from Egypt is reportedly dropped). The unilateral price hike in the crude price by OPEC during 1973 and 1974 is well-known. Since then, they have increased not only the posted price a number of times which is now rated at a little over 16 dollars a barrel against cost of production of 20 cents a barrel but also cut back production monolithically. The ITC is the oldest organisation controlling tin market, is composed of both producer and the consumer nations. India is the member of the ITC. The U.S.S.R. is also a member. On the other hand, the U.S.A. is not a member of the ITC. The international floor and selling price of tin is regulated by ITC through the International Tin Agreement, the last meeting of which was held from 11th to 14th July, 1978. The Buffer Stock Manager of the ITC located at London controls stability and price by purchasing tin from the market at a stipulated price above the floor price even if there is downward trend in the market value just to prevent slipping down the price to protect the producers' interest. The buffer organiser is authorised to release from the stock (he can hold 20,000 tonnes of tin at a time in the stock) to soften rising price for protecting the consumer's interest.

Pricing: The price quotation is made as per the commercial classification i.e. the grade. This important decision of quoting price and competing in the free market is left to acumen of the individual producers except in the socialistic countries where Government controls distribution and selling price. For specific minerals, such as copper, lead, zinc and tin, some producers are accustomed to rely on cartel or association of private firms for pricing, production and investment. In certain specific cases, it becomes necessary for a Government to fix minimum selling price or to peg the price. Formerly, the selling price of coal and coke produced in India was fixed by statutory provision under the Colliery Control Order, 1945. The price was decontrolled from 1st April, 1967 by notification dated

24.7.1967; but again subsequently controlled. At present the minimum prices of coal, barytes and mica are statutorily fixed. The F.O.B. price for the powder barytes of OCMA grade has been fixed at 57 dollars per tonne since 1978 packed in polythylene lined bituminised jute bags. If the material is to be exported in pellets, the foreign buyer is required to pay extra 6 dollars per tonne. The F.O.B. price for barytes lumps is fixed at 37 dollars minimum. A minimum selling F.A.S. (Free Along Side) price on various categories of mica meant for export was for the first time introduced on 15.2.1964. Since the introduction of the floor price, the price of various grades and qualities have been revised seven times on 1st August, 1966, 1st January, 1973, 2nd February, 1974, 15th April, 1975, 5th December, 1975, 17th April, 1976 and the last on 18th October, 1976. With the devaluation of rupee on 6.6.1966, the export duty was imposed on mica w.e.f. 15.6.1966. Such decisions are taken in consultation with the trade in a case where the country has monopolistic supply position to the world market. It also avoids under quoting and cut-throat competition amongst the domestic exporters.

Selling price is quoted at F.O.R. (Free On Rail) or F.O.R. destination. F.O.R. means the price per unit weight of minerals loaded into the wagon at the loading station. F.O.R. includes the price of the mineral at the pit/mine head plus the transport charge to the rail head, the rent of the plot for stacking at the railway yard and the loading charge into the wagon.

When a mineral is to enter the international market, prices are quoted on the basis of F.O.B., F.A.S., F.O.B.T. or C.I.F. The F.O.B. (Free On Board) price includes the F.O.R. plus the railway freight to the port, port handling charges, export duty and some incidental charges like analysis of ore. F.A.S. (Free Along Side) price is a price of the commodity on the jetty. F.A.S. price plus export duty if any and the loading charges in the ship becomes equivalent to F.O.B. Such quotations are made for the commodity which are exported in boxes or in packed form as in the case of mica. The F.O.B.T. (Free On Board Trimmed) invariably also means the price of the commodity loaded in the ship and all duties and taxes paid. Practically there is no difference between F.O.B. and F.O.B.T. except that in the case of the latter, the prices quoted include trimming charge for loading bulk commodities like iron ore, manganese ore, chromite, bauxite etc. on the ship. Trimming charge is taken just for levelling the ores on the ship. It is customary to include trimming charges in F.O.B. prices and therefore, for bulk commodities like iron ore, etc. F.O.B.T. quotation is made. The C.I.F. (Cost Including Freight) price includes F.O.B.T. plus sea freight to the port of destination including the insurance charges. The price of iron and manganese ores which enter into international market in bulk quantities are generally fixed on C.I.F. basis per unit of metal content. Different rates may be quoted for each grade. For iron ore which usually contains 4 to 5% moisture, the C.I.F. price per unit Fe for each grade is realised on dry weight, taking into account the assumed moisture content. This computation becomes necessary because sea freight is normally paid on wet tonne basis. The exchange earning, in other words the F.O.B.T. price, is worked out by deducting the sea freight from the C.I.F.

The international price of copper, lead, zinc and tin metals is governed largely by the quotations made at London Metal Exchange (LME) market. LME is a speculative market and trades both on cash (spot) and forward under 3 month contracts only. The transactions take place daily except on holidays. For tin, Penang island in Malaysia is the principal market. Penang offers spot or prompt delivery of grade 'A' tin (99.85% Sn) metal only. Other countries base their daily quotes on the Penang price taking into account the delivery cost and the financing. LME sells both Grade 'A' and Grade 'B' (standard grade, containing 99.75% Sn) tin. Grade 'B' tin is available from Europe and sources other than the Penang.

In addition to LME, the New York Commodity Market and the General Services Administration (GSA) of the U.S.A. are the two selling points. The New York Commodity Market is largely governed by the quotes of LME. GSA quotes for the selling price of various commodities for spot delivery from the stock piles, which have been sanctioned for release. Its price is also based on LME quotes. GSA receives purchase orders up to 3.30 p.m. every day except on holiday and generally executes the award on the same day.

PORT HANDLING CHARGES

Custom Duty

Export and import duties are generally known as custom duty. It is a source of revenue to the government. The export duty is charged either on ad valorem basis and/or as fixed rate on tonnage basis. Ad valorem basis is generally adopted for high priced low tonnage commodity and fixed rate on large tonnage low priced commodity. In similar way import duty is levied on commodity which is imported. In India such duties are levied under the India Tariff Act, 1934 and Customs Act, 1962.

The custom duty on various minerals as applicable during 1978 in India was as given on p. 185.

The export duty is subject to revision by the Government. It can be increased or decreased or completely waived off, as the exigencies demand, depending upon the monopolistic supply position and other international competitive factors. As for example, the export duty on mica has been adjusted several times to suit export conditions.

To study completely the competitive position of the exportable minerals in the world market one has to keep watch constantly over the F.O.B. and C.I.F. quotations of the other competing countries. The cost-oriented approach and the export pricing policy strategy should always be kept in view especially in global marketing. Much, however, depends on the capability of a country to make available the supply within specific time. A country having rich deposits but not fully developed may find difficulty in raising sufficient tonnage, available for export. Mining is not an enterprise where production can be raised in short possible time unless planning for development in relation to available reserves is made well in advance. Inter-related with the raising output a linch pin of road and rail transport as well as equipped mechanical handling capacity at the ports is of utmost importance. Ports must have good berthing capacity to allow bulk ore carriers which have generally 100,000 DWT to 150,000 DWT capacity. To facilitate berthing of bulk carrier or tanker, deep draft port is necessary. Most of the major ports in the world have capacity to load 4000 to 6000 tonnes of minerals per hour. Some have expanded capacity to load 8000 tonnes an hour and even more. In India majority of the ports can accommodate only 30,000 DWT to 35,000 DWT ships. The major ports of India are however being expanded to accommodate large ships and mechanical loading capacity improved. Except for Vizag, Bombay and Marmagao, the ore handling capacity of other ports are only about 2000 tonnes a day at present. Quicker transport and efficient port handling capacity aid much in making the commodity cheaper and competitive in the world market. Nearness of deposits to the port may help considerably in reduction of the railway freight and consequently on the F.O.B. and C.I.F.

Export Duty

Minerals	Rate of duty
1. Manganese dioxide	20% ad valorem
2. Manganese ore having more than 48% Mn	Rs. 20.00 per tonne
3. Manganese ore having 10 to 48% Mn	Rs. 12 50 ,,
4. Manganese ore having less than 10% Mn and combined iron and manganese content less than 63% (black iron ore)	Rs. 7.00 ,,
5. Lumpy iron ore having iron content 63% and above	Rs. 10.50 ,,
6. Lumpy iron ore having iron content 60% but less than 63%	Rs. 6.00 ,,
7. Lumpy iron ore having iron 58% but less than 60% iron content	Rs. 5.00 ,,
8. Lumpy iron ore having less than 58% Fe	Rs. 4.00 ,,
9. Iron ore fines including blue dust with 62% Fe or more	Rs. 4.00 ,,
—with less than 62% Fe	Rs. 3.00 ,,
10. Sillimanite	20% ad valorem
11. Steatite*	20% ad valorem
12. Mica (variable rates depending upon grade)	15 to 30 % ad valorem
Mica scraps	40% ad valorem
13. Fabricated mica	10% ad valorem
14. Mica powder	nil
15. Micanite	nil
16 Kyanite	Rs. 40.00 per tonne

*Abolished w.e.f. 27-1-79

(*Contd.*)

Minerals	Rate of duty
17. Chromite ore and concentrates	
(a) High grade fines and concentrates with Cr_2O_3 content 47% and above.	Rs. 200 per tonne
(b) Medium grade fines and concentrates with Cr_2O_3 content 35% and above but below 47%	Rs. 150 ,,
(c) Lumpy ore with Cr_2O_3 content 35% and above but below 42%	Rs. 75 ,,
(d) Low grade lumpy ore or fines or concentrates with Cr_2O_3 content below 35% and FeO content above 23%	Rs. 50 ,,
(e) Chromite concentrates produced by beneficiation of low grade chromite ore containing not more than 35% Cr_2O_3 in the natural form.	Rs. Nil

Import Duty

Minerals	Rate of duty
1. Antimony ore	40% ad valorem
2. Sulphur	10% ,,
3. Borax/Boric acid	60% ,,
4. Plumbago/Graphite*	60% ,,
5. Asbestos raw/fibre	40% ,,

*Revised to 40% ad valorem w.e.f. 27-1-79

(Contd.)

Minerals	Rate of duty
6. Fluorspar	48% ad valorem
7. Precious stones unset and imported cut	60% ,,
8. Precious stones, unset, and imported uncut	40% ,,
9. Zinc concentrates	40% ,,
10. Tungsten ore and concentrates	30% ,,

prices where a particular mineral commodity is experiencing a stiff competition in the international market.

F.O.B. and C.I.F. prices are subject to fluctuation due to several reasons, important being revision in railway freight, shipping freight, recession in market demand, release of material from the stockpile, and emergence of new countries in the world market for a particular mineral.

Ship freights are fixed and announced by different Conferences which own cargoes. Shipping Conferences are voluntary associations of multinational liner shipping companies. There are about 50 Shipping Conferences catering to India's liner exports and in equal numbers in her import trade. Three types of rates are generally announced by the Conferences. They are: (*i*) Open rate, (*ii*) General rate, and (*iii*) Specific rate. Open rate is quoted for bulk commodity like iron ore, manganese ore, etc., which is generally arrived at by mutual negotiation between the shippers and the cargo owners. General rate is the fixed rate announced by each Shipping Agency for different commodities. Each Shipping Agency may also announce independently 'specific rate' for certain commodities. Freight Investigation Bureau, Directorate General of Shipping, Ministry of Transport and Shipping, Bombay, coordinates much information in India.

11

Methods of Estimation of Reserves

THE RESERVES are known mineral assets readily available for exploitation. It implies quantity and quality available within certain dimensions. For the estimation of reserves, it is necessary to have qualitative picture of a deposit besides tonnage available within its dimensions. Qualitative picture is obtained by sampling. A careful assemblage of the sampling data helps in blocking out areas with particular grade. The degree of assurance depends both on the frequency of sampling and geological nature of the deposit.

CLASSIFICATION OF RESERVES

Reserve-estimates are placed under three categories, namely, proved, probable and possible. When the estimate is based on sufficient data, such that it will not vary much from the actual tonnage and grade when mined, the reserves thus estimated are categorised as 'proved' (or measured). For instance, in an underground mine, if a vein has been sampled at two levels, and in winzes and raises, i.e. all the four sides have been exposed and sampled, then the estimates are 'proved'. In coal limestone, porphyry copper and such other uniform deposits, boreholes drilled at sufficiently close intervals, can give adequate data for estimating 'proved reserves'. Probable' reserves (or indicated reserves) carry a lesser degree of assurance and is based on a limited data of sampling. In an underground mine, if a vein has been sampled at two levels, and if there is no data of sampling in raises or winzes, the reserves estimated from such data would be 'probable reserves'. In case of uniform deposits, probable reserves are estimated from a limited number of widely spaced boreholes. The reserve estimation when done from extrapolation of sampling data to areas where there is no data of sampling available, is termed 'possible' reserves (or inferred reserves). For example, in a property, if an ore body has been tested up to 100 metres depth, and in the neighbouring properties the ore is known to persist to a greater depth, then on geological grounds, it may be possible to assume, that the ore body will persist in this property also, to say 200 metres depth. The reserves estimated up to 100 metres depth are, say 10 million tonnes of copper ore having 2% Cu, then another 10 million tonnes of the same grade can be estimated between 100 metres and 200 metres depths. Then the total ore reserves up to 200 metres depth would be 20 million tonnes, half of which are possible reserves.

The above classifications of proved, probable and possible, equivalent to measured, indicated and inferred in estimating mineral reserves are universally accepted. The terminology under the latter category was first used by the United States Geological Survey and Bureau of Mines in 1944. It may be understood that a 'mineral reserve' is quite distinguishable from a 'mineral resource'. In geologic terms a mineral resource is a 'concentration of naturally occurring material (solid, liquid or gas) in the earth's crust in such form that economic extraction is potentially feasible. A reserve is the portion of an identified resource, estimated from geological and engineering evidences which can be mined at a profit. During the year 1974 the USGS and USBM have introduced new terminology to their earlier version of 1944 for categorisation of ore reserves and resources. This they have done to cover all possibilities for extraction now and in the future. In the new version they have introduced terms like *total resource, identified resource, demonstrated reserve* and *undiscovered*. The total resources are classified both in terms of economic feasibility (i.e. as economic and sub-economic with the latter sub-divided into para-marginal and marginal) and also is terms of degree of geological assurance (i.e. identified and undiscovered, each sub-divided into further categories). The new USGS and USBM mineral resource and reserve category definition can be visually represented by Fig. 9. The concept of measured, indicated and inferred reserves however remains the same in the revised version also. According to their new definition "total resources" will cover all the materials that have present or future value

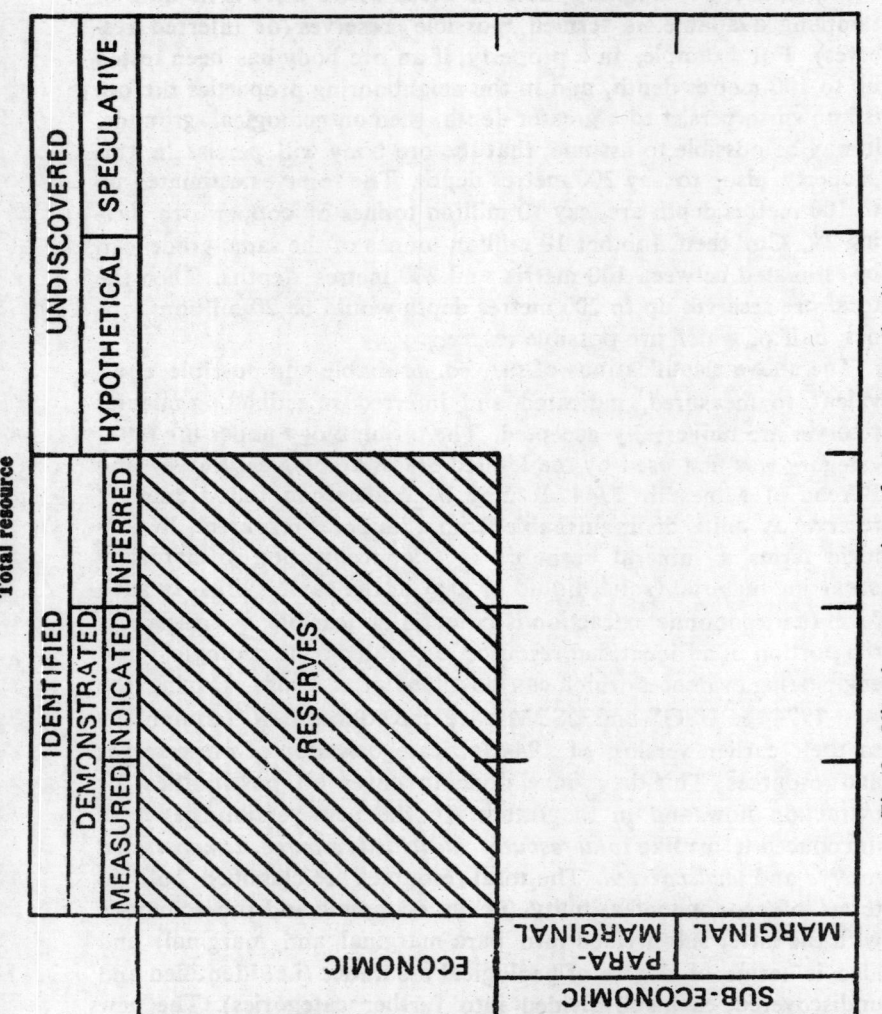

Fig. 9.   Visual representation of reserve categorisation.

and comprise identified or known materials plus those not yet identified, but which on the basis of geological evidences are presumed to exist. Following are the definitions given in the revised version of the ore reserves and resource categorisation.

Identified resources: Specific bodies of mineral-bearing material whose location, quality and quantity are known from geological evidence supported by engineering measurement with respect to the demonstrated category.

Undiscovered resources: Unspecified bodies of mineral bearing material surmised to exist on the basis of broad geologic knowledge and theory.

Reserve: The portion of the identified resource from which a usable mineral and energy commodity can be economically and legally extracted at the time of determination. The term ore is used for reserves of some minerals.

The following definitions for measured, indicated, and inferred are applicable to both the Reserve and Identified sub-economic resource components.

Measured (proved):   Material for which estimates of the quality and quantity have been computed, within a margin of error of less than 20 per cent, from sample analyses and measurements from closely spaced and geologically well-known sample sites.

Indicated (probable):   Material for which estimates of the quality and quantity have been computed partly from sample analyses and measurements and partly from reasonable geologic projections.

Demonstrated: A collective term for the sum of materials in both measured and indicated resources.

Inferred (possible): Material in unexplored extensions of demonstrated resources for which estimates of the quality and size are based on geologic evidence and projection.

Identified sub-economic resources: Resources that are not reserves, but may become so as a result of changes in economic and legal conditions.

Paramarginal: The portion of Sub-economic Resources that: (a) borders on being capable of production economically, or (b) is not commercially available solely because of legal or political circumstances.

Sub-marginal: The portion of Sub-economic Resources which would require a substantially higher price (more than 1.5 times the price at the time of determination) or a major cost reducing advance in technology.

Hypothetical resources: Undiscovered materials that may reasonably be expected to exist in a known mining district under known geologic conditions. Exploration that confirms their existence and reveals quantity and quality will permit their reclassification as a Reserve or Identified Sub-economic Resource.

Speculative resources: Undiscovered materials that may occur either in known types of deposits in a favourable geologic setting where no discoveries have been made, or in as yet unknown types of deposits that remain to be recognised. Exploration that confirms their existence and reveals quantity and quality will permit their reclassification as Reserves or Identified Sub-economic Resources.

In Russia and other East European countries the reserves are classified as A, B, C1 and C2 which correspond to measured (A and B), indicated (C1) and inferred (C2). In the above classification, the Russians sub-divide the measured reserves into A and B depending upon the degree of assurance. They put the reserves in A category only when it is blocked out or in case of open pit benches are exposed.

GSI in its report published in 1981 further elaborated the identified resources into four sub-divisions viz., developed, proved, probable and possible. These are equivalent to A, B, C1 and C2 of Russian classification. According to national classification evolved by GSI the permissible error for developed reserves has been given from 0 to 10%; 10 to 20% for proved; and 20 to 50% for probable as against 15 to 20%; 20 to 30%; and 30 to 60% allowable in Russian classification. Under possible reserve, on permissible error has been indicated under national classification; although Russian classification has provided 60 to 90% permissible error. GSI has further classified uncovered resources into: (a) prospective resources and (b) prognostic resources. These are equivalent to hypothetical and speculative of Mc Kelvey table (see Fig. 9). Prospective reserves include those reserves which have been estimated on extrapolation and projection in known areas. Prognostic resources include such type of resources which are based on speculation in unknown areas. The definitions of prospective and prognostic resources are given as follows:

PROSPECTIVE RESOURCE: Part of 'undiscovered resource'; comprising estimates not based on measurements of basic parameters but are made in known mineral districts/belts on several premises such as: (a) extrapolation beyond explored lateral and vertical limits on the strength of the knowledge of geological conditions; (b) projections made on the basis of indirect evidences such as geophysical and geo-chemical anomalies.

PROGNOSTIC RESOURCE: Speculative estimates or quantities of undiscovered mineral material suspected to exist in virgin areas with favourable geological setting as deduced from the data on crustal abundance calculations, simulated geomodelling, remote sensing and known metallogenetic history, analogy with mineral districts where similar geological conditions are known to exist.

More frequently 'quarriable' or 'mineable' reserves are used to differentiate those from the reserves *in situ*.

SAMPLING

Sampling is an art of collecting small fraction of the material, so at to represent the whole mass. It is evident that one small fraction cannot be the representative of the whole mass. A large number of samples are required to be obtained for providing satisfactory approximation to the grade and the physical characteristic of the deposit. How much and how the sample should be drawn, depend upon various geological factors, viz., nature, shape and size of a deposit and the purpose and scope for which it is required.

The samples may be drawn from outcrops, trenches, pits, boreholes or from the different sections of the quarry and underground mine. In the case of alluvial or placer deposits occurring near the surface, sampling by means of shallow pits may be adequate. In the case of simple bedded deposits and veins which are exposed to the surface or lie under a thin mantle of soil or residual product of weathering, the samples drawn from digging shallow pits and trenches would suffice. In the case of copper and other sulphide ore deposits which are leached near the surface, and exposed in the form of gossan, the samples drawn from the zone of weathering (gossan) will not be helpful. In such cases, the samples are drawn by deep pits and boreholes going below the water table. Where the deposit is not homogeneous, the samples from the deeper parts are required to be drawn either by means of drilling or exploratory mine. In such cases, the samples may have to be obtained from the drift, cross cuts, winzes and raises. It is left to the judgement of experienced field geologists who take into account all these geological factors and decide how best the samples are to be drawn and how the number of samples should be kept at a minimum. Taking of large number of samples may prove helpful but may be regarded unnecessary, entailing unwarranted expenditure It is, however, not the idea of minimising the expenditure on sampling at the cost of reasonable accuracy in the valuation of the mineral property. It is essential that in the property containing metals of high value like gold, tungsten, tin, cobalt, nickel, etc., where the incidence of mineralisation is small and erratic, large number of samples are required to be drawn at close intervals in contrast to homogeneous and uniformly laid bedded deposits.

The techniques of sampling from the outcrop, pit, trench, mine, etc. may be grouped into following:

(a) grab sample or chip sample,
(b) channel or groove sample,
(c) borehole sample,
(d) bulk sample, and
(e) car or wagon sample.

Grab sample is the random collection of broken chips from the exposed surface of an outcrop, from the mine working or from the stacked material. The material from the stack can be obtained by a small hand shovel or scoop. The sample thus collected, may be of one piece or few pieces and weigh, so to say, one to two kilogrammes or even less. Grab sample is obtained generally during the preliminary reconnoitory operation. It is also termed as picked up sample. Essentially, it is an unbiased collection of the specimen. The grade of the deposit cannot be relied upon from the assay value of such sample. It gives only an idea of what the grade is likely to be.

Channel or groove sample is collected from grooves cut systematically across the exposure of the ore body. This method is usually applied in sampling of trenches, pits, underground mines—drifts, winzes, raises, shafts and stopes. The purpose of cutting a groove and drawing a sample is to ensure that uniform quantity of material is drawn over the entire width of the ore body. Normally, the out-crops, exposures in the pit, trench or the mine working are not uniform and have irregular surfaces. It is for this purpose that a channel is cut with a width of about 10 cm and 2.5 cm deep. If the surface is very irregular and weathered, the depth has to be increased so that the fresh surface is smoothly exposed over the entire expo-sure. After cutting a groove across the ore body parallel to the true width, sample is drawn by further deepening the groove means of a chisel to a uniform depth, and collecting the broken material either in a pan, canvass or any suitable container. The amount of sample drawn is generally of the order of one kilogramme for 30 cm of groove length.

Though it is always desirable to cut the channel across the ore body parallel to the true width, it is not always convenient to do so. In trenches which expose a somewhat gently dipping ore body, it is convenient to have a horizontal groove instead of the usual inclined groove. When the ore body consists of alternate bands of richer and leaner ore, the groove is subdivided so that, each type is separately sampled.

In the underground drives, true width groove can only be cut in

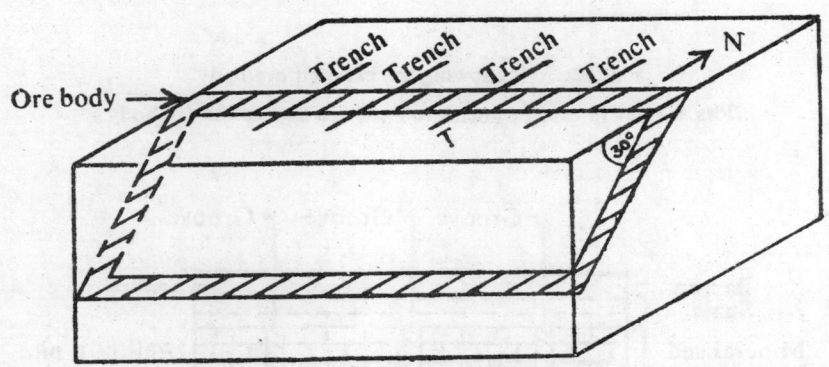

Fig. 10.   Method of cutting trenches.

(Ore body strikes E-W and dips say 30° South. Trenches are put at regular intervals across the ore body N.S. The eastern and western wall of the trenches are sampled by channels. See Fig. 11).

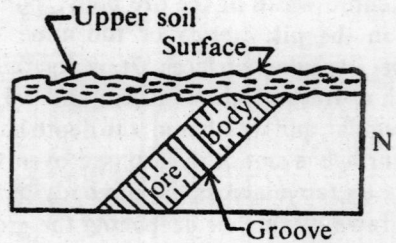

Fig. 11.   Method of cutting groove.

(Section along T.T. in Fig. 10 showing the exposed ore body and the groove)

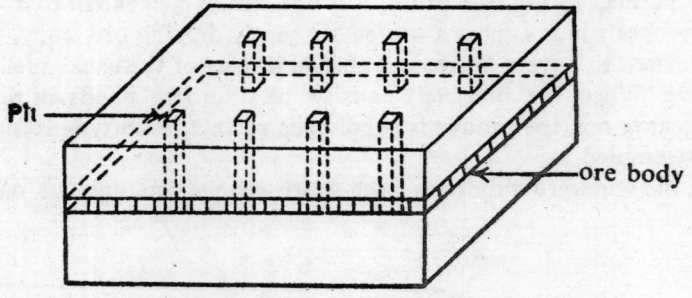

Fig. 12.   Pit showing the exposed ore body.

(Pits are put in case of horizontal beds to expose the ore body.)

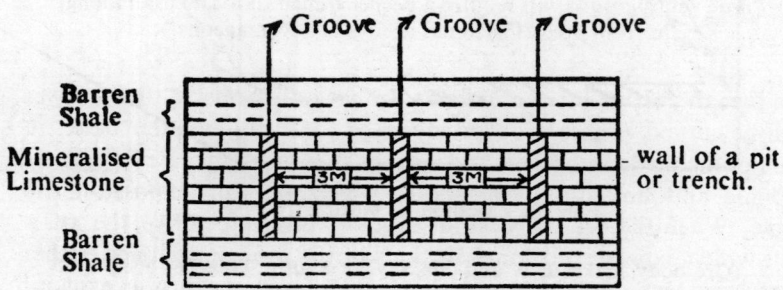

Fig. 13.   Wall of a pit or trench showing groove.

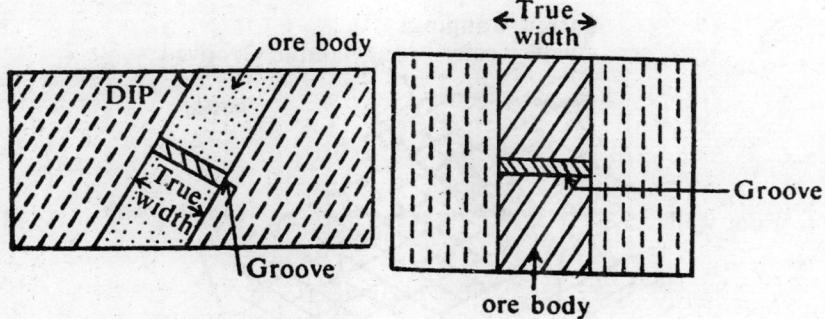

Fig. 14.   Inclined groove in a   Fig. 15.   Horizontal groove in
dipping ore body.   vertical ore body.

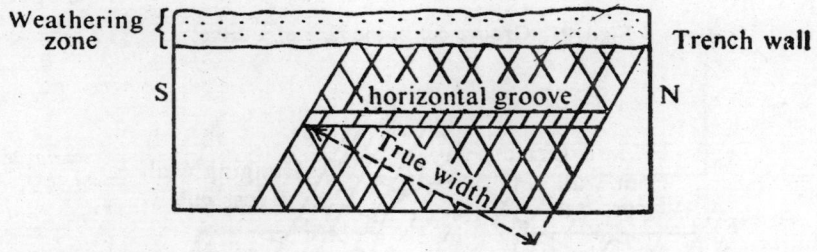

Fig. 16.   Horizontal groove in an inclined ore body.

(True width groove will require a deeper trench to locate it all along
in fresh ore. Horizontal groove is advantageous.)

the face of a drive or in cross-cuts. In back of the drive, the grooves
will be curved. These grooves cover the foot wall and the back, in
case of inclined ore bodies, and one of the walls in case of gently
dipping and horizontal ore bodies which are fully exposed in the
drive. When the ore body is not exposed fully in the drive, the cross-
cuts which expose additional sections of the ore are sampled either
by cutting true width grooves or horizontal grooves, just as explained in the case of trench samples and combined with the average
values of drive samples to obtain the total width and grade.

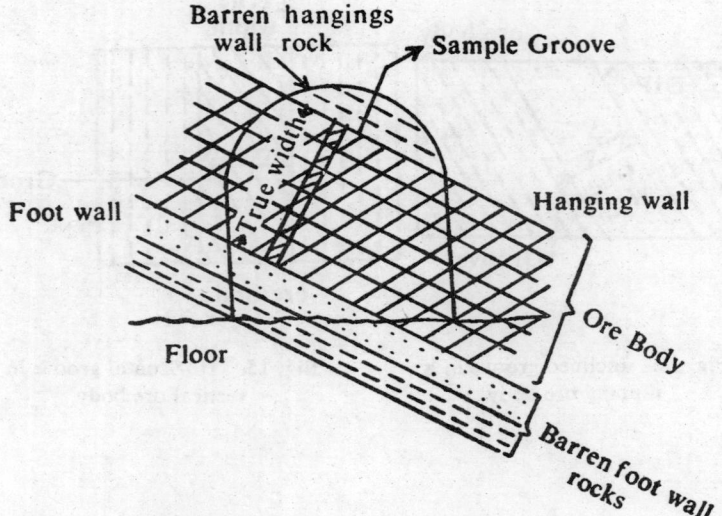

Fig. 19.   Groove cut in the face of a drive.

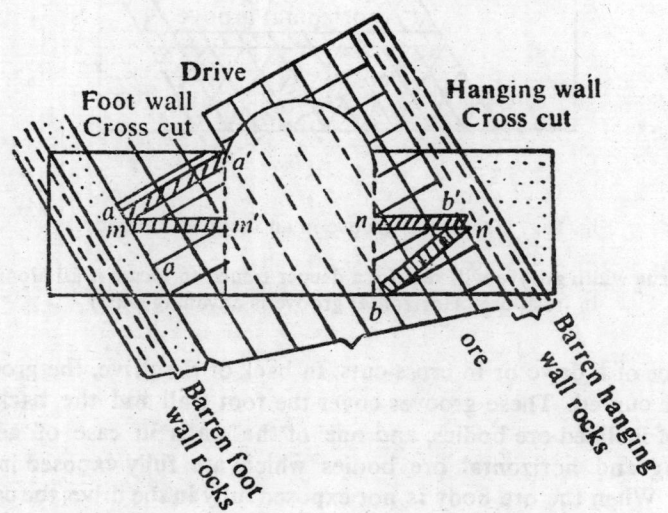

Fig. 20.   Groove cut in the cross-cut.

(In the cross-cut either true width grooves (a-a'; b-b') or horizontal
grooves (m-m' : n-n') are cut.)

Fig. 17. Trenching operation in manganese ore deposit at Balaghat, Madhya Pradesh

Fig. 18. Trenching operation in iron ore deposit at Rajhara Pahar, Madhya Pradesh

In Figure 21, 1 and 8 are barren, 2 to 7 are ore bodies. The groove is cut from x to y in the foot wall and back. Hanging

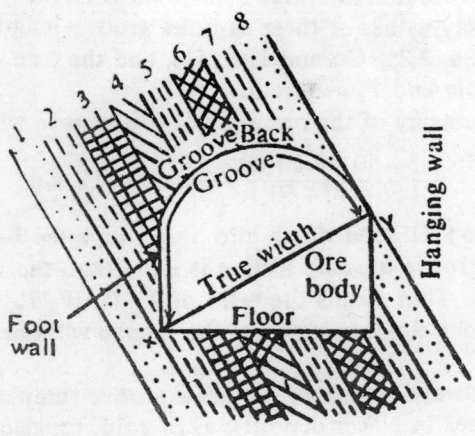

Fig. 21.   Groove cut in the drive having several ore bodies.

wall is not sampled as the rock in it is covered by back sample. Each type of rock in the ore body viz., 2, 3, 4, 5, 6 and 7 are sampled separately.

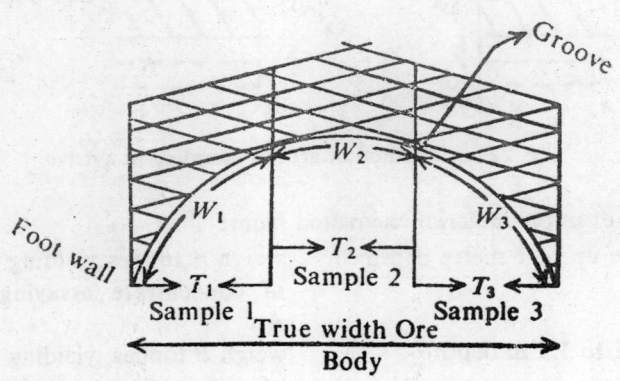

Fig. 22.   Method of groove sampling in a drive.

In the drive which exposes the ore body as shown in figure the groove x-y is subdivided for its curvature into three samples. The samples are drawn separately and each is assayed separately to arrive at the correct assay value as explained under.

Say, the assay values of three samples groove length W_1, W_2 and W_3 are 1.3% Cu, 2.2% Cu and 1.4% Cu, and the true width $T_1 = 55$ cm, $T_2 = 40$ cm and $T_3 = 60$ cm.

The average assay of the ore body at this groove will be

$$\frac{55 \times 1.3 + 40 \times 2.2 + 60 \times 1.4}{55 + 40 + 60} = 1.57\% \text{ Cu}$$

The groove x-y is subdivided into three samples because of the curvature of groove in some section is more than the true width in other sections. That means the ratio of W_1/T_1, W_2/T_2, W_3/T_3 are not uniform and one single sample for the groove will not be representative.

It is not always necessary to draw groove sample to determine the assay value. In placer deposits, as of gold, tungsten ore, tin ore, etc., the assay value is estimated by excavating material from the pit and evaluating the assay value of concentrates by panning, as illustrated below.

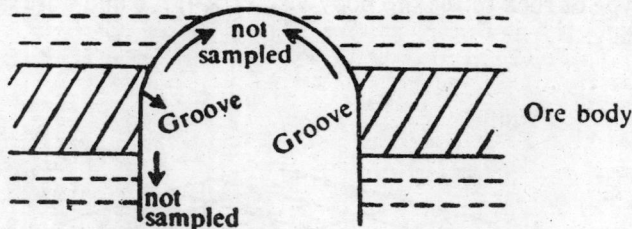

Fig. 22 (a). Method of groove sampling in a drive.

For example material excavated from:

surface up to 1 metre depth	weigh A tonnes yielding 'a' gm of concentrate assaying 0.01% Au
from 1 to 1.5 m depth	weigh B tonnes yielding 'b' gm of concentrates assaying 0.02% Au

Fig. 23. Core splitter

| from 1.5 m to 2.2 m depth | weigh C tonnes yielding 'c' gm of concentrates assaying 0.03% Au |
| from 2.2 m to 3.6 m depth | weigh D tonnes yielding 'd' gm of concentrates assaying 0.04% Au |

Then average grade is calculated as:

$$\text{for } A = \frac{a \times 0.01}{A} = \text{call it } g_1\%$$

$$\text{or } = \frac{a \times 0.0001}{A} = \text{gm per tonne of } A$$

$$\text{for } B = \frac{b \times 0.02}{B} = \text{call it } g_2\%$$

$$\text{or } = \frac{b \times 0.0002}{B} = \text{gm per tonne of } B$$

$$\text{for } C = \frac{c \times 0.03}{C} = \text{call it } g_3\%$$

$$\text{or } = \frac{c \times 0.0003}{C} = \text{gm per tonne of } C$$

$$\text{for } D = \frac{d \times 0.04}{D} = \text{call it } g_4\%$$

$$\text{or } = \frac{d \times 0.0004}{D} = \text{gm per tonne of } D$$

$$\text{The average grade of pit} = \frac{Ag_1 + Bg_2 + Cg_3 + Dg_4}{A + B + C + D}\% \text{ Au}$$

$$= \text{call it } G\% \text{ Au down to a depth of 3.6 metres}$$

In some cases, the average grade is calculated by finding out the incidence of mineral disseminated per tonne of rock. It can be described by two examples as it was followed in the case of finding incidence of diamond per tonne of gravel bed in Panna, Madhya Pradesh; and quantity of phosphate nodules obtainable per tonne of gypsiferous bed mined in Tiruchirapalli district, Tamil Nadu. In the former case the actual tonnage of gravel mined from different pits were treated for the diamond recovery, and the quantity of diamond recovered was calculated for one tonne of gravel. Similarly, the size and the quality of diamond can also be determined. Based on such inference, the reserves over the entire length, breadth, and thickness of gravel bed are calculated. In the latter case, the quantity of phosphate nodules obtainable per tonne of rock mined was ascertained by actual trial, the grade of the nodules being fairly constant, the reserve was easily calculated over the entire belt for a given thickness. It requires only the judgement and intelligence in the field to ascertain how best the method should be followed in each case.

Borehole samples are taken out by drilling. It is the most modern method of visual examination of the mineralisation underground. It helps both in delimiting the lateral as well as the vertical extension of the ore bodies. Drill holes are suitably spaced to cover the area. Depending upon the type of deposit and the degree of mineralisation, the whole mineralised portion of the core may be taken as one sample or subdivided into two or more samples and assayed separately.

Bulk sample is obtained, which may be of the order of few tonnes, either from the trench, pit, channel or from the run-of-mine. Car or wagon sample is obtained by taking a predetermined quantity of run-of-mine from each car load (underground) or from each wagon load (surface siding) in a particular order and assembled together separately for sampling. Such samples are mainly taken for ascertaining the physical properties and amenability to beneficiation by pilot plant test. In iron ore, the bulk sample may be taken to examine the ratio of fines (below 12 mm size) to the whole mass of ore mined. In case of coal, the bulk sample is taken to study the washability, the by-product obtainable under low and high thermal carbonisation and similar other tests. Such study is necessary to establish the economic feasibility of the deposit, and the need of setting pelletising sintering or briquetting plants.

SPACING OF TRENCHES, PITS, BOREHOLES AND GROOVES

It is difficult to prescribe any set method for spacing of trenches, pits, etc. because such spacings are fixed entirely on the modes of occurrence of the deposits. It is again left entirely to the judgement of the geologist in the field, to finalise the pattern. Some generalised suggestions can, however, be made. For bedded and gently dipping deposits like coal, limestone, iron ore and unconsolidated gravel, the sample locations or grid pattern prove advantageous. The distance between the grids is kept at 60 metres interval in case of iron ore and limestone, which provides fairly good idea of the deposit. In case of coal, the borehole is spaced at 300 metres interval for reconnaissance and at 150 metres interval for detailed exploration. Additional boreholes at close intervals may be needed to decipher structural features like fold and fault. For inclined bedded deposits exposed to the surface, the trenches across the ore body are put at suitable intervals depending upon the nature of the deposit. Inclined boreholes may be put for steeply dipping strata. In the drift, winze and raise, the distance between the grooves is kept at 3 metres interval in case of uniform low grade ores, 2 metres for copper ore and one metre for gold ore. In case of stopes, the distances between grooves are kept at 5 to 15 metres depending upon the value of the ore. The Russians, by and large, require drilling holes to be put at intervals of 25 to 50 metres for proving A category of reserves; for B category borehole intervals should be at 50 to 100 metres; for C1 at 100 to 500 metres and for C2 at intervals of 500 to 1000 metres. In case of complex deposits especially where distribution of mineral/metal values is not persistent, they suggest the drilling interval for all categories of reserves should be considerably reduced.

Since drilling is singularly very costly operation the borehole spacing less than 50 metres interval should not be resorted to unless in the case of extreme exigencies like deciphering structural features. In actual practice putting boreholes at 50 metres interval have been found to provide satisfactory data for calculation of reserves under measured category, supplemented by underground sampling data in the case of non-ferrous deposits.

PREPARATION OF THE SAMPLES FOR ASSAY

The samples obtained from the grooves are first crushed separately to convenient sizes of 12 to 15 mm. They are then reduced by a method of coning and quartering to about a kilogramme, half quantity of which again by further crushing, coning and quartering is sent for analysis.

The core samples are split into two halves by means of core splitter (see Fig. 23) so that one is the mirror image of the other. The one half portion is crushed, coned and quartered for the preparation of the sample for assay. The other half is generally kept for cross check and the preparation of thin slides to be tested under the microscope. It is always not necessary to do so and in that event the whole mineralised portion of the core is crushed for the preparation of the sample.

PROCESSING AND INTERPRETATION OF SAMPLING DATA

The assay value obtained from the numerous pits, trenches, crosscuts, drives and boreholes samples are studied step by step to arrive at the average value for a particular block or section of an area or the whole mass of the mineralised block as the case may be. If the grade is variant in different blocks or sections, in that case the reserves with different grades are estimated blockwise. It is more scientific to indicate reserves according to the grade available in different sections. It is invaluable from mining point of view.

Borehole samples: A borehole when drilled to intersect the ore body in depth, the entire section covered by drilling may not recover the full core length and thus the data from drilling map remain incomplete. It requires then careful interpretation to arrive at correct grade. Let us say, that a borehole was drilled to intersect a lead ore body at a depth of 210 to 215 metres. Between these depths only 3 metres of core was reconverted and another 2 metres was lost. Under the circumstances, it is very essential to ascertain the grade and other characteristics of the section lost in coring. It is achieved by collecting the cuttings (sludge) from the return water during drilling. If the deposit has uniform mineralisation over its entire width, the section lost in coring will be having the same grade as core recovered. In this case the grade of core can be taken for the entire zone, through which the borehole has intersected. Take another example of gold bearing quartz vein enclosed in soft schists. Drilling recorded mineralisation from 150 metres to 153 metres. The core recovery was 50% and the core was mostly of quartz. In this case, it is clear that the core loss is mostly in the soft schist which can be further confirmed by examining the sludge. Here the unrecovered portion is likely to be barren and the width of the core recovered and its grade may be taken as such without extrapolation. In another case, let us take the example of ore consisting of richer bands alternating with leaner bands, and a borehole has intersected 4 metres of this ore with a recovery of 80%. In this case, it is possible that the core loss has taken place either in the richer section or in the leaner section, or both. In such cases the assay of the sludge has to be suitably combined with the assay of the cores, to get the average grade. One point to be borne in mind is that the sludge represents not only the sections lost in coring but also the cuttings obtained from the borehole sides. When several runs show mineralised zone, the results have to be combined weighing for the respective widths, to get the total width and the average grade.

The borehole intersecting the lode perpendicularly, the drilling thickness will be equal to the true width. When the borehole is not normal to the plane of the lode, drilled width will be more and has to be reduced to the true width by multiplying by the sine value of the angle between the lode and the borehole (see Figures 24 and 25).

Mine samples: In the underground several grooves samples are drawn representing different rock types and ore types. To find out the average of each groove it is necessary to know whether the lode

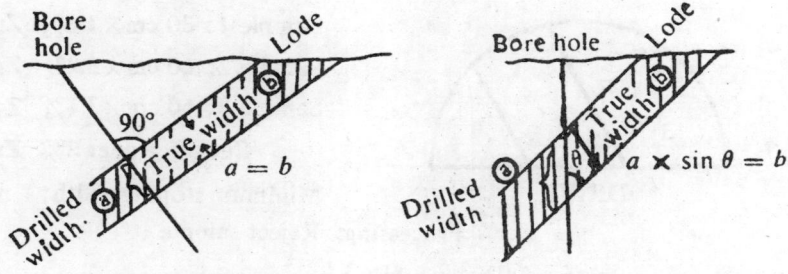

Fig. 24. Borehole intersection perpendicular to the lode.

Fig. 25. Borehole intersection at an angle other than the right angle to the lode.

is fully exposed or not. If the lode is fully exposed and narrow, say only 50 cm in width, it is obvious the width has to be enhanced to facilitate actual mining. The minimum width required for mining is known as the 'minimum stoping width'. The minimum stoping width depends upon the exact method of mining and the type of machinery to be used. In conventional method, the minimum stoping width is calculated on the basis that a man should be able to work comfortably in a stope which should have a vertical clearance of 1.50 metres and, therefore, the minimum stoping width depends upon the dip of the lode and in any case should not be less than one metre.

EXAMPLE 1:

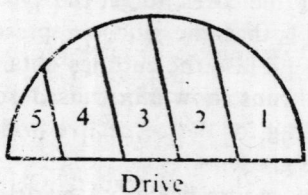

Drive

Data supplied

Sample 1:   10 cm × 0.1% **Pb**
Sample 2:   30 cm × 4.3% **Pb**
Sample 3:   40 cm × 5.1% **Pb**
Sample 4:   30 cm × 2.1% **Pb**
Sample 5:   20 cm × 0.6% **Pb**
Minimum stoping width: 1 m

Cut-off grade: 3% Pb

Processing: Reject samples 1 and 5
(Average for 2 to 4)

$$\text{Average grade} = \frac{30 \times 4.3 + 40 \times 5.1 + 30 \times 2.1}{100} = 3.96\% \text{ Pb for } 100 \text{ cm}$$

EXAMPLE 2:

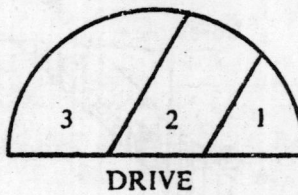

DRIVE

Data supplied

Sample 1: 30 cm × 1.6%   **Zn**
Sample 2: 60 cm × 3.2%   **Zn**
Sample 3: 60 cm × 3.1%   **Zn**
Cut-off grade: 2%   **Zn**
Minimum stoping width: 1 m

Processing: Reject sample 1

$$\text{Average grade} = \frac{60 \times 3.2 + 60 \times 3.1}{120} = 3.15\% \text{ Zn over 120 cm}$$

EXAMPLE 3:

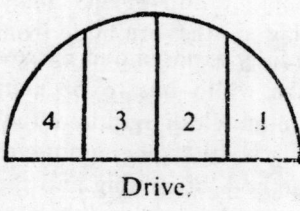

Drive.

Data supplied

Sample 1: 30 cm × 4.0%   **Zn**
Sample 2: 10 cm × 0.2%   **Zn**
Sample 3: 80 cm × 2.8%   **Zn**
Sample 4: 20 cm × 3.0%   **Zn**

Cut-off grade: 2.0%   **Zn**
Minimum stoping width: 140 cm

Processing: Do not reject any sample.

$$\text{Average grade} = \frac{30 \times 4.0 + 10 \times 0.2 + 10 \times 2.8 + 20 \times 3}{140} = 2.90\%$$

Zn over 140 cm

EXAMPLE 4: *Data supplied*

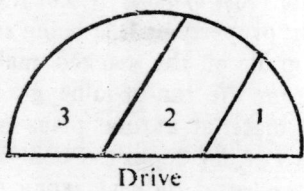

Drive

Sample 1: 25 cm × 3.2% Cu

Sample 2: 45 cm × 2.2% Cu

Sample 3: 10 cm × 3.1% Cu

Cut-off grade: 2.0% Cu

Minimum stoping width: 1 m

Processing: Average grade $= \dfrac{25 \times 3.2 + 45 \times 2.2 + 10 \times 3.1}{80}$

$= 2.6\%$ Cu over 80 cm or
2.1% Cu over 100 cm (1 m)
minimum stoping width
(grade reduced assuming
that 20 cm barren wall rock
is to be mined)

In all these cases, it is essential to know the ore-waste boundary or the cut-off grade. This can be fixed from economic considerations or in the absence of reliable data, two or more cut-offs can be assumed and reserves calculated separately.

Very often for providing better significance to the quality of the reserves available and obtainable when exploited, further sub-classification of grade is made is: (*i*) effective, (*ii*) run-of-mine, (*iii*) pithead, (*iv*) mill, and (*v*) cut-off grade, in addition to computed or sampled grade (see also Chapter 9 for grades).

Computed grade is arrived by the computation of the analytical results as obtained by actual sampling of the ore body from the available surface data, pits, trenches, borehole intersections and underground workings, and forms the basis of obtaining reliable qualitative picture of a deposit. The nature of computation varies from one property to another depending upon the extent of development, modes of occurrence and other geological factors. *Effective grade* is the grade which is available from a property after actual mining. For instance, suppose in the case of iron ore deposit, a reserve of 10 million tonnes with 63% Fe has been estimated. If the specifications are such that ore having less than 12 mm size is not acceptable and in the course of mining 50% of material produced is of this size, it will have to be rejected, thereby reducing the reserves to 5 million tonnes. If the grade of the rejected material is 64% Fe, then the effective grade of the mineable ore produced will be only 62% Fe as against computed grade of 63% Fe. *Run-of-mine* (*r.o.m.*) *grade* represents the grade of ore in certain property as it is being raised from the mine. *Pit head grade* is the grade of the stacked material near the mine site. It may differ from the run-of-mine grade in actual analysis, because the stacked material at one place might have been brought from different levels of the deposit. Such things happen generally in the case of manganese ore and many other minerals. *Mill grade* is the grade of the ore which is fed in the mill. For operating a mill efficiently, it is necessary that the ore of uniform quality is fed continuously and the degree of fluctuation in the grade must be kept within stipulated limits. The mill grade is generally used by the plant-engineers. *Cut-off-grade* is the lowest grade of the ore, which if mined can meet the direct cost of the mining. This grade makes a dividing line between the ore and the waste; the ores assaying below the cut-off grade are not taken into account while estimating reserves.

RESERVE CALCULATION

The estimation of reserves consists of finding out the total volume and converting it into total tonnage. When the computation is in the metric system (C.G.S.) the total volume in cubic metres multiplied by the specific gravity of the mineral gives directly the tonnage in metric tonnes. In British system (F.P.S.) the total volume in cubic feet is to be divided by the 'tonnage factor' which represents the number of cft in a tonne of ore. This tonnage factor (T_f) is determined by simple formula:

$$T_f = \frac{2240}{62.5 \times g} \text{ cft}$$

where g is the specific gravity of the ore.

Once the tonnage factor is determined, the specific gravity of the ore/mineral assuming constant, the total reserves of an area is determined by dividing the volume by the tonnage factor. While estimating the reserves, one practical point is always to be remembered that in nature, the weight of material may not exactly be equal to theoretical formula. It is because the ore body might be porous or might have fractures and cavities. For this reason, the specific gravity in C.G.S. and the tonnage factor in F.P.S. are required to be suitably modified for rock porosity, etc. by actually excavating a known volume of material and weighing it instead of assuming empirical factor. This correction for porosity etc. is, however, not significant for compact deposits.

For ascertaining the reserves correctly the deposit is usually sub-divided into a number of ore-reserve blocks on a plan with each sample point (boreholes, pits, etc.) as a unit. For each of the ore reserve block, the area of influence is determined and this area is multiplied by the true width of the ore body at the concerned point, to obtain the volume in case of horizontal beds or gently dipping ore bodies. In steeply dipping ore bodies, a vertical longitudinal section is used. The area in this section is converted into the area in the plane of the ore body by multiplying it with dip factor i.e. the dip length of the ore body for unit vertical depth. In an ore body dipping 45°, the dip factor is 1.41. However, in each case it has to be calculated or read directly from the chart made for the purpose.

The tonnage in the individual blocks are added up to get the total tonnage. The average grade of the deposit is obtained by the additions of the tonnage, multiplied by grade of each block and by dividing it by the total tonnage. For example, there are 10 ore reserve blocks in a deposit, each having tonnages of T_1, T_2, T_3, etc. and grades A_1, A_2, A_3, etc.

$$\text{The average grade } A = \frac{T_1 A_1 + T_2 A_2 + \ldots T_{10} A_{10}}{T_1 + T_2 + T_3 \ldots T_{10}}$$

The area of a block can be determined by various methods and broadly placed under two groups, namely, (1) geometric method, and (2) graphic method.

Under the geometric method the common methods used are: (a) included area method; (b) extended area method; (c) triangle method; and (d) polygon method.

Under the graphic method the common methods used are: (a) use of isochore maps; (b) use of stratum contours; and (c) use of transverse sections.

Before using any of these methods, the mineralised portions are first demarcated from the geological knowledge of the area in respect of ore continuity, pitch, folds and faults. Included Area Method is used when the boreholes/pits sunk are in a grid pattern. In Figure 26, pits have been put at 25 m intervals in a grid pattern. All the pits are mineralised. Assuming that total reserves can be estimated only inside the big square area enclosed by the sample points, the area of influence of each of the sample points is shown in Figure 26. It is seen from the figure, that the area of influence of the nine central pits is 25×25 square metres each; the side pits 12 in number, each have half the area of influence and the four corner pits 1/4th the area of influence in comparison to the central pits. In other words, the area of influence and hence the weightage for the width encountered in the central pits are twice and four times the

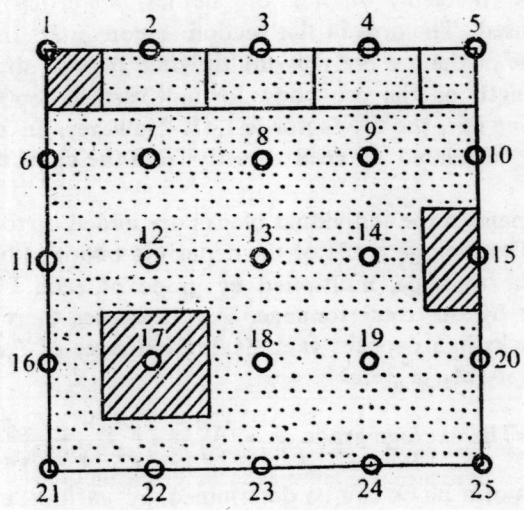

Fig. 26. Calculation of reserves by including area method.
(The area of influence by each of central, corner
and side pits has been shown by hatching.)

side and corner pits. Accordingly, the volume of the ore (V) of the area enclosed by the 25 pits can be obtained as follows:

$$V = 9 \times 25^2 \times W_1 + \frac{12 + 25^2}{2} \times W_2 + \frac{4 \times 25^2}{4} \times W_3 \text{ cu m},$$

where, W_1, W_2, W_3 are the average of the widths encountered in the central, side and corner pits. Once the volume is evaluated, the reserve can be estimated by dividing by the tonnage factor.

Similarly, the average grade (G) of the deposit can be arrived at as follows:

$$G=\frac{1}{V}\Big(9\times25^2g_1W_1+\frac{12\times25^2}{2}g_2W_2+\frac{4\times25^2}{4}g_3W_3\Big)\%$$

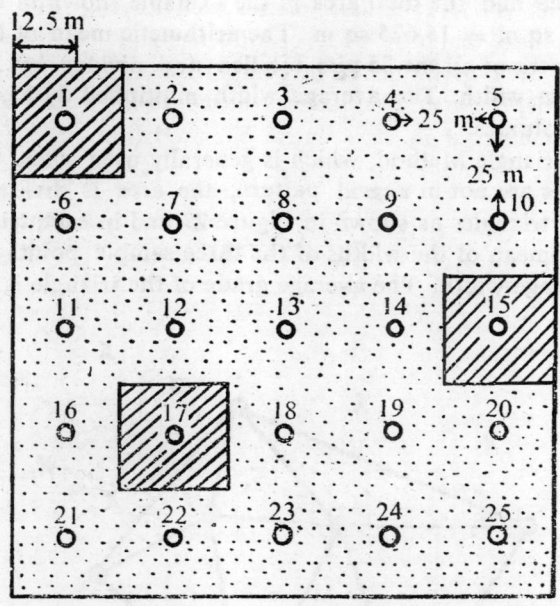

Fig. 27. Calculation of reserves by extended area method (in squares).
(The area of influence is taken as same for each pit or borehole.)

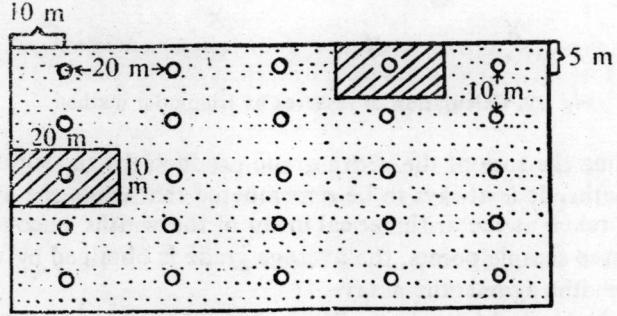

Fig. 28. Calculation of reserves by extended area method
(in rectangles).

The above calculations are taken into account when the thickness and the grades (g_1, g_2, etc.) of the ore body are varying in different pits.

In the Extended Area Method, all the blocks have equal areas of influence and the total area in the example shown in Figure 27 is 25×25^2 sq m $= 15,625$ sq m. The arithmetic mean of the width i.e. the widths of all the 25 pits totalled and divided by 25, gives the average width. The average width muitiplied dy 15,625 gives the total volumes.

In the Triangle Method, which is generally used when the sampling points are not in a grid pattern, the area is divided into a number of triangles as shown in Figure 29 and in each triangle the arithmetic mean of the widths of the three sample points is taken as the average width. The average grade of the triangle is obtained

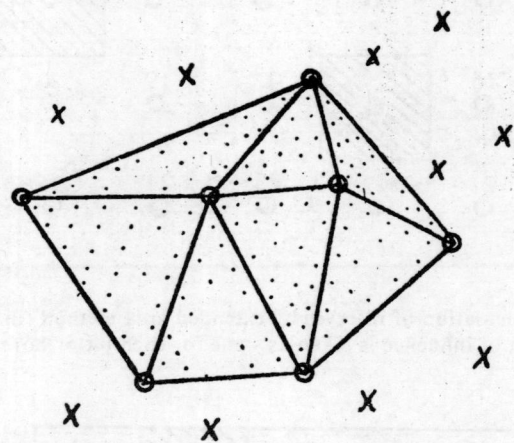

Fig. 29. Calculation of reserves by triangular method.

by dividing the sum of the width-grade products by the sum of the three widths. It is always to be remembered that while the average width is taken as the arithmetical mean of the widths encountered at the three sample points, the average grade is obtained by weighing the widths against the assay.

In the Polygon Method (see Figure 30) polygons are constructed by drawing perpendiculars to the lines joining each of the pits with the surrounding pits. It should be noted while in the extended area method the sample points have influenced half the way to the sample points on the sides and also those above and below, in the polygon method, the influence is taken as half way to all the sample parts. In each polygon, the width of the sample point around which the polygon has been constructed is used for calculating the volume.

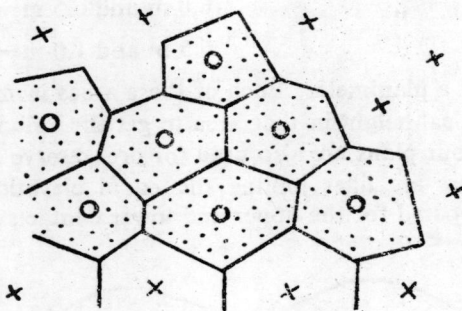

Fig. 30. Calculation of reserves by polygon method.

In the Graphic Method, isochores plans (Figure 31) are used in deposits of more or less uniform grade e.g., in coal deposits. Isochores are lines joining points of equal vertical thickness in a bed (isopaches are lines joining points of equal true thickness). In Figure 31 isochore lines having values 5 m, 5.5 m, 6 m and 6.5 m

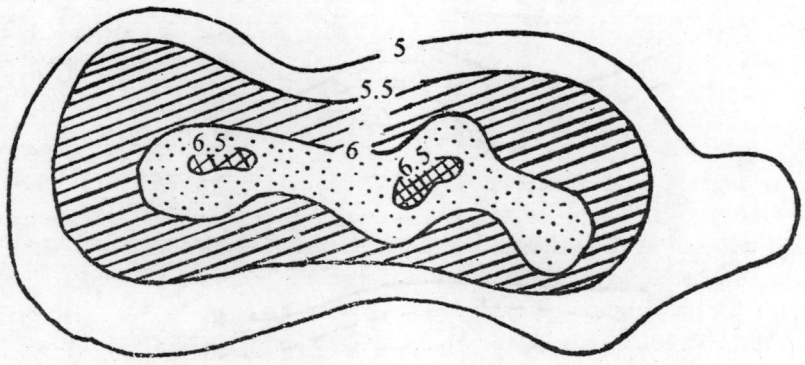

Fig. 31. Calculation of reserves by isochore plan.

are shown. The area between the contours 5 m and 5.5 m will have an average vertical thickness of $\dfrac{5.0+5.5}{2}=5.25$ m, and so on.

These areas (A_1, A_2, A_3, etc.) are measured separately:

Area between isochore   6 m and 5.5 m $=A_1$

,,   ,,   ,,   ,,   5.5 m and   6 m $=A_2$

,,   ,,   ,,   ,,   6.0 m and 6.5 m $=A_3$

,,   ,,   ,,   ,,   6.5 m and 1.0 m $=A_4$

with the help of a planimeter. Each of these areas is multiplied by the average vertical height in that area to get the volume.

Stratum contour plans are also used for ore reserve calculation. Stratum contours are lines joining the equal elevation of a bed. They can be prepared for the upper and lower contacts e.g. the roof

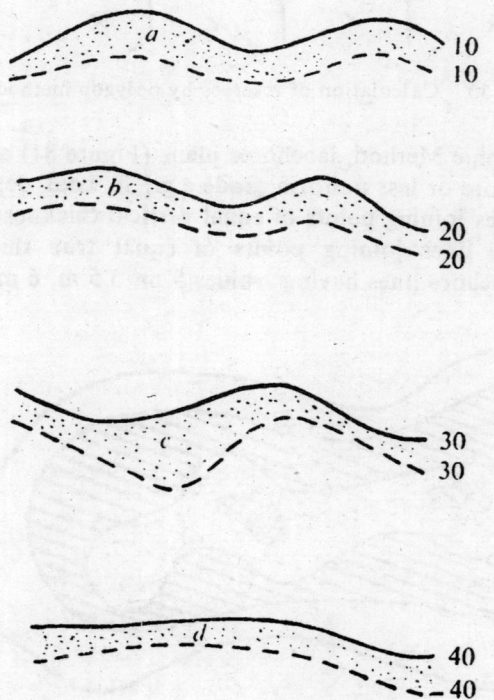

Fig. 32.   Calculation reserves by stratum contour plan.

contours and floor contours of coal beds. In Figure 32 the floor contours are shown as firm lines and the roof contours as broken lines.

The contour interval is 10 metres, '*a*', '*b*', '*c*', '*d*', are the shape of the ore body at 10 m, 20 m, 30 m, and 40 m depths. Each of these horizontal areas of the ore body can be measured separately and when multiplied by 10 m, i.e. the vertical influence of each set of contours, gives the volume at each level.

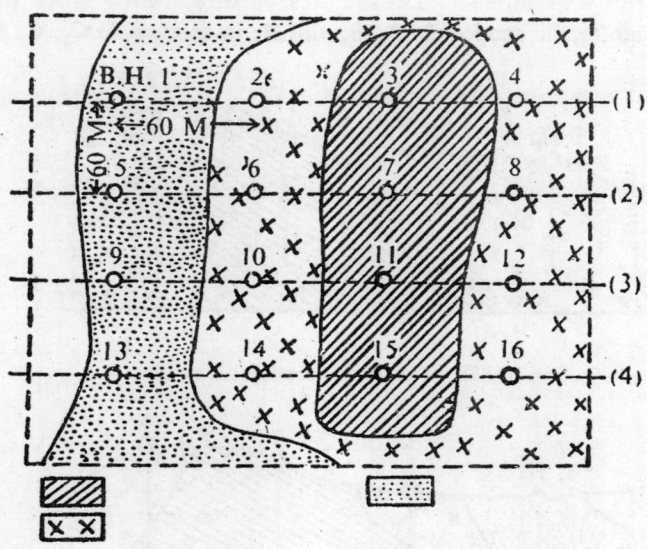

Fig. 33. Calculation of reserves by transverse sections. Plan showing an iron ore body.

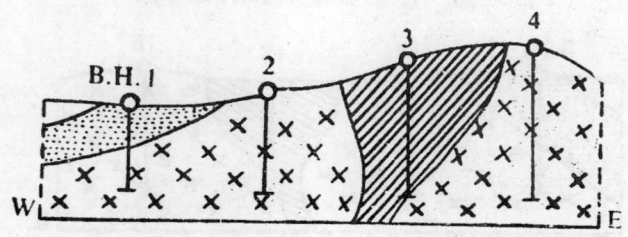

Fig. 33 (*a*). Transverse section along (1).

Transverse sections are also used in estimating reserves of limestone, iron ore and other bedded deposits. In the following example (Figure 33) of an iron ore body, there are 16 boreholes sunk in
a grid pattern at intervals of 60 m. There are three types of ore—
massive steel grey, laminated hematitic and limonitic. The reserve
of each types is estimated separately. Four transverse sections are
drawn using the data of drilling, the shapes of these types of ore
body are constructed. In each section, the area of each type is
measured by a planimeter. Let the area of these three types be—
A_1, A_2, and A_3, in section 1: B_1, B_2 and B_3 in section 2; C_1, C_2 and

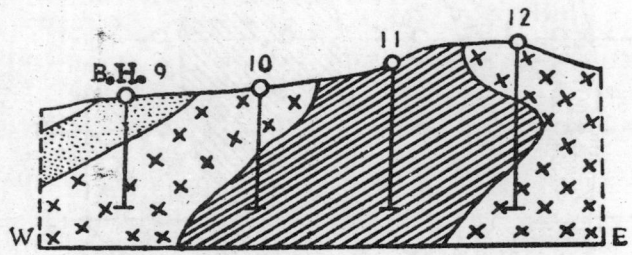

Fig. 33 (b).   Transverse section along (2).

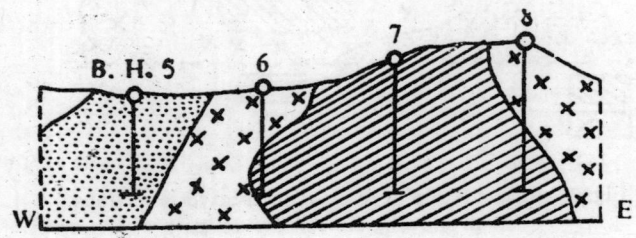

Fig. 33 (c).   Transverse section along (3)

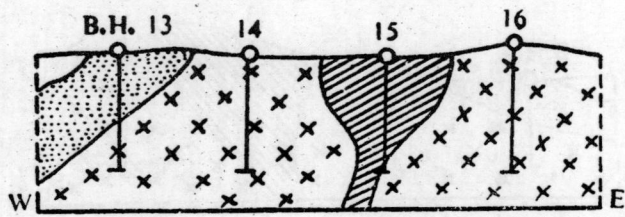

Fig. 33 (d).   Transverse section along (4).

C_3 in section 3 and D_1, D_2 and D_3 in section 4. Then $(A_1+B_1+C_1 +D_1) \times 60$ m will give the total volume of the massive steel ore; $(A_2+B_2+C_2+D_2) \times 60$ m, the volume of laminated hematitic ore and $(A_3+B_3+C_3+D_3) \times 60$ m, the volume of limonitic ore.

ESTIMATION OF RESERVES BY EXPLORATORY MINING

In the vein type deposits like copper, lead-zinc ores, gold, etc., it is not possible to estimate reserves with accuracy from the boreholes, pitting and trenching data. They may provide only the lateral and vertical extensions of the ore bodies but not their continuity and variations in tenor. In each such cases, adequate underground development work is necessary to arrive at measured and probable reserves, zone of concentration of richer and leaner shoots, level-wise average grade and all other related data. These when put together provide a clear picture of the mineralisation underground which aid ultimately in making qualitative and quantitative assessment of deposits. Such assessment work is proceeded systematically level-wise. Now let us take the results of a level in an underground development work in a zinc mine as illustrated below. Here seven

Location	Width (W) in metre	Grade (A) % Zn	$W \times A$
0 metre from the mouth of a drive	1.8	2.6	4.6
at 5 m from the mouth	1.6	1.8	2.88
at 10 m ,, ,, ,,	2.0	2.5	5.00
at 15 m ,, ,, ,,	1.5	3.0	4.50
at 20 m ,, ,, ,,	1.9	4.0	7.60
at 25 m ., ,, ,,	1.6	3.5	5.60
at 30 m ,, ,, ,,	1.8	3.5	6.30
Total	12.2		36.56

samples have been drawn at regular intervals. The width of the ore body is measured at each sample points. The average width is obtained by totalling all width and dividing by seven. The average grade is obtained by dividing the sum of width-grade products by the sum of the widths.

$$\text{Average width} = \frac{\Sigma W}{7} = \frac{12.2}{7} = 1.74 \text{ m.}$$

$$\text{Average grade} = \frac{\Sigma W \times A}{\Sigma W} = \frac{36.56}{12.2} = 3.00\% \text{ Zn}$$

When the intervals between sample grooves are not uniform, suitable influence of each groove is taken as in the following examples:

Location		Influence (L) (in metre)	Width (W) (in metre)	Grade (A) % Zn	$L \times W$	$L \times W \times A$
0 m	from mouth of drive	5	1.2	3.2	6.0	19.20
10 m	,,	10	1.4	2.6	14.0	36.40
20 m	,,	8	1.6	4.1	12.8	52.48
26 m	,,	6	1.5	3.3	9.0	29.70
32 m	,,	7	1.8	2.5	12.6	31.50
40 m	,,	9	1.9	3.0	17.1	51.30
50 m	,,	10	2.1	2.5	21.0	52.50
60 m	,,	5	1.6	3.2	8.0	25.60
Total		60			100.5	298.68

$$\text{Average width} = \frac{\Sigma LW}{L} = \frac{100.5}{60} = 1.67 \text{ m}$$

$$\text{Average grade} = \frac{\Sigma LWA}{\Sigma LW} = \frac{298.6}{100.5} = 2.98\% \text{ Zn}$$

When the ore body is not fully exposed in width, the additional average width from the cross-cuts is added to the average of the drive values as in the following example:

Location	Influence (L)	Width (in metre) (W)	Grade (% Zn) (A)
Drive data			
0–60 m	60 m	1.67	2.98
Cross-cut data			
FW, cross-cut at 10 m from the mouth	25 m	0.50	2.60
HW, cross-cut ,,		1.10	2.80
FW, cross-cut at 40 m from the mouth	35 m	0.40	2.70
HW, cross-cut ,,		1.15	3.00

$$\text{Additional width from } FW \text{ cross-cut} = \frac{25 \times 0.50 + 35 \times 0.40}{60}$$

$$= \frac{12.50 + 14.00}{60} = 0.44 \text{ m}$$

$$\text{Grade of the above width} = \frac{12.50 \times 2.60 + 14.10 + 2.70}{26.50}$$

$$= 2.65\% \text{ Zn}$$

Additional width from HW cross-cut

$$\frac{25 \times 1.10 + 35 \times 1.15}{60} = \frac{27.50 + 40.25}{60} = 1.111 \text{ m}$$

$$\text{Grade of the } W \text{ width} = \frac{27.50 \times 2.80 + 40.25 \times 3.00}{67.75}$$

$$= 2.96\% \text{ Zn}$$

	Average width (W)	Assay value (A)	W×A
Drive	1.67	2.98	4.98
FW, X-cut	0.44	2.65	1.17
HW, X-cut	1.11	2.98	3.31
Total	3.22		9.46

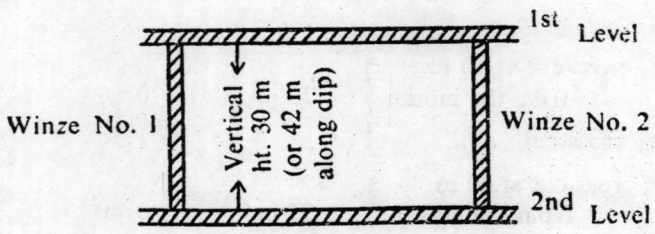

Fig. 34. Method showing calculation of average grade and tonnage by sampling in drive, raise, winze and stope.

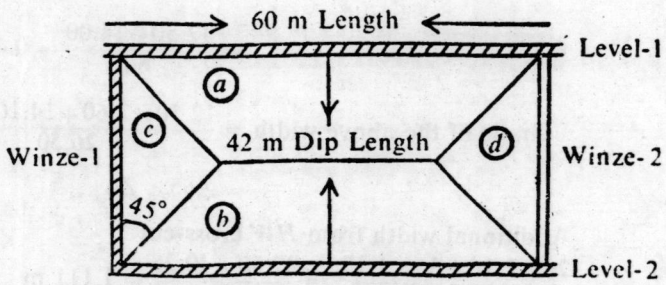

Fig. 34 (a). Method showing calculations of average grade and tonnage by sampling in drive, raise, winze and stope.

Average value at the level for full width

$$= \frac{\Sigma WA}{\Sigma W} = \frac{9.46}{3.22} = 2.94\%$$

In the same way average values are computed for all drives, raises, winzes and stope walls. When all the four sides are sampled the total tonnage and grade can be estimated either weighing the samples in the levels, raises, winzes only for their lengths or for the area as illustrated in Figures 34 and 34(a).

	Length	True width	Grade % Zn
1st level	60 m	3.22 m	2.94
2nd level	60 m	3.20 m	3.00
Winze 1	42 m	3.00 m	1.91
Winze 2	42 m	2.80 m	4.00

Average width $= \dfrac{64 \times 3.32 + 60 \times 3.2 + 42 \times 3 + 42 \times 2.8}{60 + 60 + 42 + 42}$

$= 3.08$ m

Average grade

$$= \frac{(60 \times 3.22 \times 2.94) + (60 \times 3.2 \times 3) = (42 \times 3 \times 1.91) \times (42 \times 2.8 \times 4)}{(60 \times 3.22) + (60 \times 3.2) + (42 \times 3) + (42 \times 2.8)}$$

$= 2.62\%$ Zn

Total volume $=$ Area in the plane of ore body $\times$ True width

$= 60 \times 42 \times 3.08$ cubic metres

Total tonnage $= \dfrac{60 \times 42 \times 3.08}{T_f}$ metric tonnes

In the case [see Figure 34 (a)] each level and winze has its area of influence, a, b, c and d. Each of these is weighed for its width to obtain the volume and then for its grade to obtain the average grade.

ECONOMIC CONSIDERATIONS IN RESERVE ESTIMATION

The reserve estimation has two aspects—geological and economic. In any mineral property total reserves *in situ* can be estimated assuming some cut-off grade, depth and other factors. The reserves so estimated may or may not have any economic significance at a given time except in times of national emergency when consideration of profit is of little significance. For instance, reserves estimated for a low grade tungsten ore in a remote locality may not have any economic value, but such work is necessary for assessing the country's strategic mineral potential for urgent needs. Leaving aside such critical eventualities all expenditure and labour expended on proving reserves are profit motivated and done to find out immediate possibility of opening a mine. The following important economic considerations are taken into account in locating a new deposit and opening of a mine.

(a) the price of the mineral in market;
(b) the method of mining contemplated;
(c) the treatments such as beneficiation, smelting required;
(d) distances of haulage from the mine site to the place of treatment and marketing;
(e) approachability to deposits;
(f) climatic conditions;
(g) availability of labour, power and water; and
(h) taxes, levies, interest on the investment.

The grade of ore that will meet the cost of mining will thus widely vary. A copper ore deposit analysing as low as 0.5 to 0.6% Cu can be mined by surface quarrying, while for underground mining, grades of 1.0% Cu to 2.0% Cu are required to meet the mining costs. Thus low priced deposits like cement grade limestone which have a pit's mouth value of Rs. 7 per tonne cannot be considered for underground methods of mining. Based on these considerations, the mineral deposits can be divided into two groups for purposes of estimating mineable reserve. These are:

(i) Deposits for quarrying and other surface methods of mining.
(ii) Deposits for underground mining.

Quarriable reserves: The quarrying and other surface methods are some of the cheapest methods of mining. Low priced minerals can be worked by surface methods provided the ore to overburden ratio is favourable, which is of great significance when considering exploitations. Take an example of the limestone quarry where the cost of mining per tonne is Rs. 5 and the selling price at the pit's head is Rs. 7. It can be seen that practically no waste rock can be mined since the profit accruing from sale after paying the cost of excavation, royalty, cess, surface rent, etc,, does not provide for dead work. For a given pit's mouth price, a particular limestone deposit may not be quarriable and loses its economic significance if the ratio of overburden to limestone exceeds the economic limit. In another case, a coal mine owner is selling coal at a pit's mouth value of Rs. 30 and the cost of mining by open cast is Rs. 6. In this case, up to 5 tonnes of total material can be mined @ Rs. 6 per tonne of which one tonne is coal. That is for every tonne of coal 4 tonnes of waste rock can be mined. The coal to overburden ratio in this case is 1 : 4.

In case of steep dipping veins of comparatively small width, it can be readily seen that surface methods are useful only if the material is of very high value, and is not amenable to underground methods of mining. Let us say, the ore body can yield 5 tonnes of ore per metre strike length for every 1.5 m depth. The return of ore at different phases of mining and the excavation needed are tabulated below (see Figure 35):

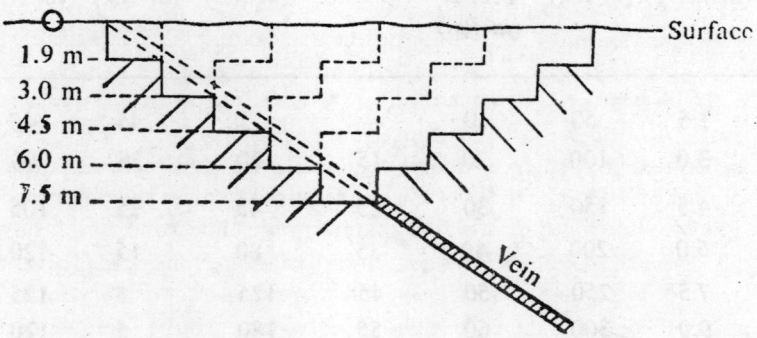

Fig. 35.   Illustration of quarrying narrow and steeply dipping view.

Thus the return is uniform for every 1.5 m depth, while the expenditure increases enormously with depth. If the value of ore is Rs. 10 per tonne and the cost of mining for every 2.25 sq m in cross-section is even kept uniform at Rs. 5 the financial position assuming a profit of Rs. 2 per tonne will be as under; the return for every 1.5 m depth remaining constant at Rs. 50.

Depth of quarry (in metre)	Total return of ore (in tonne)	Additional excavation (in cubic metre)	Total excavation (in cubic metre)
1.5	5	—	2.25
3.0	10	6.75	9.00
4.5	15	11.25	20.25
6.0	20	15.75	36.00
7.5	25	19.75	56.25
9.0	30	24.75	81.00

Depth (in metre)	Total return (in Rs.)	Total profit expected (in Rs.)	Additional expenditure (in Rs.)	Total expenditure (in Rs.)	Profit of that depth (in Rs.)	Overall profit (in Rs.)
1.5	50	10	—	5	45	45
3.0	100	20	15	20	35	80
4.5	150	30	25	45	25	105
6.0	200	40	35	80	15	120
7.5	250	50	45	125	5	125
9.0	300	60	55	180	5	120
10.5	350	70	65	245	−15	105
12.0	400	80	75	320	−25	80
13.5	450	90	85	405	−35	55
15.0	500	100	95	500	−45	Nil

'It will be seen that the project can be worked for the expected profit up to a depth of 12 metres even though the working after 7.5 metres depth is on loss for every 1.5 m depth. In the above estimation it has been assumed that the cost of mining for every 2.25 sq metre cross-section is Rs. 5, but in practice this cost increases depending upon the variation in the nature of strata, intake of water and such other factors, information on which may be available from adjoining mines. As the cost usually increases with depth, it may be necessary to work up to a depth of 10.5 metres only. If the information from adjoining mines reveal that the cost increases more sharply, the depth of the project will have to be decreased suitably. A careful planning is needed to see that the deposit is mined to the maximum possible depth where the total expenditure does not exceed the return balance. It is an important consideration from conservation and mining economics point of view.

It must be emphasised that the reserve calculations for purposes of mining require a dynamic approach with full knowledge of geology, mining costs, mineral prices, otherwise misleading conclusion will be drawn. A case in point is the often quoted reserves of phosphatic nodules at Tiruchirapalli, where a large reserves of 8 million tonnes have been estimated with recovery of 25 kg of nodules for every 4 tonnes (100 cft) of rock mined, which means, to produce one tonne of nodules about 160 tonnes of excavation is needed. One tonne of nodules carrying 20 to 22% P_2O_5 will not earn more than Rs. 80 to 100 at the present price of phosphate rock. It can be mined only if the cost of excavation is as low as Re. 0.50 per tonne. Otherwise, the estimation is not at all of any commercial value.

Mineable underground reserves : It is not possible to recover the entire ore-reserves *in situ* because part of it has to be left as pillars. Thus an 'extraction factor' is to be used to reduce the computed reserves *in situ* to the actual reserves that can be mined.

Also the grade of the ore computed will have to be reduced for possible dilution during mining and the error that may be in sampling and analysis. Thus the computed grade is reduced by what is known as an 'Assay Plan Factor' to get the run-of-mine grade. This Assay Plan Factor can be determined in course of mining by comparing the estimated and produced grades. In new deposits, data from adjacent mines can be utilised in the initial calculations.

In underground mines the cost of mining varies considerably depending on the dip of the lode, strength of the wall rocks and width of barren partings. When the lodes are steep, wide and the host rocks are strong, the stopes can be opened and bulk mining is possible at lower cost. When the rocks are weak and the dip is gentle, the stopes require timber or other form of support and the ore extracted is required to be removed from faces with the aid of manual labour and/or by some mechanical means. In such cases the cost of mining may be as high as 3 to 4 times the former cost. The cut-off grade to be fixed in any deposit will therefore be determined by the possible costs of mining of a particular deposit.

ESTIMATION OF PRIMARY OIL RESERVES

The method of estimating petroleum reserves is quite different from that of estimating metallic and non-metallic or solid fuel reserves. This is due to the unique nature of petroleum reservoirs which are always bound by certain dynamic and physico-chemical conditions underground.

In the case of oil and gas reserve estimates, the factors most commonly determined are : (i) the gross volume of the reserve underground, (ii) the recoverable reserve and the recovery factor, (iii) quality or grade, and (iv) recoverable byproducts. These terms are common with all types of mineral estimates. But apart from these there are certain other factors which are required to be determined while estimating gas and oil reserves. They are: (i) drive i.e., the mechanism that is operative during primary production, and (ii) dynamic characteristics of the hydrocarbon fluids. Petroleum reservoirs generally contain fluid and/or gas. The physical properties of the fluid or gas change with the changing temperature and pressure caused during production of hydrocarbons from the reservoir. Naturally, this leads to the fact that the reserves of oil and gas have to be estimated periodically, as for example: (1) just after drilling and well completion, (2) after regular production from the field at least for one year, (3) after few years when the production has declined, and (4) when the petroleum reservoir has reached the stage of depletion.

Estimation of Oil Reserves—Three principal methods are adopted for estimation of oil reserves. They are: (i) Volumetric Method,

(ii) Material Balance Equation Method, and (iii) Decline Curve Method.

VOLUMETRIC METHOD

The volume of hydrocarbon fluids present in an oil reservoir is calculated from the formula expressed below.

Reserves = 7758 $(A.h.)$ (f) $(I—S_w/B_i)$ (RF)
(in stock tank barrels of oil)

where,

7758 = Volume of one acre feet expressed in barrels
        (1 barrel = 42 gallons API)

A = Area of the producing formation in acres.

h = Thickness in feet of the producing formation.

f = Porosity (fraction of voids in total rock matrix).

S_w = Volume of the voids filled with interstitial water.

B_i = Formation volume factor (reservoir barrels—stock tank barrels)

RF = Recovery Factor (Fraction of the oil in place that may be ultimately recovered).

The volume of the net pay of a reservoir rock $(A.h.)$ is calculated from an isopach map of the petroleum reservoir. This map is specially prepared by correctly plotting the water/oil and oil/gas contacts, on the map. The exact sub-surface locations of these contacts are determined from well logging and core analysis data. The effective porosity and the percentage of interstitial water are also determined by the core analysis and the well logging data. The determination of the fracture or vugular porosity which is generally common with carbonate rocks, is rather difficult. The formation volume factor is calculated from the oil and gas production data and PVT (Pressure, Volume, Temperature) analysis of the hydrocarbon fluids under reservoir conditions. Here it is worthwhile mentioning that there is an appreciable shrinkage of crude oil as it moves from the formation to the surface due to the liberation of gas from solution, consequent to reduction of pressure.

All the above parameters can be evaluated with a reasonable degree of accuracy but the assessment of the Recovery Factor is rather problematic and uncertain. It depends upon a number of factors such as type of drive or recovery mechanism, structure of the reservoir, porosity, permeability, viscosity of the hydrocarbon fluids, operational efficiency, etc. It may be as much as 80% in a field under either water drive or expanding gas cap drive; or it may be 20 to 40% in a field under dissolved gas drive which is considered to be the most prevalent reservoir mechanism responsible for production of oil from the reservoir. Formulae with slight modifications are in use for calculation of recoverable reserves of oil by volumetric method under different producing mechanisms of the reservoir.

However, some rules of thumb for the estimation of the potential of a petroleum pool can be given: (a) in the case of solution gas drive oil reservoirs, the reserves may vary from three barrels per acre-foot per per cent porosity in fractured limestones or highly cemented sandstones with high solution gas-oil ratio oil to fifteen barrels per acre-foot per per cent of pore space in intergranular limestones or unconsolidated sands with medium solution gas-oil ratio, low viscosity oil and high connate water. An average figure that can, therefore, be used may be ten barrels per acre-foot per per cent porosity; and (b) regarding the oil reservoirs (only of the sandstone type) under water drive, reserves may vary from 6 barrels per acre-foot per per cent of pore space in low permeability sands with viscous oil to thirty-five barrels per acre-foot per per cent of porosity in highly permeable sands containing low viscosity oil and low connate water. An average figure that can, therefore, be used may be twenty barrels per acre-foot per per cent porosity.

MATERIAL BALANCE METHOD

In this method it is assumed that the oil reservoir attains equilibrium at all stages of production and as such mathematical expressions are developed to express the relation between the volumes of oil, gas and water produced, the decline in reservoir pressure, the quantity of water that may have encroached into the reservoir and the residual oil and gas content of the reservoir.

A material balance, which is balancing or an inventory of the materials in a reservoir, is usually formulated on a volumetric basis. That is, the initial oil volume in place is equal to the volume of oil that has been produced up to the time of calculation of reserves plus the volume of oil residual in the reservoir. It is more conveniently expressed basically by a balance on the gas present in the reservoir, because it is the presence of free gas and gas initially in the solution which plays the major role and has to be accounted for in the fomulation of the equation. The equation is formulated as follows:

Gas-in-place initially = Gas residual in the reservoir + Gas produced.

This can be expressed by a generalised material balance equation such as :

$$N = \frac{N_p B_o + B_g(R_c - R_s) - B_w(W_e - W_p)}{mB_{oi}(B_g - 1) + B_g(R_{si} - R_s) - (B_{oi} - B_o)}$$

where N = original oil-in-place (stock tank barrels):

N_p = cumulative oil produced (stock tank barrels);

B_o = oil formation volume factor;

B_g = gas formation volume factor;

R_c = net average cumulative gas-oil ratio;

R_s = solution gas-oil ratio;

B_w = water formation volume factor;

W_e = cumulative water influx during production of N_p stock tank barrels of oil (barrels);

W_p = cumulative water produced during production of N_p stock tank barrels of oil (barrels);

m = ratio of initial gas cap volume size to initial oil zone volume size (fraction); and

i = subscript indicating initial value or conditions.

The physical significance of the groups of terms in this equation is as follows:

$N_p B_o + B_g(R_c - R_s)$: Reservoir volume of the cumulative oil and gas produced.

$B_w(W_e - W_p)$: Volume of total water which encroached into the oil reservoir and retained by it.

$mB_{oi}(B_g/B_{gi} - 1)$: Expansion of gas cap which occurs with production of N bbls of stock tank oil.

$B_g(R_{si} - R_s)$: Reduction in the amount of gas initially present in stock tank barrels of oil at reservoir conditions of pressure and temperature.

$(B_{oi} - B_o)$: Change in the volume of reservoir oil comprising one barrel of stock tank oil as pressure in the system is lowered.

Among these parameters, N_p, R_c and W_p are obtained from production data; other depend on fluid properties and pressure and temperature. The accuracy with which these are recorded or assessed in the laboratory influences the final precision of calculation. Besides, m, N and W_e are the three unknowns. N and m are theoretically constant and should not change with time and they can be obtained from the volumetric method. W_e may then be evaluated, knowing N and m. Alternately, m and W_e can be assumed to be equal to zero and N can be calculated.

Material balance equations with slight modifications are in use for calculation of recoverable reserves of oil under different reservoir conditions and producing mechanisms of the reservoir. In general, it can be said that the volumetric method is usually accurate in the very early stages of depletion and the material balance method becomes reliable in the later stages of depletion when more precise data is available, if only collected and analysed carefully.

DECLINE CURVE METHOD

This method is very useful where production has been at full capacity without any restriction for a considerable time. It is assumed that the future production behaviour of a well will be governed by the trend established in its past performance. It is also seen that the decline of well production is on the average sufficiently symmetrical to permit projection of the recorded decline to indicate the probable production of a well in future years. A production decline curve is a graphic record of a well plotted on coordinate paper. The vertical scale is production per unit time, and the horizontal scale represents time.

There is another reliable method called 'constant percentage decline curve method' for estimating future production of individual wells from their own records. Here it is assumed that production during successive unit of time is expressed as percentage of the production during the first unit of time which is taken as 100 per cent. The time rate data can be plotted on regular coordinate or log paper. An empirical appraisal curve may be constructed from such data to predict the ultimate production as the function of initial production rate.

Yet there is another method called "Loss Ratio Decline Method" based on the theory that the amount of oil produced for equal time intervals and the amount of drop in production could be plotted as "Loss ratio". The curves plotted by the loss ratio method generally show either a constant loss ratio or a constancy in the differences of successive loss ratios. If the data are properly tabulated yearwise then the constancy in the differences of the successive loss ratios will be known and from this the production for any future year can be predicted.

ESTIMATION OF NATURAL GAS RESERVES

Natural gas consists of gaseous hydrocarbons, which occur and are transported in gaseous state. The unit of measurement is cubic foot or cubic metre at a temperature of 60° F and 14.7 pounds per square inch pressure absolute at standard conditions. The basis for estimating gas reserves is the law of perfect gases which is again based upon Charles's law and Boyle's law. At high pressures a deviation factor "Z" is introduced in Boyle's law to account for the amount of deviation from the law. The following methods are used for the estimation of the reserves of the gases.

(i) *Volumetric method*: In this the following equation is used for the volumetric of the gas reserves.

$$V_{is} - V_{as} = 43,560(A.h.)(f)(1-S_w)(R.F.)$$
$$\left(\frac{T_s}{P_s.\,T_f} \right) \left(\frac{P_i}{Z_i} - \frac{P_a}{Z_v} \right)$$

where V_{is} = volume of gas originally in place (SCF — Standard Cubic Feet)

V_{as} = volume of gas at abandonment (SCF)

43,560 = number of cubic feet per acre-foot or number of sq ft per acre

T_s = base temperature (520° R)

P_s = base pressure (psia)

T_f = reservoir temperature (°R)

P_i = original reservoir pressure

P_a = abandonment pressure (psia)

Z_i = compressibility factor at P_i and T_f

Z_a = compressibility factor at P_a and T_f

The other parameters are the same that are referred in similar equation for oil reserves and are determined in the same manner. The recovery factor in the case of gas reserves is much higher and may amount to 75% to 80% or even more.

(ii) *Pressure Production Decline Method* : This method is used where the withdrawal of gas has been sufficient to cause the decline in the reservoir pressure. The following formula is used for the calculation

$$V = \frac{W(P_2 Z_2 - Z_3)}{(P_1 Z_1 - P_2 Z_2)}$$

where V = Volume of recoverable gas reserve before abandonment.

W = Gas production during period of decline from P_1 to P_2 at standard temperature and pressure.

P_1 = Original formation pressure per sq inch absolute.

P_2 = Formation pressure following the period of withdrawal of gas volume, W.

P_3 = Formation pressure at the time of abandonment.

Z_1 = Deviation factor at P_1.

Z_2 = Deviation factor at P_2.

Z_3 = Deviation factor at P_3.

(iii) *Solution gas* : Solution gas is the gas which occurs in dissolved state in the oil reservoir. When oil is produced, the pressure falls below the bubble point pressure. Knowing the total volume of the oil in place (N) and oil still to be produced (ΔN_p) the volume of the solution gas produced can be found out, in the case of constant volume oil reservoir below the bubble point pressure, by the following equation.

Solution Gas Reserves = $N(R_{sl} - R_s) + N_p R_s$

where R_{sl} = Amount of gas in solution at original reservoir conditions (SCF/STB)

R_s = Amount of gas in solution of production of N_p (STB).

In case of the oil reservoir above the bubble point, the volume of the gas produced up to the bubble point would be given by the following simple equation:

Volume of gas produced = Volume of oil produced + Gas/Oil
Ratio (GOR)

In all other cases material balance equation is used for the calculation of total volume of the gas produced.

Estimation of gas condensates : Certain hydrocarbon mixtures which are in gaseous phase at high pressure get converted into liquid phase as the pressure is lowered. This phenomenon is known as retrograde condensation. In such cases, volumetric method is used usually for estimating the amount of condensate in a reservoir

$$V_{ls} = 43{,}560(A.h)\,(f)(1-S_w)\,(T_s/P_s T_f)\,(P_i - Z_i)$$

All the terms are already explained above.

TABLE 14

Classification of Petroleum Reserves

Energy source	Degree of proof	Development status	Producing status
1. Primary Reserves	1. Proved	(a) Developed	(a) Producing
Recoverable commercially by conventional methods and equipment by the natural energy inherent in the reservoir.	Proved to a highdegree of probability by commercial rate of production or by successful testing, supported by core analysis and log interpretation.	Recoverable through existing wells.	Being produced from existing wells having completion intervals open to production.
	2. Probable	(b) Undeveloped Present is undeveloped blocks which would probably produce hydrocarbons when drilled for.	(b) Non-producing Behind the casing or below or above the producing intervals but will be produced from existing wells.
	Not proved by commercial rate of production and based upon limited evidence of produceable hydrocarbons within the geological limits of a reservoir which have to be proved by further drilling and testing.		
	3. Possible		
	May be in existence, but available data does not support a higher classification.		

II. Secondary Reserves

Recoverable commercially by supplementing the natural energy of reservoir by artificial methods such as pumping, gas lift, water flooding, etc., as the physical properties of reservoir fluids change.

1. Proved

Proved to a high degree of probability by success full pilot full-scale operations in the same reservoir or by analogy from contiguous reservoir.

2. Probable

Considered to be in existence not by operational data but from core, long or reservoir data or from a case history.

3. Possible

Considered to be amenable for secondary recovery but available data does not support a higher classification.

(a) Developed

Recoverable through existing wells.

(b) Undeveloped

Will be produced upon the installation of secondary recovery methods or by drilling additional wells.

(a) Producing

Being produce from existing wells in the portion of reservoir subject to secondary recovery operations.

(b) Non-producing

To be produce from existing wells upon enlargement of secondary operation in the remaining portions of the reservoir.

12

Feasibility Study

IN THE FIELD of mineral exploration three stages are recognised. These are: (1) feasibility stage, (2) detailed project planning stage, and (3) quality control stage. These stages are required to be scrutinised step by step, at appropriate level and time, in the minutest details before a deposit can be regarded as economical for exploitation. The feasibility stage offers a basis for arriving at an investment decision while the detailed project planning stage provides all essential details for preparing the complete project report. The grade of ore output, rather the quality control, emerges as a result of the first two stages. Thereafter, mining can be planned to conform to the milling requirements or to the specifications of the buyers.

Definition

A feasibility study can be defined as an evaluation of a mineral deposit to determine if profitable mining is plausible. The exploratory work can be said to have reached the feasibility stage when it is possible to obtain a conceptual idea of the deposit, about its grade and reserve and its future prospects with certainty. This decision making moment on the desirability of making the required investment for exploitation of the deposit is indeed very crucial. It becomes essential at this stage to critically analyse all the basic data made available as a result of exploratory and related laboratory investigation work in all its ramifications to arrive at the economic viability of the deposit. It calls for an evaluation of the grade and tonnage likely to be available (reserves), beneficiation and metallurgical characteristics, other mineral and metal values, price trend, method and cost of mining and milling operations, optimum level of production to be maintained, investment on machinery and plant, investment on infrastructure and such other expenditure with a view to assess the overall profitability in relation to the total investment. The period within which the entire expenditure is to be amortized is also an important factor.

The feasibility studies are most essential today for operating mines on a large-scale primarily because of the lower and complex grade of the deposits which of necessity must be mined to meet the growing shortages. The exploitation of such deposits requires sophisticated processes of mining, beneficiation and smelting to recover all the co-products and by-products and consequently involves a high capital investment.

IMPORTANT CONSIDERATIONS OF A FEASIBILITY STUDY

The estimation of mineral reserves is the first and foremost consideration for a feasibility study. Since the assessment of mineral reserves is costly both in terms of money and time, the question arises what should be the initial input of work needed to build up an optimum level of confidence in a deposit. In other words, what should be the minimum amount that must be expended in the feasibility investigations. The answer to this basic question emerges chiefly from the results obtained in the preliminary prospecting. It will also depend on the size and the type of deposits and their mineral and metal values. The initial input to a feasibility study must always form a well coordinated exploration programme providing a two-way communication between the envisaged programme and the interpretation, assessment and evaluation of the observations at each stage of the work right from the beginning.

The approach to a feasibility study should always be viewed with an economic bias. A mineral prospect is taken up for evaluation only when preliminary prospecting has substantiated the possibility of finding the mineral in an economic quantity. Such preliminary work includes geological reconnaissance, mapping on a scale of 1:100,000 or 1:50,000, geo-chemical and geo-physical surveys, few test drilling of the geo-physical and geo-chemical anomalies and laboratory tests both chemical and petrological. In the case of flat-bedded or exposed deposits like limestone, iron ore and bauxite, geological traverses, analysis of a few samples from selected areas from pits or trenches, estimation of thickness and the extent of the bed on broad geological evidences serve the purpose of preliminary prospecting. Areas so covered and prospected with promising indications are put under the category of "mineral resource"

Only the area containing a "mineral resource" is subjected to detailed probing work leading to a feasibility study. The work required for the feasibility study can be suitably phased, to introduce an element of economy, depending upon the extent and nature of deposit and also upon the nature of assignment. The object of the assignment may be varied and an entrepreneur may assign the work to an exploration geologist or to a consulting firm, (i) to prove the deposit and ascertain its marketability only or (ii) to prepare the complete mining plan with ore transport schemes for export purposes, or the complete mining plan with milling and smelting capacities or similar other assignments. The quantum and nature of work, therefore, will vary in each case. The feasibility work may be completed in three phases in the case of non-ferrous and precious metal deposits and in two phases in the case of all others. Since no two mineral deposits are exactly identical, no rigid norm of mineral exploration can be laid down. It is, however, possible to make a tentative estimate of expenditure for the quantum of work involved to complete the first phase in each case. Work requirement as well as cost for subsequent operation can be worked out on the basis of results achieved in the earlier phase or phases. Details will have to be worked out well in advance of the investment which involves a significant element of risk and the entrepreneur apprised correctly of all implications so that the required resources could be mobilised by him.

Some economic parameters also need to be carefully considered during the feasibility work. These are the size of boreholes and concept of optimum drilling hole spacing. Drilling equipment and the size of the hole should be selected with care because of high cost of drilling. The cost of drilling per metre in hard formation may vary anything from Rs. 200 to 350 depending upon the size of the hole. There are a number of standard sizes of drill bits of which four namely NX, BX, AX and EX sizes are most commonly used. The cost of drilling in size BX will be approximately 2/3rd of the size NX for the same formation. The cost of drilling reduces considerably with size. The exploration geologist should indicate which size of the bit gives better core recovery, serves his purpose best in interpreting the geology and other analytical work. During the third phase close spaced drilling is called for to define clearly specific mineralization, geological structure and tonnage. At this point concept of optimum drill hole spacing can be introduced. At some stage of the exploration the improvement in information obtained by extra drilling will be minimal. Also there is a little value in most cases of close spaced drilling of the block/area that will not be mined for 20-30 years. Adequate drilling should confine to blocks to be mined in the first 5-10 years. Once the production begins detailed drilling work can be extended to other promising blocks. During the second and third phase some techniques of statistical analyses can be utilised to define and provide further information about the deposit. Trend analysis method has been found most useful in calculation of reserves as well as directing the search for increased reserves tonnage. Likewise, the area of influence, of assays can serve as a control for enlarged drilling programme, and sample volume investigation may ensure that the assay result obtained are logical and that the samples taken are of adequate volume to fit the mineralogical characteristics of the deposit under study. It should be noted that the statistical valuation serves only as a guide and it is no substitute for the conventional method of reserves estimation.

Phase I (Non-ferrous and Precious Metals)

Objective

The main object of Phase I is to carry out a limited exploratory work with a view to bring "mineral resource" into the inferred category. The work also aims at delineation of suitable blocks in the area under investigation so that it can be subjected to a further detailed study to obtain information about the size, shape, attitude, grade and overall potentialities in the selected blocks and in the deposit as a whole.

Scope

The work and expenditure incurred during Phase I may frequently be sufficient to reject a project. It is, however, seldom adequate to a positive acceptance of it. The Phase I work generally describe a hypothetical installation and is seldom a basis of conceptual design. It may indicate, however, the desirability of extending the work to Phase II and Phase III.

Work involved

The work involved in this phase is as follows: Construction of approach road, camping facility, construction of temporary hutments for workers and staff, construction of godown for store and magazine for explosives, laying of electric lines, if necessary, or installation of generator and such other infrastructure.

The preparation of geological map on a scale of 1 : 2000 or larger, e.g., 1 : 1000 or 1 : 500, depending on the extent of the area under investigation is frequently influenced by folds and faults. The map is required to be prepared with contour arrivals at 3 or 5 m. In the case of a flat country, the available enlarged topographical map may serve the purpose. Detailed geo-chemical and geo-physical work including the radiometric surveys on a closely spaced grid may be required to be carried out to decipher geo-chemical and geo-physical anomalies. Drilling has to be carried out at widely spaced intervals of say 300 to 500 m in addition to test drilling on the above anomalies. Trenching and pitting operations, by and large, do not serve the purpose in non-ferrous deposits. Such operations are good for bedded deposits like iron ore, limestone, manganese ore, china clay and for bauxite and chromite. The Geological Survey of India has specified that the total amount of drilling should not normally exceed 2000 m per 1000 m strike length for the nature of work involved under Phase I.

Chemical, petrological and spectrographic analysis of core-split samples should be carried out to determine the nature of association of valuable numeral components in the rock ore and to find out the presence of other metal values. Preparation of geological cross-section based on the interpretation of borehole data showing the attitude of mineralisation is found helpful in extending the work to the next phase.

Phase II

Objective

The object of this phase is to carry out exploratory work with a greater intensity in the selected block/blocks with a view to bring the inferred reserves into the indicated and measured categories and to obtain precise information about the size, shape, attitude and grade of the ore bodies in different blocks. It also attempts to delineate suitable blocks with sufficient reserves to last for 5 to 10 years at the envisaged target rate of annual production.

Scope

This phase attempts to assess the data about the minerability of the ore zone, conditions of hanging wall, foot wall and host rock, ground water, etc. and to design a process flow sheet for the recovery of metals. Phase II estimate provides a concept of the installation that might be required. This estimate cannot, however, be taken for budgeting the project.

Work involved

Closely spaced drilling, say at intervals of 50 to 100 m depending upon the structural disturbances, exploratory mining to corroborate the grade and attitude of mineralisation is indicated by the borehole data; systematic sampling work in drives and cross-cuts should be carried out to ascertain the correct grade and establish the measured and indicate reserves. Split core-samples and bulk samples should be subjected to tests for amenability to beneficiation and metallurgical characteristics so that a flowsheet could be evolved. A number of samples should be subjected to tests for determining the other valuable metals. The report of beneficiation and metallurgical test should be made available at this stage. The cost of element involved in mining and beneficiation should also be more correctly established.

Phase III

Objective

The object of this phase is to bring at least 30% of the indicated reserves into the measured category in the selected blocks. According to the Russian classification a minimum of 30% reserves should be brought into A and B category before a deposit is considered suitable for exploitation.

Scope

The results of work input of Phase III are generally suitable to determine the feasibility of the project and assist the management in budgeting for it. Phase III estimate generally describes the installation which will be built, and forms the basis for the detailed project report.

Work involved

More underground exploratory mining by development of more

than one level and cross-cuts is carried out to establish the continuity of the ore bodies. Underground drilling is also implemented to find out the extent of mineralisation. The other items of work involved are the estimation of level-wise and bench-wise reserves, preparation of level-wise geological plan and traverse sections, structure contour plan, isocore or isopach map, drawing more bulk samples for beneficiation and metallurgical test on pilot plant scale, finalisation of the process flow sheet and determination of the capacities of various items of machinery such as grinding mills, flotation cells, preparing specifications for the mining equipment and other machinery and estimating the cost involved in their purchase and installation charges after consulting and drawing quotations from the various firms.

An important aspect of this concluding phase is the estimation of the profitability profile. This involves the identification of consumer markets for the product and also to make adequate arrangement for mobilising resources for the installation of the plant. Feasibility report as a rule must provide investment analysis by describing likely design of the mine, mill, smelter and ancillary requirements using carefully selected values of controlling factors such as mineral/metal prices, equipment cost, operating cost, etc. From this stage the job requirement for the preparation of the feasibility report involves the work of interdisciplinary interests where a team of individuals with expertise in a specific field should be able to take the specific responsibility in preparation of the whole report. There are a number of techniques for investment analysis. These are largely based on: (i) Company's Investment Policy, (ii) Pay Back Method, (iii) Discounted Cash Flow Method, and (iv) Net Present Value Method, of which the last two are the most commonly used statistical methods of financial analysis.

13

Principles and Methods
of Mineral Dressing

THE ECONOMIC MINERALS and ores are rarely found in the degree of purity and form which ultimate users demand. They are usually associated with gangue, the worthless material, which is required to be separated out from the valuable mineral contents and thus render them usable. This separation is effected by adopting certain processes. It is achieved by taking advantage of the difference in physical properties likely shape, colour, specific gravity, magnetic susceptibility, electric conductivity, physical behaviour on heating and surface properties *i.e.* affinity of valuables and waste minerals to flotation reagents. The sum total of the treatments to which minerals are subjected in order to separate and discard their worthless fractions is known as mineral dressing, ore dressing or beneficiation. The methods of mineral dressing are essentially physical where the identity of mineral is rarely destroyed during the process. But as we shall find later they also take a recourse of chemical action during leaching process, the latest technique developed under mineral dressing, for concentrating metallic content in an ore, and at times by bringing change in the identity of minerals.

APPROACH TO MINERAL DRESSING

There are several processes involved in mineral dressing which may be regarded as from simple operation to complex treatments. It is evident that some selective processes required to be adopted for a particular mineral sample, depending upon the physical properties of gangue and their nature of association with one another. Practical knowledge of the physical properties of various mineral constituents in the sample, counts much in deciding the process to be followed in individual cases. To arrive at the initial conclusions, it is necessary to identify clearly under microscope the mineral constituents and the nature of intergrowth between the gangue and valuables. After identification of all the valuables and gangue minerals, the next step is to examine at what optimum crushed size the liberation of gangue from the valuable takes place. This an experienced mineralogist can indicate with good approximation. In laboratory investigation, the sample is normally crushed to 4 mesh and is allowed to pass through succeedingly finer sieves (see Figure 36). The mineral retained on sieve No. 1 would be of size + 4 mesh; on sieve No. 2 would be of −4 + 10 mesh, on the next sieve −10 + 20 and on the last sieve + 100 and on the pan would be all −100. The samples retained on different sieves are again examined under microscope to determine at what size there is an optimum liberation. If it is found, for instance, that the liberation is effective at −28 mesh + 35 mesh, then the sample is crushed to −28 mesh, and subsequent treatment for the separation of individual minerals is carried out.

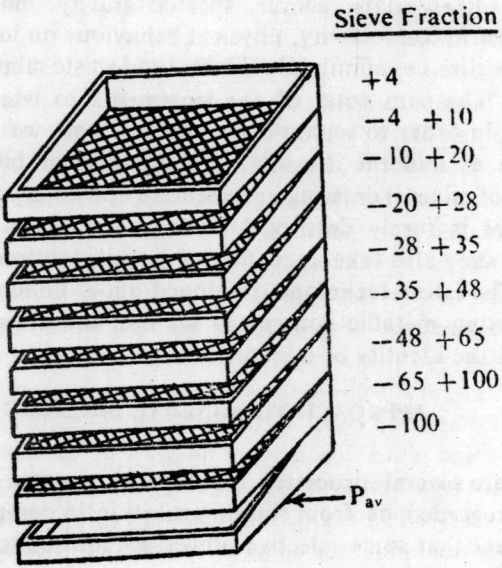

Sieve Fraction

+4

−4 +10

−10 +20

−20 +28

−28 +35

−35 +48

−48 +65

−65 +100

−100

← P_{av}

Fig. 36.   Laboratory sieve.

The processes that are chiefly adopted for beneficiation include starting from hand picking and chipping to crushing, grinding, screening, classification, air separation, washing, jigging, tabling, heavy media separation, magnetic separation, electrostatic separation, flotation, and combination of one or more of these processes suitable for particular minerals. The choice of the process depends upon the mineral constituents present and their physical properties. Filtering and drying are the common intermediary processes employed. For uniting finer material resulting from crushing and grinding for making them usable in lump form, the process of agglomeration has become the common practice under mineral dressing. It should, however, be clearly understood that the mineral dressing processes are not effective in upgrading ores where the impurities are present in a chemically combined form. It is the limiting factor in mineral dressing. Improvements can, however, be brought about by metallurgical treatments like hydro-metallurgy, pyrometallurgy, etc.

PREPARATION OF THE MINERAL SAMPLES FOR DRESSING

Comminution and sizing: Comminution means crushing and grinding to desired size of liberation. Comminution and sizing precede generally all other beneficiation methods because these operations make products suitable for application of the latter. Comminution is done essentially to liberate gangue from the valuables which enables the liberated particles to move freely relative to each other, when subjected to various treatments. There are a number of machines of different makes for crushing; namely jaw crusher, gyratory crusher, cone crusher, hammer mill for coarse grinding and ball, rod and tube mills for fine grinding. Sometimes, it is necessary to put two or more crushers in the circuit to crush ore to proper size for subsequent grinding. Such crushers are termed in the circuit as primary, secondary and tertiary. It may be necessary sometimes to put grizzly before the run-of-mine is fed to primary crushers. Gizzly is a sieve of thin steel bar (rod) strong enough to withstand load impact of run-of-mine when discharged on it. The purpose of grizzly is to allow to pass only sized materials receptible to primary crushers. Grizzly serves to provide another important function. It sieves out the finer material not required to be crushed thus improving the capacity of the crushers.

Jaw, gyratory and cone crushers are particularly suitable for hard minerals. Hammer mills are used for comparative softer minerals like bauxite, phosphorite, limestone, etc. Jaw crusher and the gyratory crusher can give trouble with sticky minerals whereas hammer mill can crush and eject anything fed into it particularly if equipped with moving plates. For crushing hard and sticky minerals it is necessary to wash them before discharging into crusher to remove fines which cause stickiness.

There has been some new development in grinding since 1964. It is in autogenous grinding by aerofall mills. Autogenous grinding means grinding ore with ore itself. It consists of a thick steel cylinder of large diameter approximately of 6 metres long, kept horizontally, in which ore is packed and rotated. By a mere fall of mineral on mineral and, with abrasive action on each other, mineral is ground to powder which is removed by suction. By regulating the air discharge mineral of any size can be obtained.

The use of screen is to size the crushed and ground material. It is an appliance having uniform square or round openings (apertures). Screens of various mesh sizes are available which are all standardised. Mesh denotes number of holes per linear inch. When we call 100 mesh or 200 mesh, it means 100 or 200 openings per linear inch. The large the mesh number the finer is the product passing through it. They are manufactured according to the specifications of Tyler, British, American, French or German standards sieves series. BSS, ASTM and Tyler series sieve follows closely the Rittinger series that is the ratio of opening of larger siever to the opening of next immediate smaller sieve is $\sqrt{2}$: 1. It means, for an example, the opening of 6 mesh designated sieve is $\sqrt{2}$ times larger than the next Tyler sieve designated as 10 mesh. Similarly, 10 mesh opening is $\sqrt{2}$ times larger than the opening of next sieve of 14 mesh. The Indian standard series follow the Rittinger series only approximately, covering the requirements of corresponding sizes in all these series. There is very little difference in the mesh size of one standard corresponding with another. Tyler series is the most common and universally adopted. The base of the Tyler series is 200 mesh with aperture equal to 0.0029 inches i.e. 74 microns.

Sieves up to 400 mesh are generally employed. Further analysis is not possible with sieves due to very fine size of the particles. Analysis range below 400 mesh is accompanied by methods such as elutriation infra-sizing, etc.

METHODS

Gravity separation: The process of gravity separation takes into consideration the difference in specific gravity between associated minerals i.e. the valuable and the gangue. It is effective only when the difference is more than unity, although theoretically it is possible to separate them even when the difference is less than unity. Heavy media, jigging and tabling are the common methods employed under gravity separation.

Heavy media separation: It is adopted for separating minerals of coarse size generally from lumps to $+10$ mesh. It is of two types, drum and cones types. In this process a heavy media is created by suspension of heavy minerals and alloys like galena, barytes, magnetite and ferro-silicon in water. The specific gravity of these suspended materials are:

Barytes	—	4.8
Galena	—	7.5
Magnetite	—	5.18
Ferro-silicon	—	between 7.5 and 7.1
(85% Fe, 15% Si)		

By mixing any one of the above materials in pulverised form in suitable proportion with water, a medium of predetermined density can be created. This suspension when kept slightly stirred behaves like a heavy liquid. The density of the medium is normally maintained between the specific gravity of valuable and waste so that the waste separates as float and the valuable as sink product.

Of all the materials used in making heavy media, ferro-silicon is regarded as the best because of its high specific gravity and easy recovery from the sink product by magnetic separation. The slurry containing ferro-silicon thus obtained is recycled into the plant making the whole operation a continuous process. Galena although has high specific gravity is not preferred as suspending material, because its recovery from the sink product is not so simple. It can be recovered only by froth flotation. Also it produces lot of slime which causes inconvenience in the process. Ferro-silicon has completely replaced galena nowadays. Out of the other two, magnetite and barytes, the former is preferred again for its easy recovery by magnetic separation. Therefore, for lighter minerals whose specific gravity is between 1 and 3 generally magnetite is used and for minerals heavier than 3 and less than 5, ferro-silicon is used. Atomised ferro-silicon, that is ferro-silicon particles having rounded size, is most commonly used as it has been found to have better dispersion. This permits operating at higher specific gravity 4 when compared to maximum 3.3 with ground ferro-silicon.

In coal washing sand, felspar and barytes are used but nowadays these are being replaced by magnetite in the preparation of heavy medium for the very reason of its easy recovery by magnetic separator.

Jigging: Jig consists of a framed sieve. There are several types of jigs from hand operated to those having mechanical device for giving pulsation in the fluid. The process separates valuables from the gangue when the mixture of the two is allowed to settle in a rising current of water. The heavier minerals, which are usually the valuables, settle at the bottom while the lighter ones rise with the current and are removed from the top. Jigging is usually applicable to size range of particles between minus 8 and plus 28 mesh. There is no rigidity about the lower limit of mesh size. Even the finer materials are separated by jigs depending upon the nature of the valuables and the associated gangue. For beneficiating tin and tungsten ores finer mesh can be treated with advantage by jigs.

Hand operated jigs are commonly employed for beneficiating low grade manganese ores by the mine owners in India for the removal of associated quartz and the country rock.

Tabling: The method of tabling is usually applied when the size range is minus 28 plus 200 mesh. Again the lower limit of size range in this case also is not restricted. Ores of very fine size can be

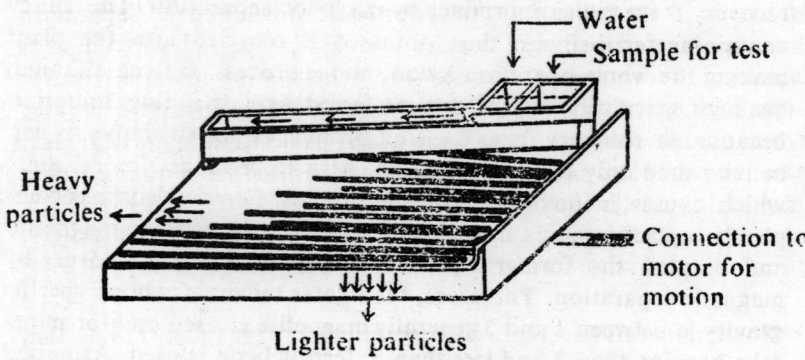

Fig. 37. Laboratory table.

treated depending upon the nature of association and the valuables. For the treatment of finer sizes separate decks (slime deck) are available.

A table consists of a deck kept at a slightly sloping angle. Deck is fitted with riffles which act as sinking tanks. When table is given reciprocating motion at right angle to the flow of fluid, generally water, the heavier particles settle down in the riffles because of the forces acting upon them (see Figure 37), and carried along diagonal line of the table and collected as shown in the illustration below.

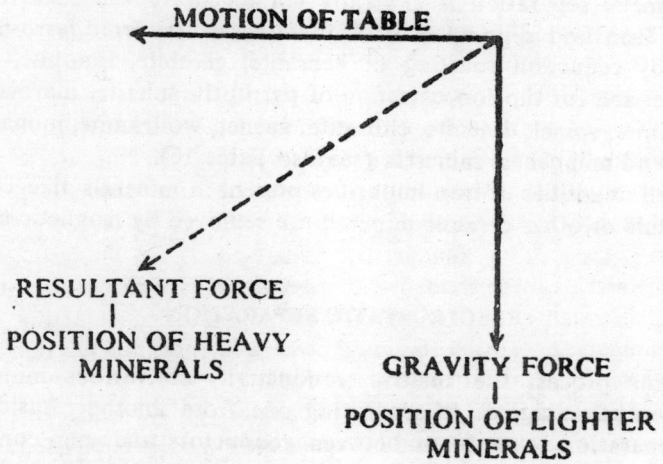

The lighter particles are acted only by gravity forces as they do not settle in the riffles and are collected as shown above. Separation is thus effected between heavier and lighter particles. Tabling is conveniently employed usually for beneficiating chrome ore for separating serpentine and talc. Mica can be conveniently separated from quartz although both have nearly same specific gravity. Here the flaky shape of mica helps in being separated from quartz. Mica moves down with the gravity force.

MAGNETIC SEPARATION

Some minerals have natural susceptibility to magnetism i.e., they are attracted by magnet. The most natural magnetic mineral is magnetite. In this method, use is made of varying magnetic susceptibility of minerals. Table 15 indicates the magnetic susceptibility of some common minerals compared with iron. By varying the magnetic field of separator, minerals of different magnetic susceptibilities can be separated from one another. With low intensity the highly magnetic minerals first separate out and then by increasing the intensity of the magnetising current, the minerals lower in the series, having feeble magnetic property can be separated. In case where separation is to be effected in finer sizes, wet magnetic separation is used, provided the minerals are highly magnetic.

Magnetic separation is generally employed for the removal of tramp iron and separation of iron minerals rendered ferro-magnetic by reduction roasting of hematite, goethite, limonite, and siderite, and for the concentration of pyrrhotite siderite, marmatite, franklinite, spinel, ilmenite, chromite, garnet, wolframite, monazite, rutile and manganese minerals (see also Table 16).

Small quantities of iron impurities present in minerals like china clay, talc or other ceramic minerals are removed by magnetic separator.

ELECTROSTATIC SEPARATION

In this process the relative conductivity of various minerals is taken advantage of in separating one from another. Basically, the separation takes place between conductors and non-conductors. Minerals are capable of momentarily acquiring electrostatic electricity when brought under the influence of electric field. Some are capable of discharging the charge instantaneously when brought into contact with conducting material. Such minerals are called conductors. The other minerals which do not leak off charge when in the field are called non-conductors. Generally, metallic minerals are conductors, and non-metallic minerals are non-conductors. Examples can be given of magnetite, ilmenite, rutile, silicate, etc., as conductors and quartz, sulphur, garnet, apatite, silicate minerals as non-conductors. The principle and working of electrostatic separation can be explained by Figure 38. In the equipment an electric field is applied between the wire electrode and the rotar. When the mixture of conducting and non-conducting particles is fed through the field, the conducting particles as they pass on the charge through the rotar remain unaffected and follow the tangential path Non-conduct-

TABLE 15

Relative Magnetic Susceptibility of Common Minerals

Strongly magnetic

Iron	100
Magnetite	40.18
Franklinite	35.38
Ilmenite	24.70

Weakly magnetic

Pyrrhotite	6.69
Siderite	1.82
Hematite	1.32
Zircon	1.01
Limonite	0.84
Corundum	0.83
Pyrolusite	0.71
Manganate	0.52

Feebly magnetic
or
non-magnetic

Garnet	0.40
Quartz	0.37
Pyrite	0.23
Sphalerite	0.23
Molybdenite	0.23
Bornite	0.22
Apatite	0.21
Chalcopyrite	0.14
Fluorite	0.11
Galena	0.04
Calcite	0.03
Witherite	0.02

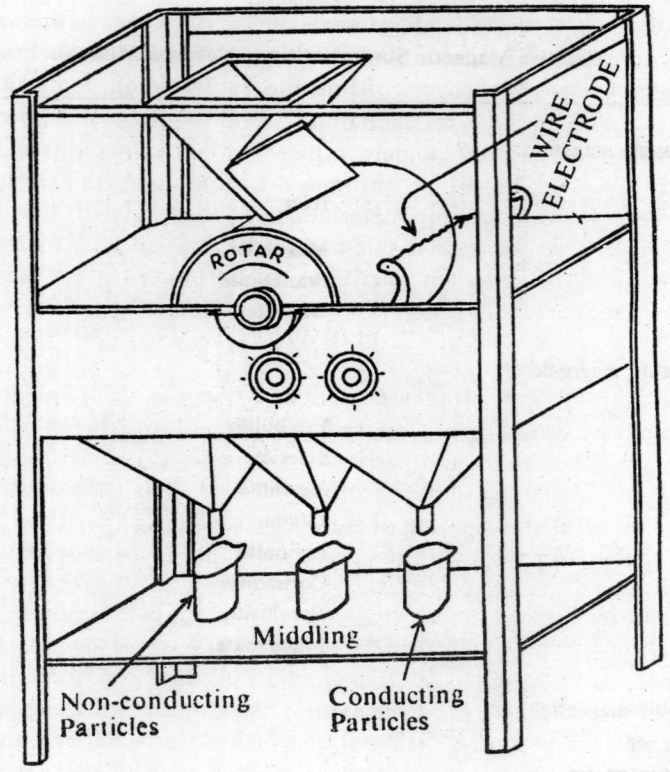

Fig. 38. Electrostatic separator.

ing particles retain charge and remain pinned down on rotar in the field. The particles which are neutral, rather poor conductors, fall vertically down as middling. One precaution must be kept in mind that while using this process the ores should be dry and preheated to remove moisture.

Table 16 provides the separation characteristics of common minerals depending upon specific gravity, magnetism, and conductivity.

Starting with a mixture of any of the above, it can be determined whether or not, they can be separated by electrostatic (high tension), magnetic or gravity methods and whether any one or a combination of methods are required. The minerals appearing in different columns can be separated by high tension and or magnetic alone. Two or more minerals appearing in the same column can be separated by gravity concentration if they have sufficient difference in their specific gravity. The combination of magnetic and electrostatic method is employed for the recovery of ilmenite, rutile, zircon, monazite, etc., from the beach sands of Kerala and Tamil Nadu coast. Electrostatic separator is nowadays widely utilised in African countries for the recovery of diamond. A chrome ore containing talc, serpentine and magnetite as gangue material can be beneficiated first by gravity separation to remove talc and serpentine and then by magnetic separation to remove magnetite to obtain concentrate of chromite.

FLOTATION

Flotation is the most important and complex technique amongst the mineral dressing processes requiring an intimate knowledge of the physico-chemical properties of minerals. Whenever flotation is spoken it generally implies froth-flotation. It is essentially a process of selectively floating of the desired mineral particles in pulp by their attachment with the rising air bubbles with the aid of suitable reagents. Mineral particles possess an inherent property of having preferential wettability. They may be either hydrophilic or hydrophobic. A hydrophilic particle is one which is easily attracted to water. A hydrophobic particle under the same condition tends to repel water. There are some natural floatable minerals such as, coal, graphite, sulphur and diamond on which when water is poured are not wetted as they possess non-wettable or hydrophobic surfaces. In the flotation technique these hydrophobic and hydrophilic characteristics of minerals are taken advantage of in separating one from another. Such property on the mineral surfaces can also be achieved or modified with the help of suitable reagents as discussed later which forms the basic principles of flotation. The principle of flotation was first applied in 1906 to sulphide minerals only. Subsequently, its application was enlarged to cover oxide and non-metallic minerals with the use of anionic and cationic reagents. Its application is fastly growing and now nearly all minerals have been covered and rendered possible for concentration including coal.

This process is generally applicable when the size of the particles is finer than 48 mesh. For efficient operation, it is necessary that not only all mineral particles are free from interlocking but also the size of each particle should be such as to adhere to the rising air bubble.

TABLE 16

Separation Characteristics of Common Minerals

Non-conductor (High tension pinned)			Conductor (High tension thrown)			
Magnetic	*Weakly magnetic*	*Non-magnetic*	*Highly magnetic*	*Magnetic*	*Weakly magnetic*	*Non-magnetic*
1	2	3	4	5	6	7
Monazite (5.3)		Scheelite (6)			Wolframite (7.5)	Gold (19)
					Columbite tantalite (6.2)	Copper (9)
					Samarskite (5.7)	Galena (7.5)
						Cassiterite (7)

Hematite (5.2)

Magnetite (5.18)

Pyrite (5)
Molybdenite (4.8)

Zircon (4.7)
Barytes (4.5)

Xenotime (4.5)
Garnet (4.5)

Chromite (4.5)

Ilmenite (4.5)

Chalcopyrite (4.3)
Rutile (4.18)

Corundum (4)

Celestite (3.9)

Brookite (3.8)

Siderite (3.8)

Staurolite (3.7)

Limonite (3.6)

Kyanite (3.6)

Diamond (3.5)

Topaz (3.5)

Epidote (3.5)

Sphene (3.4)

TABLE 16 (*Contd.*)

1	2	3	4	5	6	7
—	Olivine (3.3)	Sillimanite and Fluorite (3.2)				
—	Apatite (3.2)					
—	Hornblende and Tourmaline (3)					
		Anhydrite (2.9)				
		Mica and Beryl (2.8)				
		Felspar and Calcite (7)				
		Quartz (2.65)				
		Gypsum (2.3)				
		Sulphur (2.07)				Graphite (2.2)

Note. The figures in brackets are specific gravity of respective minerals.

From experience it has been found that the mineral particles coarser than 48 mesh (about 295 micron or 0.295 mm in diameter) cannot be effectively floated.

The entire mechanism of flotation is governed by a group of three reagents, namely, collectors, frothers and modifiers. The essential mechanism of flotation and the processes involved in it can be summarised as follows:

(1) Grinding the ore to a size sufficiently fine to liberate the valuable minerals from one another and from the adhering gangue material.

(2) To prepare pulp in water in a ratio usually of 25% solid and 75% water. The content of solid can be raised to 35%, provided the desired viscosity of the pulp is not interfered. In practice, ratio of solid to water is kept at 1 : 3 or even less.

(3) Addition of reagent in accurate quantity for adherence of the desired mineral particles to the rising air bubbles. Adherence is achieved by reagent called collector.

(4) Preparing sufficient surface tension so that the air bubbles

Fig. 39,   Flotation cell.

do not break at the surface of the pulp and loose the minerals. It is achieved by use of reagent called frother.

(5) Creating conditions favourable to collector for adhering
particular mineral to air bubbles and depressing other which
are not required to be floated. This condition is created by
reagent modifier.

(6) Creating a rising current of air bubbles in the pulp.

(7) Forming a mineral laden froth on the surface of the pulp.

(8) Removal of the mineral laden froth, the concentrate.

The above operations are carried out in flotation cells. In mill
operation, there may be a number of cells in circuit, depending
upon the plant capacity. It should be clearly understood that all
processes involving flotation are either selective or differential such
that one mineral or a group of minerals is floated away from
accompanying gangue. The term 'bulk separation' is referred to
the separation of unlike minerals, such as sulphides from non-
sulphides. The term 'differential flotation' is applied to operation
when separation of a mineral is to be effected from similar other
mineral types, for example, the concentration and subsequent successive removal of galena, sphalerite, chalcopyrite, pyrite and similar
other sulphide minerals from the single ore. It is now possible by
means of flotation to single out accurately each mineral from the
group of minerals such as sulphides from non-sulphides; silicates
from non-silicates; even one copper mineral, say, chalcopyrite from
calcocite; apatite from rock phosphate and so forth. Such differential or bulk separation is achieved by the use of a particular 'collector' aided by conditioning agents called 'modifiers'. The above
separation work on the principle of making a particular mineral or
a group of minerals desired to be floated hydrophobic and others to
remain hydrophilic. The differentiation in behaviour rather pulp
conditioning is induced by modifiers. After conditioning, when air
bubble is blown through the pulp, the hydrophobic particles are
carried up by the collector with the rising bubbles and stabilised
at the surface of the pulp in a mineralised form which is scooped
out as concentrate. The use of modifiers is a must in case of differential flotation.

How a particle becomes hydrophobic or hydrophilic is a subject
matter of great detail and for this a reference to standard books on
physical chemistry is suggested.

REAGENTS

In the following paragraphs it is intended to describe the functions of various flotation reagents and how they work in the operation. These are required in minute quantities. Actual requirements are determined by laboratory performance tests.

COLLECTORS

The collector plays a most important role in flotation mechanism. Collectors are broadly classified as either cationic or anionic. A collector is anionic if the part which imparts water repellency to mineral surface carries a negative charge. Or conversely, if this part of the collector is positively charged, the collector is cationic. When a particular collector is added to the pulp it produces hydrophobic film on the mineral particles required to be floated. Each collector molecule contains a polar and non-polar group. The polar part gets attached to the mineral particle and the non-polar part to air bubble and moves up. There are a number of reagents used as collector for sulphide, acid, basic and oxide minerals. The most widely used collectors in sulphide flotation are the xanthates and dithiophosphates. Xanthates are potassium and sodium salts of xanthic acid, an organic sulphydril compound. Xanthates are very powerful collector for sulphides. A number of xanthates are known, namely, potassium and sodium ethyl xanthate $(C_2H_5OCS_2K/Na)$, potassium and sodium isopropyl xanthate $(C_3H_7OCS_2K/Na)$, etc. Xanthates are dissociated by strong acids and are usually added to pulps having a pH of 7 or more, i.e. when the pulp is having alkaline condition.

Dithiophosphates are the products of phosphorus pentasulphide with phenol, alcohol, thio-alcohol, amines and nitrites. These products are sold under the name of 'aerofloats'. They are weaker collectors than xanthates. Xanthates or dithiophosphates are ineffective on oxide or hydroxide ores like siderite, hematite, magnetite, limonite, regardless of the amount of agent added.

Fatty acids and soaps are used for floating non-sulphide minerals. The common reagents are oleic acid and sodium soap or oleate. Oleic acid and its sodium or potassium soaps are used in the flotation of oxides, halides, fluorite, apatite, limestone, gypsum, etc.

Amines are cationic collectors used for floating silicate minerals.

Good separation can be achieved between silicates and oxides or carbonate minerals by a proper use of collector and depressant. Silicates respond to cationic collectors, while the basic minerals attract the anionic fatty acids.

FROTHERS

These reagents are used to create froth. It is well known that pure water (whose pH is 7) does not froth whatever amount of air is blown. Add to it a small quantity of soap-water and then blow. Air bubbles will form on the surface. The same function is done by frother in the pulp. It reduces the surface tension of water and thus helps in forming the bubbles, which is the characteristic of all frothers. It does not, however, possess the property of a collector. Organic substances are best frothers. There are a number of known frothers such as eucalyptus oil, cresylic acid, M.I.B.C.* and Pine oil. These are regarded as the best for both commercial availability and cheapness. Other frothers commonly used are alcohol, aniline, xylidine, pyridine, etc. Most of the inorganic substances raise the surface tension and hence are not used. Frother has a great practical value in flotation. Froth is dependent upon the pulp density. Saturated pulp will not give desired quantity of the froth which consequently affects the recovery. Hence the ratio of solid to water in flotation cells is not generally allowed to increase to more than 1 : 3. Frother is added in a very small quantity. In flotation of sulphide ores, 75 gm of frother is sufficient for one tonne of ore.

Control on the quantity of frother is not generally as critical as on the collector; however, the choice of proper frother is extremely important. The proper use of frother should be established by the results of laboratory experimentation coupled with actual mill testing.

MODIFIERS

The list of modifying or conditioning agents used in flotation is long and includes all reagents whose principal function is neither collecting nor frothing. Modifiers are used to bring out latent hydrophobic property at appropriate stages of treatment. It prepares surfaces on mineral to which collector gets attracted.

*Methyl isobutylcarbinol.

The functions of the modifying agents are many, the important ones may be grouped as under:

(*i*) pH regulator
(*ii*) depressant or wetting agent
(*iii*) activator
(*iv*) dispersant
(*v*) resurfacing
(*vi*) precipitant

By far the most important conditioning factor is the pH regulation of the pulp. Most minerals have a range of pH within which they float best and above which it is impossible for any collector to carry them up with the bubbles. The effectiveness of the flotation reagents depends on the degree of alkalinity or acidity of the pulp solution. A solution becomes acidic or alkaline when the pH is less or more than 7. Therefore, in all flotation processes the maintenance of correct pH of the pulp is sine-qua-non. It is easy to determine by pH meter. Soda ash, caustic soda and lime are the common compounds for increasing alkalinity. Sulphuric acid is used if the pulp is to be made acidic or for reducing the alkalinity. All sulphide minerals with the exception of pyrrhotite float in alkaline pulp using potassium ethyl xanthate as collector. The maximum pH at which some sulphide minerals can be floated is shown below. A particular mineral ceases to float above given pH.

Mineral	Max. flotation pH
Chalcocite (Cu_2S)	14.0
Bornite (Cu_5FeS_4)	13.8
Covellite (CuS)	13.2
Chalcopyrite ($CuFeS_2$)	11.8
Pyrite (FeS)	10.5
Pyrrhotite (Fe_nS_{n+1})	6.0

The pH modifier thus works effectively in differential flotation. It is however not enough and the help of other modifying reagents is also necessary. Depressant is used to make the desired particles hydrophilic. Activator on the other hand, enhances the hydrophobic character of the particle. Frequently, several functions may be performed by a single modifier. For example, soda ash may serve as an activator and a depressant as well as a pH regulator. Lime also acts as a depressant for galena and pyrite. Sphalerite floats weakly with xanthate. Copper sulphate activates its flotability. Sodium chromate and dichromate are used to depress galena from Pb-Zn-Cu sulphide ores by the formation of an insoluble lead chromate. Sodium hypochlorite is used to depress copper ore from molybdenite. In lead-zinc sulphide ores, sodium sulphide, sodium cyanide or zinc sulphate is used as depressant for sphalerite. Air is depressant for pyrrhotite. Cyanides are used for depressing copper sulphides. Sodium silicate and dextrin are depressant, for quartz (siliceous gangue). A comprehensive list of depressants, and activators and the guideline for their use under different circumstances can be found in the standard books on mineral dressing. While using these reagents a thorough knowledge of minerals being treated and the effect of reagents on them are essential. This helps the mineral dressing engineer in knowing correctly what is coming out as concentrates and what remains in the tailings. The ore-microscopy finds the most useful role at this stage.

Sometimes, it is necessary to disperse slimes which mask the surfaces of the particles and thus prevent collector to act effectively. Dispersants include dextrin, lignin sulphonates, quebracho and starch. In fact, these organic colloids might also be considered as depressants which when used in excess, can prevent all flotation with anionic or cationic collectors. Their principal application is most effective on ores admixed with carbonaceous shales or clays. With ores containing such minerals, the organic colloids often reduce collector consumption as well as improve dispersion and depression of the gangue slimes. Sodium silicate is used as a dispersant for quartz slimes for removing them from sulphide surface.

The action of the resurfacing agents is to reprepare the surface which has been modified by the action of other reagents during the operation for selective flotation afresh. In a broad sense, this action of resurfacing agents is either activation or depression. Generally all metal ions might be considered as resurfacing agents but only a few are commonly used for the purpose. Copper, one of the most important, finds use in the flotation of sphalerite which is unaffected by xanthate. Copper ions are added to the pulp as copper sulphate; copper is absorbed on the sphalerite surface by replacing zinc in the mineral lattice, forming a copper sulphide film on sphalerite. Sphalerite, therefore, behaves as a copper mineral which is easily floated with xanthate. Frequently, the presence of naturally occurring resurfacing ions may interfere with, rather than aid, flotation separation; where such interference is encountered, it becomes necessary to precipitate or complex these ions to render them inactive. The use of precipitants may also inhibit the formation of the resurfacing ions

Chemical processing: In the mineral dressing techniques, a significant development has taken place towards chemical methods of recovering valuable metallic contents by amalgamation, leaching, dissolution and precipitation. A number of new ideas of solvent extraction are being experimented. As a matter of fact these methods form a part of hydrometallurgy and are being usefully extended within the mineral dressing parlance.

Gold forming amalgam with mercury is well known. This process is used to recover metallic gold from the gangue. The amalgam process has been replaced by cyanide process. In this process the cyanide (KCN) is allowed to flow through the crushed ore and leach out gold. As the weak cyanide solution circulates through the mill, it becomes richer and richer in gold content. The precipitation of gold from the pregnant solution is effected by mixing it with zinc in excess of the gold content, after removing as much oxygen as possible, because gold remains dissolved in presence of oxygen. Since zinc and gold lie on opposite sides of neutral point in the electro-chemical series, zinc easily and most effectively precipitates gold. The cyanide process is being extended to copper and other ores. American Cyanamid Company has developed a process for extraction of copper by cyanide solution from sulphide and oxide minerals, flotation concentrates and tailings. The leaching is done with alkaline cyanide solutions at atmospheric temperature and pressure. The dissolved copper is recovered by precipitation through the addition of acid and sulphide and the cyanide by a stripping operation using air or steam. Cyanide requirement is reported to be low. At the new Mountain Pass, California, U.S.A., plant of Molybdenum Corporation of America, a solvent extraction method is used for separating the rare-earth group elements. Europium is recovered from a very dilute aqueous solution (0.002 of a pound per gallon) with a very high yield. A hot froth flotation process give a 60% rare earth oxide concentrate which is roasted and leached after which solvent extraction is employed Now the application of bacterial leaching in hydrometallurgical process is also being mentioned. Commercial application of bacterial leaching is being done on copper, and more recently on uranium; though experimental work on bacterial extraction of nickel and zinc from sulphides and molybdenum and tin sulphides has also given favourable results.

New techniques: These days photoelectric sorting is expending its use where marked colour difference exists. As the sized particles fall through two or more beams of light, the reflection is transferred via lenses to a sensing cell, together with the background illumination. If the colour differs from that of the selected background, a jet of air from a high-speed value controlled by the cell deflects the particles. Concentration experiments have been carried out with talc, felspar, magnesite and limestone.

The National Coal Board of Britain has developed an electro-pneumatic machine for separating coal from stone on the basis of transparency to x-rays. Material, sized between 3 and 6 inches, is monitored as it falls to a row of 16 nylon fingers, each 14 inches long covering a space 22 inches wide. Coal is deflected, but stone, which has only a quarter of its x-ray permeability, causes the appropriate finger to retract so that the stone falls through. A finger operating cycle is 0.1 second and sensibility can be varied to suit changes in the feed. Dirt coating does not appreciable affect efficiency.

In South Africa, much use has been made of the 'optical separator machine' for the recovery of diamond. In this machine, the operating principle is that a diamond will reflect and transmit light that strikes it, while the associated material is usually comparatively opaque. An intense light beam is diverted onto the material to be separated. The diamonds reflect this light which is picked up by sensitive photo-electric cell and the minute current generated in the photo-electric cell is multiplied through an electronic circuit and used to trigger a divider gate which diverts diamonds from the stream of other material.

AGGLOMERATION

In the foregoing pages it has been described how the minerals and ores are beneficiated to make them usable, which require crushing to some fine size to make the whole mass amenable to beneficiation process. Generally, the concentrates thus obtained are required to be made into lumps before they are put into the furnaces either for recovery of the metal content or used as an aid in recovering the metal content from other concentrates. It is absolutely necessary especially in case of concentrates required in iron and steel metallurgy; though it may not be so in case of non-ferrous metallurgy. During the course of mining iron ore, also large quantities of fines result, the ratio of lumps to fines may vary anything from 70 : 30 to 40 : 60 depending upon the friability of the ore in a deposit. Therefore, for effective utilisation of fines ($-\frac{1}{2}''$ size) and conservation of the mineral resource, it is necessary to make them usable by combining the fine sizes into a suitable size receptible to b'ast furance.

The process of combining the smaller particles into large hard mass retaining the identity of original particles in broadly called agglomeration. Different names have been assigned to agglomeration depending upon the technique involved in uniting the particles. These are: (i) briquetting, (ii) sintering, (iii) pelletising, and (iv) nodulising.

Briquetting is the process of agglomerating the fine particles by application of pressure. Some binder is usually added. Common binders are starch, lime, sodium silicate, asphalt and cement. Sometimes, molasses, magnesium chloride and sodium chloride are also used as binders. The briquettes may be prepared in the forms (shapes) of tables, balls, pillows, barrels, eggs, double pyramids or rods.

The most common machine for briquetting is a double roll machine where the rolls are of alloy steel and are provided with pockets of desired shape. The briquetting temperature is about 100°C and pressure applied is around 4000 lb per square inch. Briquetting is most commonly employed for lignite, coal, limestone, manganese ore, chromite. Neyveli Lignite Corporation (South Arcot district, Tamil Nadu) prepares lignite briquetts in large quantity for domestic as well as for use in thermal plant. Iron ore fines have not found much favour for agglomeration by this process, although it is practiced in the U.S.A., Germany and the U.S.S.R. in a limited scale. The techniques of sintering and pelletising have proved advantageous over briquetting as far as iron ore is concerned, and they are fastly being adopted in each and every country.

Sintering technique involves bonding on mineral particles by incipient fusion. This technique is commonly applied for iron ore fines having size of —12 mm. It is well known that iron ore of size less than $\frac{1}{2}''$ (12 mm) is not used in the blast furnaces. Hanging and slipping of a blast furnace generally result when gas passing through the burden is restricted by an excess of fines. This condition is dreaded by all blast furnace operators. Sintering offers solution to this problem of fines which could not be utilised otherwise.

The sintering process follows three main steps, viz. (i) preparation of the feed, (ii) fusion of the prepared layer of raw materials on the sintering machine, and (iii) cooling of the agglomerated mass.

In case of iron ore the feed is prepared by mixing fines in suitable proportion, predetermined by trial, of coke breeze, limestone and dolomite fines, blast furnace fuel dust and mill scale. The size of the iron or fines is usually —12 mm and other materials like coke breeze is —3 mm. The prepared mixture is added with proper quantity of water, the whole mass is mixed thoroughly in pug mill or drums and then the mix is conveyed and laid on the moving chain of the sinter machine. It is ignited at about 1,200°C–2,300° C by either blast furnace or coke oven gas or a mixture of the two; and as the charge moves forward, burning of the mix proceeds downwards by suction applied across the bed depth. The process is completed just before the sintered mass is discharged through a breaker and a screen to the cooler. The latest trend is in the preparation of self-fluxed sinters. The amount of limestone and dolomite is added in such a quantity so as to meet the entire requirements of fluxes in the blast furnace. The manufacture of self-fluxing sinters has proved advantageous in improving the blast furnace performance. Nowadays most of the blast furnace operates on 100% sinter with exceedingly good results.

While making selection of fines for sintering, determination of their chemical constituents is very essential from blast furnace operation point of view. The iron ore fines generally contain high proportion of Ai_2O_3 and SiO_2 the ratio of which should normally be unity (up to 1 : 1.5 is tolerated), and the total of Al_2O_3 and SiO_2 should not exceed 10.5%. The total of $Al_2O_3 + SiO_2$ can be increased provided the low ash coke is used, keeping the percentage of Al_2O_3 not to exceed 18% in the blast furnace, beyond which melting temperature increases considerably, consequently requiring higher consumption of coke.

Pelletisation is an advanced system of sintering where pellets instead of a loose mix, are fired and heat hardened. It is a method of agglomerating the fine particles by rolling to form small spheres— the pellets. The green or wet pellets are subsequently heat treated to impart them required strength. The fines and the concentrates suitable for making pellets must pass 90 % through 300 mesh. Ores ground to 325 mesh are regarded most suitable. Pelletisation is the newest innovation in the process of agglomeration, and was commercialised only since 1957 as a result of extensive research work carried out by Allis Chambers Manufacturing Company, U.S.A. on the use of hematite flotation concentrates. The revolutionary invention of grate-kiln process by this company for making the pellets brought boom in mining and treating of low grade hard siliceous magnetic ore (taconite) by magnetic separation and flotation for recovering the iron ore concentrates and then using the same in the form of pellets. Taconite contains approximately 30 % Fe. It is known as itaberite in Brazil.

Pelletising involves two basic steps—balling and firing. The grind material is added with bentonite in proportion of 5 to 8 kg per tonne and with requisite quantity of water and mixed to acquire plasticity. Bentonite acts as a bonding agent. The main reasons for using bentonite are the high 'green' strength resulting from its plasticity, and the high 'dry' strength resulting from its capacity to absorb and then give off moisture. Other binders like borax, lime and magnesia are also used. In cases where the iron ore contains higher percentage of alumina, the ferrous sulphate is used as binder. The mixture is fed to either drums or discs, also known as flying saucers. The balls or green pellets are formed as a result of the rolling action. These are then coated by rolling in a drum with finely powdered fuel such as coal or anthracite. These coated balls are next fired and heat hardened on a travelling grate-kiln at a temperature ranging between 1200 and 1370°C. Further advancement has been made in the firing technique which has dispensed with the use of solid fuel coating. In that case the pellets are fired by oil or gas fired furnaces. Pellets have proved to offer better performance than sinter in the blast furnace. Size of the ore has the greatest effect on reduction rate. Reduction rate is inversely proportional to the size of the particles. The usual pellet size are 10 mm to 25 mm.

Nodulising involves incipient fusion of the particles. This method requires continuous movement between the charge and the equipment, and is generally carried in a rotary kiln at a temperature of 1250°C to 1370°C. At this temperature spherical bodies of different diameters are formed from the fine particles. Nodulising process is used for practically any fine size but this process is nowadays more or less replaced completely by pelletisation and sintering.

14

Strategic, Critical and Essential Mineral

THE TERMS strategic, critical and essential minerals are the outcome of war emergency in relation to supply position necessary for winning a war. Only those who have witnessed war and are at the helm of affairs would realise the shooting rate at which the consumption of minerals, metals and other material rises during war munitions and destructive operations on land, sea and in air. War, whether it may be on compelling account for defence or may be offensive action, requires massive and quick mobilisation of men, money and material. Food and cloth hold important positions, men and material hold the most. Material provides the fighting power and comes from minerals alone. Most argue, why should we think of military and talk of war at all. Is it not a most haunting parable of civilisation whenever man is at war with man? Despite that war has come, and is continuing. Peace is regarded only an interval between wars. Military preparedness has thus to be assured for national security and ward off territorial infringement. Any shortfall in such preparedness would lead a nation to defeat and succum helplessly in event of confrontation. This deficiency, the shortfall in domestic resources, danger of supplies getting cut and difficulty in procurement from outside sources, was greatly felt by the warring nations during the First World War. Since then involved countries began thinking seriously on the menacing recurrence of global war and material shortages in that eventuality. It has clinched unfettered faith in the policy of all governments now in maintaining up-to-date inventories of domestic resources and take such measures of procurement and stock piling programme that would deter enemy from invasion.

CLASSIFICATION OF MINERALS FROM MILITARY
POINT OF VIEW

War requires encyclopaedic list of material for fighting power. From military point of view the minerals should be put under two groups viz (*i*) War Minerals, and (*ii*) War Supporting Minerals. War minerals will include those which are absolutely essential for warfare, munition, communication and transport. They provide the fighting power in the front. On the other hand war supporting minerals will be those which supplement the war efforts. There is hardly a mineral which can be differentiated between those required during war than those required for normal civilian needs during peacetime. All the minerals and metals that are needed for the production of arms and munitions and other equipments for fighting war are also needed for production of various machineries, plants and vehicles, etc. for peacetime use. It is difficult to make a subtle differentiation between minerals needed for warfare or civilian needs. Nevertheless emphasis in use of a particular mineral during war is quite different than peace. In peacetime economy this distinction does not have much significance. In the case of war, lack of sufficient supply position of certain minerals, required even in a small quantity, may prove a great hazard to the security of the country.

At the same time, it may be understood that all the minerals cannot be herded into war group. There are a few number of minerals whose presence or absence may not jeopardise the fighting power. These are diaspore, pyrophyllite, vermiculite, garnet, ochre, chalk, wollastonite, precious stones excluding industrial diamond, etc. In this list we can also include a group of alumino-silicate minerals like kyanite, sillimanite, and andalusite for which effective substitutes have been found out and these may not attact much military significance.

Each of the two categories namely, war and war supporting minerals can further be classified as strategic, critical and essential depending upon total inadequacy, deficiency and sufficiency in resource position. Strategic is one in which a country has more or less negligible resource available and depends upon outside sources. Critical includes those in which a country is deficient in supply position, but known occurrences are such that they can be worked as a war time measure, irrespective of the cost. Essentials includes all those for which known resources are quite large and produced in sufficient quantities. Evidently, essentials are all those which do not fall under the category of strategic and critical.

The minerals, to be categorised under each, will thus vary from country to country depending upon the resource position of individual countries. The ultimate strength of a nation depends upon how best it has developed capability of utilising its resources and processing, metallurgical and fabrication industry.

Applications of minerals in war: Iron ore and subsidiary minerals required for the production of steel are of vital importance. Steel is the industrial vehicle which keeps going all industries including its own. It is the basic material for munition and provides a galaxy of supplementary war material. It is always required in bulk and talked in terms of millions of tonnes. Sufficiently large supply position of steel and special steels is sine-qua-non for war efforts as well as for all industrial needs during war and peace. Manganese ore is essential for the manufacture of steel which is used in the form of ferro-manganese.

Tungsten is required in the manufacture of munition, armour plates and heavy guns. Armour piercing bullets manufactured out of tungsten steel have a marvellous piercing power. High speed tools contain as much as 18% W. Molybdenum is a good substitute for tungsten. It is one of the best known and most commonly used refractory metals having melting point of about 2610°C and is one of the most important alloying elements used in steel and certain nonferrous alloys. Only 0.25% of it makes the alloy steel very hard. Peacetime use of tungsten is as filament in incandescent lamps and tungsten carbide for cutting tools. Molybdenum is also used as filament. Vanadium is very small quantity is required for the manufacture of vanadium steels; 0.5 to 5% in high speed tool steel 0.15 to 1.3% in low alloy tool and special steels and in minute proportion in carbon and alloy steel for structural uses. Vanadium has a relatively high melting point (1899°C) and low density about midway between the densities of titanium and steel.

A number of simple vanadium alloys appear to beat the best titanium alloys and stainless steel in strength and creep properties in the range of 538°C–649°C. Vanadium alloys are ductile and can be welded. Its thermal conductivity is double that of titanium or stainless steel which is regarded important for aircraft and nuclear reactor. Vanadium steel, alloyed with metals like chromium, manganese and tungsten, is utilised for the production of high speed steels used in the manufacture of axle, clutch, gear, etc., where strength, hardness and fatigue are primary requisites. Ferro-chrome is essential for armour plate as well as a variety of structural steels. Titanium metal is finding increasing use in aircraft and missiles. On strength-density basis, titanium is superior to all light structural metals like magnesium and aluminium. It is harder than aluminium and approaches the high hardness possessed by some of the heat treated alloy steel. Its significant properties are its low density (about 0.6 that of steel), magnificent corrosion resistance to marine and saline atmosphere. It has good resistance to oxidation damage up to 649°C. Nickel is used for stainless steel and high nickel alloys which are generally used for crankshaft, marine, electronic, nuclear power and aerospace applications. Chrome-nickel alloy is made to throttle the flow of electricity to convert it into heat or light as in the case of tungsten and molybdenum filaments.

Cobalt is one of the useful ferro-alloy metals. It improves the magnetic qualities in iron and hence it is used in the electronic industry for producing permanent magnets. Without cobalt there could be no communication system because magnets are needed in radio, television and many other electronic devices. Cobalt steel is employed in the manufacture of jet aircraft engines, gas turbines and high speed tools. It is also used in the production of atomic energy.

War uses of antimony is confined to manufacture of antifriction metals and hardening of lead for shrapnel. Magnesium is sought after for making light metal alloys for planes, automobiles and trains. It is used as incendiary bombs for flares. Cadmium is used in the processing of uranium and in the manufacture of special steel. It imparts excellent resistance to corrosive actions. Pure zirconium is resistant to heat and corrosion and has complete inability to capture thermal neutrons. It has, therefore, attained prominence as a structural and cladding material in atomic reactors. Zirconium is alloyed with iron, silicon and tungsten. Zirconium alloyed with steel has been used to manufacture a number of equipment during the last world war. Zircon mineral as such is largely used as moulding sand for high temperature castings like heavy cylinders. Aluminium came into market in commercial quantity in 1939. Since then it has caught up rapidly as a light structural metal for military use. Its abundance has replaced copper, lead, zinc, tin and also steel to great extent in certain uses.

Copper, lead and zinc are mainly alloy metals and actively employed in the manufacture of war munition in various forms. The use of copper is not only confined to the manufacture of munitions or shells but all active fronts are connected with rear with communication wire, telephone exchange, etc., which requires miles and miles of copper wire. All armatures of electric motors require copper wire. About 400 kg of the metal go into the building of a tank, a tonne is needed for a large bomber and 1000 tonnes for a battleship and even larger amount is required for ammunition. Good amount of copper goes in the manufacture of anti-aircraft guns and aeroplanes. Copper-zinc alloys are used in the manufacture of jacket of cartridges, air bullets and shells. Both lead and zinc are used as protective non-recoverable coating on steel sheets. War use of lead is confined to the preparation of 100-octane aviation gasoline and ammunition foil. Lead paint prevents penetration of gamma rays. Radiation and blast hazards of atomic bombs can be checked to a great extent if the glass panes of doors and windows are painted with lead. Other protective measures can be the construction of underground houses. Peacetime use of lead is storage batteries, paint and pigment, cable covering, tetra-ethyl lead for high test gasoline, building construction, and various alloys chiefly solder, type metal and bearing metal. Tin is essential for food canning for air dropping military supply in the front. Silver-lead solders were reported to be used to save tin in brazing alloys for joining parts of ship, planes, tanks, guns, shells, rockets, torpedos, etc., during the Second World War.

Beryllium metal is used extensively as a moderator in atomic piles and nuclear reactions. It has a property of slowing down neutrons without absorbing them. Beryllium and copper alloys develop a high tensile strength and ability to withstand repeated stress. These alloys being non-sparking and non-magnetic find use in many special forms in parts of aeroplane, engine, motor, switches and watches. Beryllium oxide is an excellent high temperature refractory (melting point 2570°C) and finds use in aircraft spark plug and high temperature electric furnaces.

Mica is invaluable for its dielectric and insulating properties and is used in the manufacture of every electronic equipment. A number of substitutes for mica have been developed but none equals in dielectric strength compared to mica. Piezoelectric quartz crystals and ca have a great importance in radio, telecommunication and electronic industry. Quartz plates cut out from quartz crystals, prevent wavelengths from wandering and thus transmit them with correct frequency. They are used as oscillator plates, employed in radio transmitting and receiving sets, for controlling the frequency and detection devices for locating submarines. In radar and in acoustic submarine devices, they amplify a mechanically produced sound wave, transforming it into electric energy and projecting it to detect objects. The electric wave bounces back and is transformed into sound by the quartz plate. The sound is projected upon a screen indicating the approximate position of submarines.

Germanium metal has led to the invention of transistorised radio-communication. This metal made it possible to make smallest sets which can be carried even in pockets for communicating between the army in front with rear line of control.

Sulphur indirectly aids very much the war efforts. It goes in the manufacture of so many products which help to mobilise essentials of war. Because of its importance, it is placed under war minerals rather than war supporting minerals. Sulphur is used in the manufacture of explosives, refining of petroleum, sugar and in the medicinal preparations, disinfectants and fungicides. It is the starting point of several synthetic chemical industries, the most common being sulphuric acid which is called the king of all chemicals and largely used in pickling of steel, galvanising and tinning, manufacture of fertilisers, alum, processing of textile, etc. India had to face difficult time during the Second World War when the supply of sulphur was practically cut. Its absence can paralyse many industrial outputs.

War use of phosphorus is in the manufacture of incendiary bombs, tracer bullets, smoke screens. Peacetime uses are in the manufacture of match stick and fertiliser.

Lithium and its compounds, after the Second World War, have become important for war and peace uses. Lithium perchlorate is used as rocket propellants and in nuclear reactor. Lithium has two isotopes with mass numbers of 6 and 7. The former isotope is used for the production of tritium, extra heavy hydrogen, an intermediate explosive in the manufacture of hydrogen bombs. Lithium-7 meta-phosphate and pyrophosphate are valuable constituents in fused salt nuclear breeder blankets because of their relatively high thermal stability and low absorption of thermal neutrons by phosphorus compounds. Lithium hydride and lithium borohydride are used for war purposes. One pound of lithium hydride yields 22.5 cu ft of gas and that of lithium borohydride 66 cu ft. Peacetime uses of lithium minerals and compounds are in glass and ceramics, refrigeration, as a bleach in laundries and disinfectant to swimming pools. It also finds small use in alloying element where it is reported to impart toughness and tensile strength. Lithium is also used in signal flares and tracer bullets.

Strontium compounds have many war time applications. Strontium oxalate finds use in tracer bullets to control the burning rate of the trace composition. Strontium nitrate is used for military flares, rockets and shells. In war, generally every fifth bullet fired is a tracer bullet.

Graphite can be divided into two classes, one natural and other artificial. The artificial graphite is manufactured by heating anthracite coal or petroleum coke with small quantity of sawdust and quartz. This product has replaced natural graphite, used widely in industry for lubricant, dry battery and electrodes. The natural graphite is used as foundry facing, in the manufacture of crucibles, pencils and also as lubricants. Graphite lubricants are ideal for machinery working at high temperature where ordinary lubricants like grease tend to melt out quickly. Pure graphite is used as 'moderator' in showing down the action of neutrons of uranium. It was a primary material used for research leading to the manufacture of atom bomb in 1940.

Discovery of uranium, as a most destructive power and possessing tremendous power for peaceful use as well, was established 1946. Nobody could imagine before the first bomb was dropped in Nagasaki and Hiroshima, Japan, in 1946, that these tiny isotopes of uranium keep locked up such a vast amount of energy. The atoms of radioactive elements when made to split up generate tremendous amount of heat which can devastate anything in a twinkle of a second. The energy released by the fission of a single uranium atom is 200 million electron volts. Radioactive elements are classified as fissile and fertile. Uranium and thorium are two important fissionable elements which are in plentiful supply and used in atomic weapons and energy. Thorium (Th_{232}) is known as fertile element for the production of U_{233} in a breeder cycle. But U_{233} is less useful from the fission point than U_{235}, a most powerful fissionable element. Natural uranium is the source of two desirable fissionable fuels U_{235} and U_{239}. Natural uranium contains the most common isotope U_{238}. U_{235} is present in proportion of 0.7% maximum to common isotope. U_{239} (plutonium) is produced in breeder reactor by striking U_{238} by the neutrons emitted by U_{235}. Mass of pure U_{235} is utilised in the preparation of atom bombs. Separation of U_{235} from other isotopes is highly complicated and expensive process, and is kept secret. In the breeder reactor, U_{238} is used as a mass. In the reactor, one of the neutrons resulting from U_{235} is utilised in sustaining chain reaction, and the remaining ones are captured by U_{238} resulting in the production of U_{239}. It can be separated chemically and used as fuel as well. Instead of U_{238}, Th_{232} can also be used in conjunction with either U_{235} or U_{239} as fuel, the U_{233} thus produced is separated out and reutilised as fuel in generating nuclear heat. One kg of U_{235} releases heat equivalent to 3700 tonnes of coal. In other words its heating value is 30,000 times more than coal in terms of B.T.U.

Military dependence on coal, coke and petroleum and its products is unlimited. Coal and coke are inner strength of nearly all metallurgical chemical and fabrication industries; a source of thermal heat and electricity and locomotion. Petroleum provides the fighting power on land, sea and air. Petroleum jelly is used in the manufacture of napalm bombs, much heard of in Vietnam War, which, when exploded, go on consuming oxygen of air thus suffocating the lives.

Industrial diamond is needed for cutting hard substance like steel plates for the manufacture of aircraft, armour plates, war equipments and in the manufacture of diamond dies for drawing microscopically thin wires of copper, etc. required in the manufacture of intricate instruments. It is indispensable to modern industry as well as for war efforts.

Fluorite has emerged as a mineral of immense value. It is chiefly utilised in the manufacture of synthetic cryolite which in turn is used in the extraction of aluminium from alumina melt where cryolite acts as an electrolyte. Natural cryolite having exhausted, fluorite has rather rescued aluminium metallurgy. Other important uses of fluorite are as flux in cryolite bath where it helps in balancing the fluorine content and in steel metallurgy where it reduces viscosity. Atomic bombs production requires several fluorine chemicals.

New uses developed of borax have brought this mineral considerable military significance. For this reason it can as well be put under 'war minerals' rather than under 'war supporting minerals'. Borates have been found good anti-knock agents in gasoline. Boron hydride has potential value as rocket fuel developed in recent years.

In the field of nuclear energy, boron (B^{10} isotope) protects personnel from the harmful effects of reactors. Boron has the unique property of absorbing neutrons produced by nuclear reactions without the emission of harmfull secondary gamma radiation. Ammonium pentaborate is used as a 'poison charge' in atomic submarines. Borax is widely used in glass and ceramic industries, in the manufacture of optical glass, as flux in the manufacture of artificial gem and refining of gold and in several chemical industries.

Mercury has uses but both in war and peace its use is mostly confined to pharmaceuticals. Army uses large quantities of mercurial preparations for prophylactics and antiseptics. Mercury fulminates are used for detonating explosives but other substances have substituted them. Large use is found in electrical apparatus, thermometers and soda ash industry.

Asbestos especially of chrysotile variety is invaluable for heat and electrical insulations. It is largely used as insulating material, as clutch plate and brake lining of heavy motor vehicles. Asbestos cement sheets are used by army for quickly erecting barracks and storage godowns. Asbestos cloth is used during the fire fighting.

Mineral intelligence and the war: War intelligence requires a deeper probe not only of own country's resource position but also of hostile countries, the sources from which they obtain supplies and the stocks accumulated by them. In a long-drawn war, mineral supply plays a dominant role, for on it depends a country's industrial might and on its supply the sinews of war. The defeat of Germany and Japan in the last war was largely due to their inability to maintain mineral equilibrium. Both Germany and Japan had spectacular successes in the beginning, because both had built stockpiles of minerals of which their indigenous resources were either limited or non-existent. Further, bold thrusts made by Germany in France and Japan in Indonesia (formerly the Dutch East Indies) obtained for them various minerals needed for their war commitments. The Germans drive to the Ukraine in 1941 was in part the result of a desire to control Caucasian oil field, and to secure the manganese field of Nikopol. Germany got huge stocks of petroleum products and the iron and steel resources of northern France while in East, Japan got oil of Indonesia. As their stocks started depleting, and as Germany failed to reach by force of arms, the rich oil and manganese ore fields of Russia; and Japan the oil and iron belts of eastern India made it possible for the Allies to maintain mineral superiority, so that the defeat of Germany and Japan became only a question of time. The supply of petrol, vital for quick movement by land, air and sea, became so limited that both Germany and Japan were often unable to undertake sorties except for the most essential purposes. The Allied proved superior and very shot fired by the Axis Powers was returned a hundred-fold by their opponents. The Ruhr and the Silesian coal fields of Germany were heavily bombarded which paralysed completely the German's mining and metallurgical activities. The synthetic petrol production plant based on coal was destroyed which crippled Germany, reducing it to military smithereens. In Japan, the mineral supply was so critical that there was little cement for fortification; the army was provided with shell cases made of some dull grey substitute metal, for there was no more brass; some military units were issued even bamboo spears. Japan even today is a weak nation so far as domestic supply position of some of the basic minerals and ores are concerned. Japan depends for 98% of her petrol requirements on imports. The fall of Italy also became imminent immediately after her petrol supply line was cut. Eventually, each country has to make its own resource survey and take stock of the situation from time to time. A country may have a potential of rich mineral wealth but, if it is locked underground, may expose it to same hazard, as if it were devoid of mineral wealth and thus the consequent fighting power.

India's status in resource position: India enjoys unique position in possessing sufficiently large reserves of many useful war and war supporting minerals. During a period of last one decade, as a result of intensification of geological works, many new discoveries have been made and reserve positions of several known deposits improved thus bridging the gap between the demand and supply. New reserves have been added to copper, lead, zinc, nickel, tungsten and uranium ore deposits, bauxite, asbestos, pyrite, phosphorite and petroleum. Presence of cadmium, molybdenum, gallium, germanium and selenium have been found associated in recoverable quantities in several base metal deposits. Raw material for iron and steel industry are abundant. Rich deposits of iron ore containing over 60% Fe exist in Bihar, Orissa, Madhya Pradesh and Karnataka. There are six major iron ore hill ranges, namely Bonai range stretching for a continuous length of 51 km from South Singhbhum in Bihar to Keonjhar with offshoots in Mayurbhanj, Cuttack and Sundergarh districts of Orissa; Bailadila range having 14 detached hills covering a length of 58 km in Bastar district and Dalli Rajhara hill in Durg district, Madhya Pradesh; Bellary-Hospet hill range extending for 16 km in Bellary district, and Chicknahali and Chikmagalur hill ranges Chikmagalur district, Karnataka, and the Goa hill ranges in Goa. There are several detached yet large deposits, viz., Ari Dongri in Bastar; Kudremukh in North Kanara and Bababudan hills in Chikmagalur, Karnataka; Kanjamallai in Salem district, Tamil Nadu; Chandrapur and Ratnagiri districts, Maharashtra and at several places in Andhra Pardesh, Karnataka, Haryana and Rajasthan. The Bonai hill range is estimated to contain 8000 million tonnes of ore, and worked extensively at several places which forms the source of iron ore to five major steel plants in the eastern sector. The total reserves in Bailadila deposits are estimated at 3000 million tonnes with rich iron content. The deposits numbering 14, 13, 11, 10, 5 and 4 have proved reserves of 150 (including 20 million tonnes of blue dust), 100, 215, 130, 206 and 215 million tonnes respectively. The grade of the deposit number 5 is $+67\%$ Fe and of remaining $+65\%$ Fe. The estimated reserves of Kudremukh deposits are placed at 1000 million tonnes with 33% Fe. Reserves in Goa are placed well over 500 million tonnes. The Kudremukh and Salem deposits are magnetite whereas the remaining hematite type. Goan ores are mostly of blue dust containing 58 to 61 % Fe which pelletise well and find good export market. The total inferred reserves of iron ore in the country are placed at over 21,000 million tonnes with proved reserves of 12,000 million tonnes. Production of iron ore is in exportable surplus. Major deposits of manganese ores are found to occur in an arcuate belt for a length of 210 km extending from Chindwara in Madhya Pradesh to Nagpur and Bhandara in Maharashtra to Balaghat again in Madhya Pradesh which contain about 77% of the total reserves in the country. Other major deposits are located in Orissa, Karnataka, Andhra Pradesh, Bihar and Gujarat in order of importance. Deposits of smaller magnitude are found in Ratnagiri, Maharashtra; Goa and Banswara in Rajasthan. The special characteristics of ores of Maharashtra and Madhya Pradesh belt are high in Mn content, low in iron content and in general containing a little high phosphorus of 0.15 to 0.25%. Ores with low phosphorus containing 0.1% or little less are found in selected areas like Bharveli and Ukwa in Balaghat. Float ore contains usually low phosphorus. This belt is the major producer of ferro-manganese grade, manganese ore which is obtained by blending ores from different mines. Ores of Orissa and Mysore are low in phosphours as well as in Mn content and high in iron. Gujarat ore is high in phosphorus although Mn content averages 44 to 48%. Other deposits are of medium to low grade ferruginous type mostly suitable for blast furnace charge. Total reserves of manganese ores in the country are estimated at 104 million tonnes, of which Maharashtra and Madhya Pradesh belt contains 42 million tonnes; Bihar and Orissa 44 million tonnes; Karnataka 17 million tonnes, Rajsthan 2 million tonnes, Gujarat 2.5 million tonnes, Goa one million tonnes and Andhra Pradesh 0.5 million tonnes. The resources for alloy metals like chromium, titanium, vanadium, zirconium except tungsten are quite large. Chromite suitable for metallurgical, chemical and refractory are mined in Bihar, Maharashtra, Karnataka and Orissa; the Sukinda area of Orissa State being the major producing centre accounting for 98 % of the total production. Total reserves of chromite are placed over 30 million tonnes of which Orissa alone contains 90% of the reserves.

Sizeable reserves of nickeliferous laterite containing 0.5 to 1% nickel have been established in 3 blocks in Sukinda area extending from Sukinda in Cuttack to Simplipal basin in Mayurbhanj, Orissa. A large part of the reserve is situated in Kansa block of Cuttack district. Nickel is associated with cobalt in the range of 0.06%. A feasibility study has been completed to produce 4,800 tonnes of nickel and 200 tonnes of cobalt per annum from this area. Nickeliferous chromite deposits have also been established in Manipur State where production has commenced in 1976.

In ilmenite, an ore of titanium, the country has abundant supply from the beach sands of Kerala, Tamil Nadu and Orissa. The reserves of beach sands containing 65 to 70% ilmenite are estimated over 250 million tonnes. Beach sands are also a rich source of zircon, rutile, monazite, sillimanite, garnet, etc., besides ilmenite. The average composition of the Chavara, and Manavalakuruchi beach sands are as follows:

	Chavara (Kerala) in %	Manavalakuruchi (TN) in %
Ilmenite	60–70	45–55
Rutile	4–7	2–7
Zircon	5–8	4–6

Monazite	0.5–1	3–4
Sillimanite	4–8	2–3
Garnet	—	7–14
Leucoxene	0.5–1	0.5–1
Quartz, Shell, etc.	4–25	10–25

The chief source of vanadium is the vanadium bearing titaniferous magnetite deposits found in a belt, stretching from the southern part of Singhbhum district of Bihar to Mayurbhanj district of Orissa. The reserves are estimated at about 22 million tonnes. The vanadium oxide (V_2O_5) content in the deposit has been found to vary from 0.5 to 0.48 %, going up to 8 % in some cases. It is not worked at present. The known reserve position of tungsten ore is small and production inadequate. Two deposits are worked at present, one at Degana in Nagaur district, Rajasthan and other at Chendapathar in Bankura district, West Bengal. In the former case only the alluvial deposits are mined and concentrated to obtain are analysing 63% WO_3. In Chendapathar, wolfram occurs in quartz veins transversing phyllites and quartzites. The concentrate produced in this area analyse 66% WO_3 or more. One more wolfram deposit occurs at Agargoan in Nagpur district, Maharashtra. Mineralisation is however erratic. Recent work done on the tailing of Kolar and Hutti gold mines has shown presence of scheelite ($CaWO_4$). Laboratory work has shown that a concentrate analysing 67% WO_3 can easily be obtained. A pilot plant has been established for recovering tungsten from the dumps. Prospecting work done in Kolar gold mine has shown concentration of scheelite in the 48th to 57th levels. These new finds have improved the position of tungsten ore supply.

The supply position of copper, lead and zinc is strategic to critical at present but likely to achieve essential position in view of several new discoveries made. Major production of copper ore comes from Mosabani and Rakha areas, Bihar and Khetri and Darido areas, Rajasthan worked by Hindustan Copper Corporation. The chief sulphide minerals occurring there are chalcopyrite, pyrrhotite with some pyrite, pentlandite, violarite and millerite. Average grade worked during 1975 in Bihar was 1.4%Cu. The production of copper metal is small about 14,000 tonnes per annum, as against our minimum annual requirements of 60,000 to 70,000 tonnes, even after effecting sub-stitution. The reserves proved in new deposits are, however, suffi-cient to produce 50,000 to 60,000 tonnes of metal annually. Impor-tant mines under development are Madan-Kudhan, Kolihan, Satkui, Chandmari in Khetri copper belt and Dariba in Alwar, Rajasthan. Rakha mines block, Roam Sideshwar, Ramchandra Pahar and Tamapahar in Singhbhum copper belt, Bihar; and Nallakonda in Agnigundla belt, Guntur district, Andhra Pradesh. In Madan-Kudhan section of Khetri copper belt total reserves proved so far are 46 million tonnes with 1% Cu. Indicated reserves are placed at 46.6 million tonnes with 1% Cu. In Kolihan the proved reserves in the central blockstand at 29 million tonnes with 1.57% Cu. Pre-sence of gold, silver, cobalt and nickel has been established. The reserves of copper ore in Rakha area are placed as follows: Proved 41.7 million tonnes with 1.35% Cu; Probable 24.6 million tonnes with 1.33% Cu; and Possible 21.4 million tonnes with 1.28% Cu. In nearby Tamapahar block, reserves of 25.8 million tonnes with 1.2% Cu; have been estimated. New reserves have been established in Malanj Khand, Balaghat district, Madhya Pradesh. It is the single largest deposit discovered in the country. The ore reserves estimated are 65 million tonnes containing 1.3% Cu. The reserves in Agni-gundla area are estimated at 18 million tonnes. It is primarily a lead-copper deposit with lead content varying from 3 to 9.1% and copper from 0.56 to 1.71%. The production of lead-zinc-silver ores from Zawar area, Udaipur district, Rajasthan where the ore deposits occur in dolomitic rock forming seven hills of which Mochia Marga, Zawar Mala, Balaria, Bowa and Baroi are important. Mining is at present confined mainly to central portion of Mochia Marga hill. The Balaria deposit is under development. Reserves proved so far are of the order of 120 million tonnes with 7% total metal content. The average grade of ores mined contain about 3% Pb and 4% Zn. Silver is associated mainly with lead and cadmium with zinc. Silver content in the lead concentrate on an average is 774.5 gm per tonne. Zinc concentrate also contains small quantity of silver about 171.5 gm per tonne besides 0.22% of cadmium. Important minerals identified in this deposit are galena, sphalerite, pyrite, chalcopyrite, arseno-pyrite, pyrrhotite and argentite. Presence of elements like gallium, indium, cobalt, bismuth, chromium, zirconium, antimony, alumi-nium and manganese has been traced. Zawar deposits are worked by Hindustan Zinc Ltd. which has an electrolytic zinc smelter plant nearby at Debari having 45,000 tonnes licensed capacity. Another rich deposit of lead-zinc-silver ores associated with molybdenum, bismuth, arsenic, cadmium have been established in Dariba-Rajpura area, in the same Zawar belt outside the Zawar lease hold. Both areas are held by HZL, A tentative reserves of 110 million tonnes with 7 % combined metal content has been estimated down to a depth of 300 metres.

In resource position of bauxite the country is happily placed. Estimated reserves 2250 million tonnes distributed chiefly in Bihar, J & K, Madhya Pradesh, Karnataka, Tamil Nadu and Goa including the newly located deposits in Uttar Pradesh east coast covering Andhra Pradesh and Orissa. Production capacity of aluminium metal is adequate. Goan deposits contain more than 50 million tonnes. Reserves in east coast are placed at about 2000 million tonnes.

Pegmatite belts of Andhra Pradesh, Bihar and Rajasthan are the principal source of beryl and mica, in which the country has a large supply position. Production of beryl chiefly comes from Rajasthan and Andhra Pradesh. Production ranges from 1000 to 3000 tonnes a year, most of which is exported. Reliable source of piezoelectic quartz crystals has not been established as yet outside Brazil. A survey of all working mines of quartz in the country has given disappointing results. Germanium has been found in the lignite of Kashmir, Assam coal and lead-zinc ores of Zawar.

India lacks completely in the source of elemental sulphur. Efforts to recover elemental sulphur from Amjore pyrite in Bihar by well-known Orkla & Outokumpu processes have not gained ground. The grade of the ore is 40% S. A new pyrite and pyrrhotite deposit has been found at Saladipura in Sikar district, Rajasthan. The grade however is low containing only 22% S. The Pyrites, Phosphates and Chemicals Ltd., has started developing this area. Inferred reserves are 115 million tonnes down to 250 metres. Recovery of by-product sulphuric acid is being made from zinc and copper smelters. Amjore pyrite is mined to produce 400 tonnes of sulphuric acid a day at Sindri to meet the requirements of Fertiliser Corporation of India. The Madras Refinery plant recovers at the rate of 6000 tonnes of elemental sulphur per annum during the refining of petroleum. In apatite and phosphate rock, the country has improved its resource position considerably from complete deficiency to near sufficiency. New discoveries of several phosphorite deposits in Mussorie area in

Uttar Pradesh, Jaisalmer and Udaipur districts in Rajasthan and Jhabua in Madhya Pradesh have been made. The deposit, one located, at Jhamar-Kotra in Udaipur is of considerable economic importance. Here an 18 metres thick phosporite bed containing 30 to 35% P_2O_5 occur with possible reserves of 100 million tonnes. The country's resource in strontium mineral is negligible. Small deposit of celestite ($SrSO_4$) is found in Tiruchirapalli, Tamil Nadu, which is not worked, Known resources of lithium minerals are also small. Intermittent production comes from mica mines of Bhilwara and Hazaribagh. A new deposit is reported to be discovered in Bastar, Madhya Pradesh.

The position of graphite in India can be rated as critical rather than strategic. Setting up three synthetic graphite plants at Durgapur, Barauni and Bangalore has considerably eased the position of electrode supply so vital to the manufacture of aluminium and other electro-metallurgical industry like ferro-manganese. Production of natural graphite chiefly comes from Orissa and Andhra Pradesh. However indigenous graphite is unsuitable for use in atomic reactor, as high purity graphite has not been found. For use as a moderator, graphite should be low in certain neutron absorbing elements notably boron and the rare earths. The position about fluorite is critical at present. There are four known deposits of economic importance. There are: (1) Ambadungar, Baroda district, Gujarat, reserve 11.6 million tonnes with 30% CaF_2, (2) Mando-ki-Pal, Dungarpur district, reserve 1.6 million tonnes wish 18% CaF_2, (3) Chowkri-Chapoli, Sikar district, reserve 0.254 millon tonnes with 18% CaF_2, both in Rajasthan, and (4) Chandi-Dungri in Durg district, Madhya Pradesh, reserve 0.535 million tonnes with 16% CaF_2. There is a beneficiation plant about 3 km from the Ambadungar mine to treat run-of-mine for producing metallurgical and acid grade fluorspar. The plant has the capacity to treat 30,000 tonnes of ore per year.

The country has rich source of fissionable material. Uranium minerals are worked in Bihar and Rajasthan. Copper belt of Bihar is an important source of uranium containing about 0.5 to 1 kg per tonne of rock. Uranium content varying from 0.03 to 0.1% is economically extractable. Several mica pegmatite veins in Bihar, Rajasthan and Andhra Pradesh contain uranium (pitchblende), fergusonite, brannerite, etc. Singar mica mines of Nawadah and Sankara mica mines of Nellore are famous for such occurrence. The Barren

Measures of Gondwana are found to be a good source of uranium.
The beach sands containing monazite in Kerala and Tamil Nadu
are chief source of thorium and uranium. Monazite sands yield 0.2
to 0.4% uranium oxide (U_3O_8) and 5 to 11% thorium oxide (ThO_2).
These are being produced at Indian Rare Earths Factory, Kerala.
Total reserves of uranium in beach sands estimated by Department
of Atomic Energy, Government of India are about 30,000 tonnes.
Known reserves of thorium oxide are estimated to be between
450,000 tonnes and 40,000 tonnes.

The present annual production of monazite is approximately 2000
tonnes mainly employed in the manufacture of cerium compounds.
The cerium is recovered for use as atomic fuel. Monazite is the store
house of rare earths and fissionable material. It usually contains
20 to 31% cerium, 1 to 1.5% yttrium and 21 to 31% rare earths
oxides besides containing thoria, uranium, phosphate and several
rare elements. Estimated recoverable reserves of combined rare

Composition of Kerala Monazite

	In per cent
ThO_2	8.1
CeO_3	30.6
La_2O_3	15.7
Pr_2O_3	2.9
Na_2O_3	10.5
Europium, gadolinium and terbium oxides	0.7
Yt_2O_3	0.4
Dysprosium, holmium, erbium, ytterbium and lutecium	0.1
CaO	1
W_2O_3	0.3
P_2O_5	26.2

earth oxides are estimated over 400,000 tonnes. A typical composition of monazite from Kerala is given in the table on page 285. Uranium mineral is worked at Jaduguda, Singhbhum district, Bihar by Uranium Corporation of India Ltd. where daily 1000 tonnes of ore is mined and processed to recover fissionable elements.

Large reserves of coal are available in the country. The total reserves are placed at 120,000 million tonnes down to 900 metres depth. But likely reserves are many times more. In supply of petroleum the position is critical. In industrial diamond, it is strategic. Only a small production of about 1500 carats to 1600 carats is obtained from Panna diamond field against our present requirements of 265,000 carats a year which is likely to increase to 500,000 carats in a very near future. Production of chrysotile asbestos is also small. For the first time amosite deposits of small dimension have been located in Bababudan Hill area in Chikmagalur district, Karnataka. An area of 194.25 hectares has been acquired by M/s Hyderabad Asbestos Cement Products Ltd. Other raw material like limestone, dolomite, gypsum and refractory minerals, the country has plentiful supply position.

TABLE 17

Relative Positions of Various Minerals and Metals in India During 1978

Minerals			Metals		
Sufficient	Inadequate	Deficient	Sufficient	Inadequate	Deficient
1	2	3	4	5	6
Barytes	Apatite and rock phosphate	Antimony ore—totally absent	Aluminium	Ferro-boron (based on imported borax)	Beryllium—no production
Bauxite	Copper ore	Arsenic ore—totally absent	Antimony (based on imported ore)	Ferro-phosphorus	Bismuth—absent
Beryl	Fluorspar	Asbestos	Cadmium	Ferro-tungsten (mainly based on imported tungsten)	Cobalt—no production
Bentonite	Graphite	Borax—totally absent	Ferro-chrome	Ferro-vanadium (mainly based on imported vanadium pentoxide)	Germanium—no production
Calcite	Lead-zinc ore	Industrial—diamond	Ferro-manganese	Chromium	Gold
China clay	Lithium ore	Sulphur—totally absent	Ferro-silicon	Copper	Mercury—absent
Chromite	Petroleum	Mercury ore—totally absent	Ferro-titanium	Lead	Molybdenum—no production
Coal and Lignite		Strontium ore	Iron & Steel	Magnesium	Nickel—no production
Corundum		Tin ore—nearly absent		Phosphorus	Niobium—no production
Dolomite					Platinum—absent
Diaspore					Selenium
Felspar					Silver
Fireclay					Strontium—no production
Fuller's earth					Titanium—no production
Garnet					Tin—no production
Graphite					

TABLE 17 (Contd.)

1	2	3	4	5	6
Glass sand		Wolfram		Zinc	
Gypsum		Piezoelectric		Special steel	
Ilmenite		quartz crystals.		Vanadium	
Iron ore					
Kyanite					
Limestone					
Magnesite					
Manganese ore					
Mica					
Monazite					
Ochre					
Pyrite					
Pyrophyllite					
Rutile					
Salt					
Sillimanite					
Talc					
Uranium ores					
Vanadium ores					
Wollastonite					
Zircon					

15

Conservation and Substitution

CONSERVATION means making the most of what we have. It is an effort to utilise in best possible way, the resources available depending upon the industrial needs and changing technological requirements. It is a misconception that keeping minerals underground for future use or use of the future generation is conservation. The mineral asset is just like deposit in a bank but with a difference that there is no interest accruing to it; with every cheque or draft drawn, the reserves continue to diminish. Therefore any effort that we can make to increase the life of a deposit, without sacrificing our own needs, is an attempt towards conservation. The use of pellets and sinters prepared from fines, which were till recently thrown as waste, has increased many folds the reserve positions of iron ore, manganese ore, limestone and many other minerals Also reuse of scrap has prolonged the life of the metallic mineral deposits considerably. The efforts of conservation are perceivable at all stages. It can be achieved by (*i*) efficient method of mining to take out every possible tonnage locked underground, (*ii*) utilisation of leaner and unmarketable ores by innovations in ore dressing and metallurgical practices, (*iii*) use and reuse of scraps, (*iv*) recovery of all associated elements as co-product or by-product, (*v*) economy in use of minerals, and (*vi*) by substitution.

Minerals, for the purpose of utilisation, are placed under two classes; namely expendable and non-expendable. Expendable minerals are those which have practically no salvage value and once consumed are lost for ever. These include coal, petroleum, gas and most of the non-metallic minerals excluding precious stones. Non-expendable minerals are those of metallic group. Metallic minerals have a good salvage value as scrap and are recycled into normal run of consumption in course of time, except those quantities that are lost due to corrosion and abrasion, destroyed and littered in war and drowned in sea or otherwise. Metals entering into chemicals have little salvage value. Non-expendable value of minerals has greatly aided conservation and thus become a continuous exercise of research.

Common metals like iron and steel, aluminium, copper, lead, zinc, tin and antimony have good recoverable value as scrap. Precious metals like gold, silver and platinum are seldom lost. Gold is most stable. It does not rust and hence our treasury vaults go on piling up new productions year after year. These metals, except for inadvertent loss, remain preserved.

Metallurgical and fabrication industries produce considerable quantities of mill scrap or factory scrap resulting from splashing, cutting, trimming and finishing the products. In steel industry alone, considerable amount of scrap in the form of spillages, trimming and cuttings results while producing finished steel from ingot. These are recovered, remelted and put into use again, and form a useful source of secondary metal. Recovery of metals from a secondary source is true to every metallurgical and fabrication industries with variations in quantities recoverable. Besides these, scraps result from outlived machinery, plants, wagons, coaches, rails, automobiles and several other things including household articles. They form huge source of secondary metals. It may be remembered that during the Second World War even the horseshoe nails, ordinary nails and other broken steel structures were collected in India to meet the shortages of steel.

It is difficult to say precisely what quantities of secondary metals are recovered in the country every year from scraps. There is no organisation to collect and study such statistics as existing in the U.S.A. and other advanced countries. Relatively small production of primary metals has been one reason for not initiating such study as yet.

In industrially advanced countries, scrap has become an industry by itself. There are a number of furnaces working on scraps only. In the U.S.A. about 70 % of the total steel production and 32% of the total aluminium production come from secondary sources. Huge piles of scraps flow like big rivulets into the factories. The following statistics will provide the magnitude of scrap industry in the U.S.A.

METALS SCRAP USED IN U.S.A. DURING 1976

Steel	89.9	million short tonnes
Aluminium	1.74	,,
Copper	416,000	short tonnes
Lead	1003,108	,,
Zinc	306,508	,,
Tin	11,161	,,

According to survey made by the United States Bureau of Mines, the primary copper comes into circulation as scrap in about 25 years and quantity recovered is 60%. Copper goes into use as it is, and also in the form of alloys such as brass, bronze, etc. In addition to the considerable amount of scraps originating from the smelters themselves as a result of croppings and trimmings of billets, bars tubes, there is also a flow of scraps from other allied manufacturing industries. In modern brass works, 60% or more of the material fed into the furnaces is scrap. Copper, zinc or other metal is added to the extent necessary to make the particular alloy. Copper wire used in telephone, telegraph, electric lines is hardly lost; when new lines are laid, old wires are rolled up, sent to mill where they are remelted and rolled into wire again.

Lead and zinc metals entering into white pigment and lithopone and chemicals are generally regarded to be lost, although they can be recovered. Seventy-five per cent of the quantity of lead metal used in storage batteries and printing metal, is likely to be put into recirculation in 3 to 4 years. In case of its use in the electrical cables the period is expected to be 15 to 20 years and only 50% of the material can be expected to reappear. In other words, there would be recovery at the rate of 2.5% annually of the total metal consumed after a period of 15 years. Similarly in the case of zinc, it has recovery value from the scrap of dry cells and alloys like brass containing zinc. While separating zinc from brass, considerable loss of zinc occurs. Recovery of the zinc from alloys is estimated to be 30% only.

Economy in use, to say it more clearly, putting mineral of limited resource to specific and indispensable uses only and substituting it by a different mineral wherever possible, is a measure of conservation. To illustrate this viewpoint, the substitution of sulphuric acid, in which sulphur is the most important ingredient, by hydrochloric acid can be cited. Recent experiments have shown that in pickling of steel for galvanising and tinning purposes, sulphuric acid can be substituted by hydrochloric acid. This measure can greatly help in economy in the use of this scarce element for India, and divert the quantity thus saved for other indispensable uses. It has been possible also to partly substitute sulphuric acid by hydrochloric acid in the manufacture of triple-superphosphate.

Substitution is an exercise of 'get going' in event of material shortage by finding out another material, a substitute, which could serve the purpose. It is again an effort to find out an alternative to material in short supply. Substitution results as a matter of expediency. Efforts of substitution and conservation in most cases are supplementary to each other. A substitute can have only a nearly similar properties to that which is substituted and its availability cheaper and easier. If a new product is invented or a new mineral or metal having better performance is found out than the mineral of metal in use, in that event it cannot be categorised under substitution. It can rightly be placed under 'new invention', a new source of raw material. This idea can be made clear from the example of natural cryolite and nitre which have been replaced by synthetic cryolite and nitre. These synthetic products were invented by Germans during the two world war periods. The synthetic cryolite is universally used in place of natural cryolite, the reserve of which is practically exhausted. Synthetic nitre has become a strong competitor of Chilean nitre which is regarded as the largest single source in the world. A number of synthetic abrasives and refractories are now manufactured which have a better performance than some of the common abrasive and refractory minerals. They all form a new source of raw material and thus cannot be classed under substitutes. Examples of some of the substitutes developed and how they help in conservation can be found from the following paragraphs.

For mica, a number of substitutes have been developed which are in market and sold under different trade names, like reconstituted mica, samica, filmica and ceramics which are steadily substituting natural mica in capacitors and becoming popular among the design engineers for electrical equipments. For mica it can be rightly said that mica has substituted mica. In other words scrap mica in the form of reconstituted mica has replaced splittings and blocks in a number of uses.

Electricity has substituted, where its supply is cheap and abundant, much needed coke in blast furnace to the extent it is required to generate heat. Dieselisation and electrification of many rail routes have helped much in conserving good quality coal. Hydropower has saved consumption of coal in thermal stations. Nuclear energy has also become a good competitor of thermal energy in an area where the supply of coal is to be obtained from over 600 km. Hydroelectricity is the cheapest of all but its power cannot be transmitted to every nook and corner unless sufficient hydel potential is available. Between thermal and nuclear, in certain areas, the latter has an advantage. Cost of generating nuclear power per unit at Tarapur in Gujarat and Ranapratapsagar in Rajasthan was estimated at 2.8 paise as compared to 3-1/4 and 5-1/4 paise respectively of thermal power.

Worldwide shortage of copper and relative abundance of aluminium has made aluminium substitute coppor wires in high transmission lines, although aluminium has only 70 % capacity to allow flow of electricity compared to the same volume of copper. All underground cables are manufactured of aluminium instead of copper since 1962 in India. This has resulted in saving about 3600 tonnes of copper metal annually. Possibility of substituting zinc and tin, in galvanising and tinning of steel by aluminising it, has also become a reality It can also save considerable quantities of zinc and tin metals.

Tungsten carbide has been found to be a good substitute of industrial diamond in cutting tools especially in drilling. Bauxite has to a great extent substituted fluorspar, which is another scarce mineral in India, in steel metallurgy where it is used as flux and in reducing viscosity. By this, the consumption of fluorspar has been brought down from previous rate of eight kg per tonne to less than one kg per tonne of steel. During 1973 and 1974 the rate of consumption of fluorspar per tonne of steel produced by different steel plants in India was as follows: Bhilai 0.1 and 0.03 kg; IISCO 2 and 2.4 kg; Bokaro nil and 0.39 kg; VISL 0.57 and 0.92 kg; TISCO 0.35 and 0.44 kg and Rourkela 0.13 and 0.26 kg respectively. Durgapur did not consume at all. Adequate availability of steel scrap can replace completely the use of 'Steel Melting Shop Grade' iron ore in steel making process. During 1974 in India a total of 560,470 tonnes of SMS grade iron ore was used alone.

There are very many possibilities of substitution and conservation in mineral indusry, for example use of pyrophyllite in place of soapstone, use of soapstone in place of china clay in paper industry, use of diaspore in place of kyanite refractory bricks, use of synthetic graphite in place of natural graphite in the manufacture of lubricant, synthetic diamond in place of natural industrial diamond, synthetic quartz in place of natural quartz crystals, replacement of antimony by calcium in place of antimonial-lead to calcium-lead in storage battery, etc. They are, in fact, a part of strategy to keep the wheel moving. Every strategist and mineral economist has to take note of these possibilities in planning and projecting requirements of future mineral supplies.

16

Changing Pattern of
Mineral Consumption

THE LAST one decade or so has witnessed several revolutionary technological developments and innovations in the field of mineral utilisation, extractive metallurgy, new routs to metal production, increasing use and manufacture of synthetic mineral raw material substitutes in addition to utilisation of submarginal grade ores by improved ore-dressing technology. Technological changes have shifted pressure from conventional mineral resources to resources hitherto unnoticed. New material are progressively being experimented and used displacing the conventional raw material; as a result certain minerals which were regarded indispensable till lately have now been rendered surplus. All these changes have undoubtedly made a great impact on the future resource and consumption of minerals. Cumulatively their effect can be discussed under following five headings, namely,

(1) altogether displacement of certain conventional raw material;
(2) use of altogether new and unconventional raw meterial displacing the conventional raw material;
(3) changes in the norms of consumption;
(4) impact due to synthetics and substitutes; and
(5) use of by-product recoveries.

To quote some examples, the production of synthetic industrial diamond, so vital for the manufacture of precision cutting tools, now equals the production of natural industrial diamond. Similarly, the synthetic cryolite is practically meeting the entire requirements of aluminium industry. Even the use of cryolite for this purpose is now threatend to be by-passed by new methods developed for the aluminium metal production. In the U.S.S.R., the most common ore of aluminium metal, i.e. bauxite has been displaced by nepheline-syenite and alunite. Piezo-electric quartz crystal, in which India has no resource and perhaps Brazil is the only country in the world producing such crystals, has been replaced to a great extent by synthetic quartz reported to be giving equally good performance. In the case of germanium, valued for its semi-conducting property, and largely used in transistor, is nowadays being replaced by silicon metal. By-product recovery is also making good in road and displacing natural mineral. In the case of gypsum, hydrated calcium sulphate produced during the manufacture of phosphoric acid from rock phosphate and recovery of common salt from sea water has practically threatened the mining activities of natural gypsum in the country. In India about 8 lakh tonnes of byproduct hydrated calcium carbonate is being produced per annum at present. The by-product calcium carbonate sludge obtained in Sindri plant of the Fertiliser Corporation of India is regularly being utilised for the manufacture of cement. The changes in steel technology have also given a good impact on the consumption pattern of various raw material required for the production of iron and steel. The introduction of L.D. process for steel making has practically displaced the use of chromite as a refractory bricks and its place has been taken largely by dolomite and magnesite. The changes in steel technology are also likely to relieve a great pressure on coking coal which is being substituted by natural gas or hydrocarbons produced from non-coking coal or by using 'formed coke'. There is also a significant change in the preparation of blast furnace feed from merely lumpy graded or unprepared ores to prepared ores in the form of self-fluxed sinters and pellets.

ALTOGETHER DISPLACEMENT OR CERTAIN CONVENTIONAL RAW MATERIALS

In this group three important mineral raw material, namely, chromite, cryolite and coking coal can be discussed conveniently. Chromite hitherto held an important position as refractory making material in steel industry. With the distinct swing towards L.D. process its place has now been taken over by magnesite and dolomite in the basic oxygen converters. Chromite refractory seems to give unsatisfactory results and not able to sustain the variant shocks and heat conditions operating in the L.D. process. Presently more than 50% of the world production of primary steel is produced by this process. The new capacities being added all the world over are adopting L.D. process because of economy and faster rate of production. This trend has given a great impact on the future consumption of chromite as a refractory brick.

Although magnesite and dolomite bricks have taken the place of chromite in the L.D. process, at the same time the use of natural magnesite in turn is gradually being reduced by marine magnesite. Natural magnesite generally is not found of that purity as required by steel makers. It usually contains a little high percentage of silica whereas the manufacture of refractory bricks stipulate the silica percentage within 2% to meet the requirements of the L.D. process. Preparation of high class refractory from natural magnesite generally calls for beneficiation, sintering and calcination. Initially, the sea water magnesia also fell somewhat short of L.D. specification because of their low density. Hence a method was devised whereby magnesium hydroxide extracted from sea water was first calcined at low temperature and then formed into pellets in a special rolls before dead burning. Thus, double burned magnesia proved to be better but still it was not comparable to the product made out of natural magnesite. After extensive investigations the causes were identified to the impurity contents, particularly those of lime, silica and boron. The lime and silica are introduced by calcined limestone and dolomite added to precipitate magnesium hydroxide from sea water whereas the boron originates from the sea water itself. Methods have been perfected to eliminate those impurities and to prepare magnesia of more than 99% purity of required density very suitable for refractory manufacture. By chemical treatment it has been possible to keep the phosphorus content within 0.005% as well in the magnesia prepared from sea water.

The use of cryolite in the aluminium metallurgy is well known. Aluminium is produced by Bayer and Hall processes where bauxite is digested in caustic soda to produce alumina, and in turn alumina is electrolysed in cryolite bath to recover aluminium metal, Hitherto the Hall method is the only commercial method universally adopted for the extraction of aluminium metal. Cryolite forms indispensable bath for reduction of alumina to aluminium. No other mineral or material is known in which alumina can be melted for electrolysis. Cryolite deposit was known to occur only at one place in the world, namely Ivigtut in Greenland where mine reserve is exhausted. Only stockpiles are being reworked and marketed after beneficiation by Denmark which meets only a partial requirements of aluminium industry. Now the synthetic cryolite has taken over the place of natural cryolite which has been discussed later. However, during the past few years two new processes have been invented by which not only the use of cryolite is by-passed but also provide choice of a wide variety of aluminous ores. Both these processes are feasible. In early 1972 Alcoa of the U.S.A. demonstrated a new process in which alumina is combined with chlorine to form liquid aluminium chloride, which in turn is separated into molten aluminium and chlorine by electrolysis. It is claimed that the consumption of electricity by this process is 30% less than the conventional Hall process. Alcoa is constructing a 15,000 tonnes per year pilot plant in Texas which can be expanded to 3 lakh tonnes per year if the pilot plant test proves commercially sound. In other process aluminium chloride is passed through manganese metal to produce aluminium metal. Both manganese metal and chlorine are recoverable for recycling. Pilot plant based on this process is also expected to be set up in the U.K. This process was discovered by Charles Toth.

With the fast progress towards iron and steel making by direct reduction/arc furnace routes, there has been corresponding elimination of coking coal by gaseous hydrocarbons either natural or prepared from non-coking coal. It is another significant development of great importance to many countries devoid of coking coal. Unlike the conventional blast furnace process which yields iron by smelting a mixture of iron ore, coke, limestone, etc., direct reduction processes produce iron in a solid form without smelting the ore and without use of coke and fluxing material. In a direct reduction process iron ore is fed into the reaction chamber where heated reducing gases (heated up to 2000° F) remove oxygen from the ore leaving behind pure iron in lump form. The reaction which takes place in the chamber can be shown by following formula :

$$Fe_2O_3 + CO + 2H_2 \rightarrow 2Fe + CO_2 + 2H_2O$$

The product obtained by direct reduction process is known as 'sponge iron' 'pre-reduced pellets', 'metallised pellets' or 'direct iron'. The sponge iron forms the feed material of mini steel plants instead of scraps. A number of different processes of direct reduction have been patented by different companies of which four are very important, namely, Midrex, Armo, Purofer and HYL processes. A number of units utilising the above processes are already under operation in different parts of the world. The total world installed capacity of sponge iron till 1973 was about 6.3 million tonnes. Iran is planning to build one million tonne plant using natural gas. A sharp rise in coke prices in some countries and the cost in favour of gaseous and liquid fuel in some other have led to the invention of these processes where non-coking coal, even lignite can be used in gaseous form. The direct reduction processes differ from each other in a minor way in the gas reforming unit for producing gas only. The condition prevailing in India warrants setting up several mini steel plants under operation and proposed to be erected in the near future. It is expected that very soon India will have not less than one thousand mini steel plants in the country. The availability of scraps presently in the country fall short of the requirements.

USE OF ALTOGETHER NEW AND UNCONVENTIONAL RAW MATERIAL DISPLACING THE CONVENTIONAL RAW MATERIAL

In this field the use of dolomite as refractory material and large scale use of iron ore fines in the form of pellets and sinters as well as limestone and manganese ore fines in steel industry and nepheline-syenite as an alternative to bauxite can be described. Before the introduction of L.D. process, the application of dolomite was mainly confined to as fettling material in place of costlier magnesite. Its use rose to prominence with the dawn of L.D. process. Now the high purity dolomite bricks are universally being used in L.D. process for lining the convertors and its demand has suddenly shot up.

Amongst all the developments the aluminium industry has witnessed significant change both in respect of technology and raw material. Bauxite-starved superpowers e.g. the U.S.A. and the U.S.S.R. for considerable long time were trying to find out alternatives to bauxite as raw material. In this direction the U.S.S.R. has made a major breakthrough by using nepheline-syenite as a source of alumina. The U.S.S.R. has also commercialised the use of alunite (basic potassium aluminium sulphate) for the production of alumina. The Soviet Union's first alumina plant from alunite is under operation at Kirovobad in Azerbaijan. Alunite is mined at Zaglinsky by open pit and transported 35 km to Kirovobad plant by aerial rope-way for processing it to alumina. Process appears to be similar to that developed by Kalunite Incorporated of the U.S.A. But in the U.S.A., this process has not been commercialised whereas in the Soviet Union the use of alunite is already being made for the pre-paration of alumina. The process generally involves roasting, crush-ing and leaching it with dilute sulphuric acid to obtain alum. Alumina and potassium sulphate (for use as fertiliser) are then separated from the alum by subsequent roasting and leaching.

The Achinsk alumina complex on the Chulim river in Central Siberia is the biggest plant in the Soviet Union. The nepheline-syenite for this plant is mined in the Kuznetsk-Alatau mountains about 305 km from Achinsk. There is another alumina plant based on nepheline-syenite operating in the Kola peninsula in the north-west Russia. Here nepheline-syenite is obtained as a tailing of local apatite concentrations. Siberia plant is reported to have an annual capacity of 8 lakh tonnes of alumina per year.

In the Toth process calcined high alumina clay or bauxite is chlorinated in fluidised bed reactor using chlorine gas with coke as reducing agent. This process has shown the possibility of bringing in usage raw material other than bauxite from the production of alumi-nium metal. Unlike Alcoa process, the Toth process completely eliminates the intermediate stage of alumina making thus eliminating altogether use of caustic soda and cryolite as well.

The use of iron ore fines in the form of pellets or sinters and also that of limestone and manganese ore fines has radically changed the entire position of these raw materials. The use of pellets as a feed material for the blast furnaces has opened a great possibility of using low grade ores like taconite or itabarite (siliceous iron ore containing 30% Fe) after beneficiation. It is estimated that in the world at present nearly 150 million tonnes of installed capacity of pellets is available. In India there are two plants producing pellets, one is owned by Chougule and Company in Goa and the other by Tata Iron and Steel Company at Jamshedpur. The Goa plant has got a capacity of near about half million tonnes and the Tata nearly one million tonnes, Sinter capacity is available in all the steel plants in India. The total installed capacity of sinters in the country is about 7 million tonnes. The concept of high grade lumpy ore has undergone tremendous change with the advancement of blast furnace technology, whereby it is possible to use ore of size less than 12 mm. Japan, for example, uses iron ore lumps with size varying from 6 to 25 mm. In India, generally 10 to 50/80 mm lump size is used. In future, this will be reduced to 10 to 53 mm in Hindustan Steel Ltd. plants. In Bokaro, provision has been made to use iron ore lumps of size varying from 10 to 40 mm. Manganese ore, limestone, dolomite fines and coke breeze are used for the preparation of self-fluxing pellets and sinters. The steel plants in the country use on an average self-fluxed sinters varying from 20 to 40% of the burden. The Bhilai Steel Plant contemplates using up to 60% of self-fluxed sinters in blast furnace feed. Bokaro contemplates still larger proportion of sinters which may go up to 75% of the burden. Improved furnace feed besides other factors has been largely responsible for better blast furnace performance and consequently lower consumption of coke, fluxing and lining materials.

CHANGES IN NORMS OF CONSUMPTION

The norms of consumption may be defined as the rate of inputs in terms of quantity for production of unit weight of end product. There has been recently drastic cut in consumption of certain raw material and inputs due to technological changes in design of manufacturing units and/or manufacturing process of input itself. Significant changes are perceivable in respect of following:

Coking Coal

In advanced countries consumption of coke per tonne of pig iron has been brought down to 600–650 kg as against present figure of 800–900 kg in India. The most impressive achievement has been reported from Japan which has reduced coke rate to about 375 kg. All these changes have been affected by combined efforts of several technological innovations. These can be enumerated as follows:

(i) *Substitution partially or supplementation by some auxiliary fuel injection*: It has been established that each one kg of coal powder injected replaces 0.61 kg of coke with an injection rate of 80 kg per tonne of hot metal. Such experiments are being planned in Indian blast furnaces.

(ii) *Raising blast temperature*: Though a blast temperature of 1100°C is common in most countries, in Indian furnaces this is around 900°C. Bokaro, however, envisages 1000°C. The HSL plants are planning to increase blast temperature up to 1000°C.

(iii) *Increasing pressure at the top of furnace*:   This technique involves partial restriction of the flow of outgoing gases from blast furnaces. Countries like the U.S.S.R. and Japan use top pressure of 2-2.5 atmosphere. Only Bhilai is designed for a top pressure of up to one atmosphere and Bokaro for 1.5 atmosphere.

Refractories

The consumption of refractories per tonne of ingot steel has registered decline due to efforts to improve the lining life of the steel melting shops and elimination of some of the stages in the process which requires a large quantity of refractories. Bhilai Steel Plant and Rourkela Steel Plant consume 26 kg of fireclay refractories against 42 kg by TISCO due to elimination of Duplex process which requires a larger quantity of fireclay refractories.

Improvements in manufacture of refractories have brought progressive decline in the levels of this consumption. The soaking zones of reheating furnaces now use electrocast refractories with considerable increased life. Use of high pressure techniques in manufacture of refractories have resulted in improving bond between refractory grains bringing better resistance to slag attack and erosion. As a cumulative result, maintenance requirement of refractories has dropped from 100 kg to as low as 40 to 50 kg per tonne of ingot steel.

Fluxes

A commendable reduction in consumption of fluxes has been achieved in iron and steel industry by the following techniques:

(a) *Washing of iron ore*:   It has been estimated that each 1% of extra alumina in the ore requires about 61 kg of extra fluxes. Maximum allowed alumina in the iron ore is 5% but it is brought down to 2.5±0.5 in lumpy feed to 3.5 to 4.5% in the fine ore by washing.

(b) Mixing of fluxes in the self fluxing sinter/pellets.

Impact Due to Synthetics and Substitutes

The synthetics and substitutes have made profound impact on the natural raw material. In this group we can discuss synthetic cryolite, synthetic quartz, synthetic diamond, reconstituted mica, synthetic mullite, synthetic corundum, formed coke and synthetic graphite, electrolytic manganese dioxide and use of silicon metal in place of germanium. Presently and in the coming years also the above material are going to play a dominant role in displacing the natural mineral in many fields of application. As stated before, the synthetic cryolite has already taken a dominant position in the place of natural cryolite in the aluminium metallurgy. It is estimated that nearly 90% of the world requirement of cryolite is met by synthetic cryolite produced from acid grade flourspar or utilising the by-product fluorine gases emanating from phosphate rocks during the manufacture of phosphatic fertilizer. Phosphate rock invariably contains 2.19 to 4.42% fluorine as a mineral calcium fluoride which forms very useful source for by-product recovery and can be used in the manufacture of fluorine compounds and chemicals like cryolite, sodium fluoride, etc. It has been estimated that a plant producing 474 tonnes of phosphoric acid by wet process a day can recover 10 tonnes of cryolite per day. In our country there are five units, two located in West Bengal, one each in Gujarat, Tamil Nadu and Kerala which have started utilising by-product fluorine although in a modest scale. About half a dozen more units have been licenced to recover fluorine and manufacture chemicals and compounds thereof. In addition to above production of cryolite utilising acid grade fluorspar by two units, viz., Navin Fluorine Industries, Surat and National Fluorine Corporation, Thana is already being made. One more licence has given under joint sector in Tamil Nadu.

17

National Mineral Policy

NATIONAL mineral policy is, in a sense, the attitude of a government towards development of its mineral resources. Broadly, the policy can be viewed from two angles, one domestic and the other foreign i.e. the attitude of a government towards foreign investment. Every country has its own problem and each tries to develop its known resources, make search for new ones and reduce dependence as far as possible. To achieve this objective some follow "close door" policy in which no foreign investor is allowed to participate except those of allies; other "open door" policy where foreign investment and participation is freely allowed, or have "restrictive" policy where investment and participation is allowed on minority share basis, if felt so expedient to do so. "Close door" policy is followed mainly by communist countries. This policy decision again varies from "have" countries to "have not" and industrially developed or less developed to underdeveloped countries. There are a number of countries which have advanced industrially but withoat having suffi- cient mineral resources; a mention may be made of the U.K., Japan and most of the European countries. Such countries look upon for supplies, from different and diversified sources just to maintain regular inflow of raw material even during political upheavals. The general attitude of the governments of those countries is to encour- age their own people acquiring mining rights whenever possible and to extend technical know-how in mining and get back ores in return. In our country itself, Japanese have financed for development of Bailadila hill ranges to ensure supplies of iron ore for their own country. Japan has also extended considerable technical know-how, capital equipment and finance in the iron ore mines of Goa. Such collaboration on mutual benefit has become a common feature of modern thinking. At the same time, collaboration in fabrication and metallurgy is viewed with suspicion and political motives. Such collaboration does not come forth easily unless a larger interest of the lending country is involved. It is difficult to setforth, point by point in print, a national mineral policy particularly concerning foreign investment. This attitude has necessarily to be flexible. In the case of newly independent countries, there is natural fear of foreign domination if such mineral concessions are freely granted to foreigners whereby control over the basic industries goes in the hands of a foreign investor. Coupled with this, is the anxiety to develop speedily the natural resources and so the developing countries, to which India belongs, follow a 'restrictive' policy.

The mineral policy, on the whole, is based on realistic knowledge of a country's mineral resources and its industrial status in relation to competing powers. A country's mineral resources are susceptible to progressive measurement upon new reserves added or discoveries made. The policy should, therefore, be subject to revision promptly upon new mineral evidences. In formulating mineral policy four axioms are universally recognised, these are: (i) minerals are depleting assets, (ii) the modern industrial economies are based on minerals and mineral fuels, (iii) the boundaries of countries bear little or no relation to mineral geography, and (iv) no nation has domestic resources sufficient for its industrial needs.

Keeping these factors as a base and as a guideline, further framework of policies is constructed in relation to: (i) appraisal of mineral resources, (ii) leasing system of mineral rights, (iii) mines safety, (iv) mineral taxation policy, (v) imposition of protective tariff and quota system, (vi) conservation and substitution, (vii) ore dressing and metallurgical innovation, and (viii) stockpiling programmes. A detailed discussion on these points and how they should be viewed have been made in foregoing chapters. The significant points which need mention here are discussed.

First of all, having a realistic picture of a country's resources is singularly important, for with it is interlinked an integrated policy relating to mining, setting up industries, export import and even barter deal that can be setforth. The whole policy towards mineral appraisal aims at enlarging reserves. To achieve this aim, the first and foremost task before any nation is to establish, organise and expand the geological units depending upon its needs, Besides geological reconnaissance which prelude all intensive prospecting to bring up promising areas into productive stage, the organisational set up should have such experienced staff as to undertake different stages of specialised work involved. Practically every country, big or small, has a geological survey unit. The oldest Geological Survey unit is probably of the U.K. which was established in 1835. The Geological Survey of India was established on 4th March in 1851. Thus our present knowledge of the country's mineral wealth is largely based on the classical work done by the Geological Survey of India over 125 years.

For minerals and metals in which the resources are inadequate or non-existent, there must be top consideration for assessing the situation from time to time, the sources of present supply, searching for new deposits and encouraging research into more economical methods of extraction and utilisation and even substitution. Intensive exploration, improved methods of mining, new processes of concentration, smelting and refining techniques have expanded mineable reserves, making it possible to develop new ore bodies, to utilise leaner grades, to win more metals from ores, to make a better quality product and to recover by-product elements. Likewise, the secondary metal industry has become a powerful factor in eliminating avoidable waste. Stockpiling programme of strategic and critical minerals should, as a matter of policy, be encouraged. As an instance, the Netherlands oil firms by law are required to hold stock of crude oil half of their annual refinery capacities. Another very important aspect of national mineral policy is the maintenance of an uptodate inventory of minerals resource available. Inventory should include the reserves established category-wise, viz., measured, indicated and inferred categories. Such categorisation provides a definite assurance to the country for planning and embarking upon further project and also to find out the gaps in the knowledge and the quantum of further work needed for meeting the present and immediate further requirements. The inventory should also indicate not only the reserves position but also the quality of minerals available, their suitability to different industrial use and also the beneficiation characteristics. It is a continuous process which must continue without any relaxation. The task is not so simple as it may appear to be. It requires keeping vigil over all the exploration activities proceeding on in the country, interpreting the results and reports and watch over new discoveries.

DOMESTIC MINERAL POLICY

The first guiding factor in the formulation of mines and minerals legislation; so far grant of mineral concession is concerned, was the Industrial Policy Resolution of 6th April, 1948, which included some of the minerals amongst industries requiring large investment and a high degree of technical skill and kept under purview of Central regulation and control. The Policy Resolution of 1948 was superseded by another set of Industrial Policy Resolution of 30th April, 1956. The revised Resolution was based on the socialistic pattern of the society which the Parliament had accepted in December, 1964. Under this, minerals have been put under three categories. The first category includes mining and processing of coal and lignite, mineral oil and gas, gypsum, sulphur, gold, diamond, copper, lead, zinc, tin, molybdenum, iron ore, manganese ore, chromite and tungsten ores, and minerals specified in the Atomic Energy (Control of Production and Use) Order, 1953. The second category contains all other minerals, excluding minor minerals. The third category consists of minor minerals. The resolution defines that the future development of minerals and their related industries of the first category, save wherein grant is already in the hand of private sector, will be the exclusive responsibility of the State. Minerals of second category will be progressively State-owned and only the State will, generally take initiative in establishing new undertaking but in which private enterprise is also expected to supplement the effort of the State. The minerals of third category have been left entirely for the initiative and enterprise of the private sector. To implement the policy underlined in the above Resolution, suitable provisions have been made under the Mines and Minerals (Regulation and Development) Act, 1957 and Coal Bearing Areas (Acquisition and Development) Act, 1957. Although the Resolution gave a bias in favour of the public sector, in practice, however, there is no dogmatic approach in the matter of granting licence or lease for minerals of the first two categories to any sector, as it is realised that too great a rigidity might defeat the purpose in view.

According to the Constitution, the mineral rights belong to State Governments. The rights were vested in the States by the Government of India Act in 1935. After Independence under Article 295 of the Constitution, all mineral rights previously vested in the rulers of princely States were deemed vested in reorganised States into which the former princely States have been incorporated or in the Central Government in the case of a Union Territory. But some private ownerships have remained for which the Mineral Concession Rules contain a chapter entitled "Grant of Mineral Concessions by Private Persons." The Central Government frames all rules for the grant of mineral concessions, development, conservation and safety and their enforcement is also effected by the Central Government organisations, except the grant of mineral concessions. Conservation and development other than coal are looked after by the Indian Bureau of Mines and safety by the Directorate of Mines Safety. The conservation aspect of coal is looked after by BCCL and CMAL. In the present framework of leasing system as governed by the Mineral Concession Rules, 1960, the State Governments act as a lessor. The State Governments are authorised under the Rules to grant leases directly for major minerals other than those given in the first schedule of Mines and Minerals (Regulation and Development) Act, for which prior approval of the Central Government is necessary (see Chapter 6). For minor minerals each State Government frames its own rules and it has the exclusive right for the grant of leases. As a lessor each State Government has a very onerous responsibility to perform in creating condition for attracting capital to this sector of industry. In developing countries like ours, a minimum amount of geological work necessarily at least be made available to entrepreneurs about the areas earmarked for leasing out. Along with it, the State Government may extend all facilities of roadways and transport, assist in acquiring land for mill, colony, school and hospital and also in getting railway sidings. It may, at times, be necessary for the State to physically participate in the venture, at the initial stage, if such assistance is sought and required.

Like most of the countries, except the U.S.A. and France, a tenure system is followed in India for grant of mineral rights. The tenure of lease for certain minerals is 20 years including petroleum and 30 years for others unless shorter period is asked for with a provision of renewal for equal period of the original term. In the U.S.A. and France the mineral rights pass to the party like surface estate. There are several controlling laws for development, conservation and safety. But the mining firms are so organised that hardly any necessity is felt for the Government to interfere in their activities. As a point of argument it is questioned which of the two systems, tenure or perpetuity, is better. The progressive thinking is, in the favour of tenure where a lessee is at his option to renew the lease for the period equal to original term, any number of times, if the statutory conditions have been complied. In the petroleum concession rules, however, the provision has been made for only one renewal equal to original term. It gives a statutory pressure on a lessee to work the area which has been acquired. At the same time, it provides a controlling power to the government to disallow renewal if the lessee is found wanting in development. Prior to 1949, most of the mineral concessions were just akin to perpetuity where individual leases were granted by princely States for a very large area for a period varying from 99 to 999 years but without having stipulated any condition for development. This disparity in leases has been largely modified under Mining Leases (Modification of Term) Rules, 1956, except lease granted in Goa during Portuguese regime.

Once the foreign capital admitted is treated at par with Indian capital for security and taxation, rather with little more latitude towards taxation. Remittance of profits and dividends is freely allowed subject to availability of foreign exchange. In event of any foreign enterprise being compulsorily acquired compensation is paid on a fair and equitable basis and the Government provides reasonable facilities for the remittances of proceeds.

18

Growth of Mineral Industry and the Economy

THE EARLY records of mining in India are traced back to 300 B.C. when Kautilya wrote "Mines are the source of treasure"—indicative of the fact that there existed a flourishing mining activity in those early times as well. Although there is a great hiatus in our knowledge of what happened even prior to middle of nineteenth century, evidences provide positive testimony to extensive mining activities in the past. The heap of slags obtained during iron and non-ferrous smeltings as found in several places in Rajasthan, Bihar and elsewhere are the mute testimony to this fact. India was then leading producer of diamond, gold, copper, lead, zinc and steel in the world. Till 1886, the entire diamond wealth of the world came from Panna in Madhya Pradesh and Wajra Karur in Anantapur district, Andhra Pradesh. The latter is quiscent since late eighteenth century but the former has survived through all vicissitudes of political upheavals and still forms the only field in the country where production is going on, although on a small scale. Authentic records of mineral production· in some details are available from the middle of nineteenth century when coal was the only mineral recorded to be produced. By 1900 seven minerals—coal, gold, manganese ore, mica, salt, salt petre and petroleum—were under regular production. A mention has been made about the mining of agate in Combay, rock crystal in Tanjore (Thanjavur) and beryl in Nellore prior to 1900. By 1906, chromite, magnesite and iron ore (hematite) were added and also a regular production of diamond was restarted. Production of monazite started in 1911. The two world wars gave good impetus to mining activities. A number of new minerals like ilmenite, rutile, limestone, bauxite, barytes, kyanite, sillimanite, copper ore, lead-zinc-silver ores, etc., were added to the list of minerals worked in India. Between the period 1934 and 1946, twenty-two minerals including coal and petroleum were worked. During 1948, the number of minerals worked rose to 30 excluding petroleum and atomic energy minerals. India now mines some 53 useful minerals and ores. The post-Independence era, especially from 1954 saw a great spurt in the activities of mineral exploration, mining, expansion and setting up of new industries. New organisations were established by the Central Government for the exploration and exploitation of minerals. The geology and mining departments of each State Governments were reorganised to share increasing responsibility effectively. Subsequently, the State Governments opened their own Corporations for working some of the minerals. The Indian Bureau of Mines was the first Central agency established on 1st March 1948, primarily for advising government on mineral policy matters; later in 1950 and 1953 its activities were enlarged to mineral exploration. With a policy decision coal exploration activity of the Bureau was transferred on 20.3.64 to the National Coal Development Corporation created in 1956 as a successor company to the State Collieries. The remaining exploratory wing for metalliferous minerals was transferred to the Geological Survey of India on 1.1.66. The Department of Atomic Energy was created in August 1954 to look after the interests of atomic energy minerals. The National Mineral Development Corporation was created in November 1958 for mining metallic deposits. A separate Oil and Natural Gas Commission was set up in 1956, for exploration of oil and gas; later on its activities were extended to exploitation of oil and gas also. During April 1958, Oil India Ltd., a joint enterprise of Assam Oil Company and the Government of India on 50:50 basis was formed for exploration and exploitation of oil in the areas of Moran, Noharkatiya and Hugrijan leased out to Assam Oil Company. Pyrites and Chemical Development Corporation was set up on 22nd March 1960 for mining Amjhore pyrites. The activity of this Corporation was later enlarged and the name changed to Phosphate, Pyrites and Chemicals Ltd. in 1968. In April, 1956, the State Trading Corporation was established for trading in minerals and other commodities, later on October, 1963, trade in minerals and metal was entrusted to a newly established sister concern, namely the Minerals and Metals Trading Corporation. MMTC has an exclusive right to export iron and

TABLE 18

Value of Mineral Production in India by Groups, 1960-78

(in Rs. million)

Year	Fuels (Coal and lignite crude petroleum and natural gas)	Metallic minerals	Non-metallic minerals	Total
1960	1,115	326	210	1,651
1961	1,210	311	291	1,812
1962	1,419	326	383	2,128
1963	1,690	307	420	2,417
1964	1,701	344	479	2,524
1965	1,960	382	563	2,905
1966	2,253	419	607	3,279
1967	2,605	466	628	3,699
1968	2,985	528	652	4,165
1969	3,392	542	731	4,665
1970	3,487	620	781	4,888
1971	3,476	689	869	5,034
1972	3,692	739	965	5,396
1973	3,807	871	1,043	5,721
1974	6,153	1,121	1,343	8,167
1975	7,904	1,337	1,368	10,609
1976	9,677	1,778	1,725	13,180
1977	9,689	1,768	1,720	13,177
1978	9,728	1,711	1,700	13,189
1979	14,186	2,235	1,463	17,884
1980	17,146	2,518	1,533	21,197
1981	30,227	2,968	1,788	34,983
1982	45,536	3,645	2,100	50,290

TABLE 19

Average Number of Mines Reporting Production and Average Daily Labour Employed during 1977–1982

| | 1977 | | 1978 | | 1979 | | 1980 | | 1981 | | 1982 | |
| | No. of mines | Labour employed | No. of mines | Labour employed | No. of mines | Labour employed | No. of mines | Labour employed | No. of mines | Labour employed | No. of mines | Labour employed |
1	2	3	4	5	6	7	8	9	10	11	12	13
Coal	524	497,346	505	496,610	498	498,929	484	498,872	496	513,358	516	530,216
Bauxite	42	3,815	52	4,509	49	4,269	54	4,832	48	4,317	54	4,686
Chinaclay	162	6,867	148	6,618	142	7,047	149	8,113	111	6,870	159	7,781
Chromite	18	5,485	17	4,829	16	4,512	18	5,647	15	5,345	17	5,394
Gypsum	55	2,464	65	2,652	61	2,231	47	2,029	28	1,116	35	1,377
Limestone	247	48,731	243	47,682	270	51,439	252	49,752	262	49,770	296	52,786
Magnesite	10	7,156	12	7,396	12	6,839	12	6,684	13	7,010	13	7,115
Iron Ore	224	53,155	224	51,518	213	46,205	199	43,878	205	44,933	187	47,244
Manganese Ore	176	27,830	158	26,796	162	28,401	162	26,842	155	26,534	164	28,186

TABLE 19 (*Contd.*)

1	2	3	4	5	6	7	8	9	10	11	12	13
Mica	201	7,366	175	6,848	162	5,911	176	6,662	193	6,735	181	6,599
Steatite	118	4,049	111	4,022	106	3,932	109	3,959	107	3,966	113	3,984
Total of India (including Other minerals, Oil and Stones)	2,529	747,836	2,463	741,998	2,456	742,128	2,358	749,680	2,272	749,803	2,422	779,551

SOURCE: Statistics of Mines in India, Vol. II, Non-Coal, 1977–82, issued by the Directorate-General of Mines Safety, Dhanbad.

manganese ores, except those of Goa origin. The export trade of mica except that of fabricated and powdered mica was canalised through MMTC w.e.f. 24.1.72. A subsidiary company viz. MITCO was floated by MMTC to handle export trade of mica which was registered on 28th June, 1973. But it started functioning from 1.6.75 only. The Indian Oil Ltd. was formed for the distribution of petroleum and its products. Separately, the Indian Refineries (P) Ltd. was incorporated on August 20, 1958 for refining crude oil produced, from oil's field in Assam. These two firms have now been incorporated into the Indian Oil Corporation Ltd. The Neyveli Lignite Corporation was formed in 1956 for mining and utilisation of lignite deposits of South Arcot district, TN. The project was inaugurated on 20th May and later taken over by Neyveli Lignite Corporation in December 1958. The Kolar Gold field mines previously worked by M/s John Taylor and Sons were nationalised by the Mysore Government on 21.11.56 and subsequently on 1st December, 1962, handed over to the Ministry of Finance. Government of India. This was the first nationalisation move in post-Independence era, in keeping with the Industrial Policy Resolution of 1956. In January 1966, the lead-zinc-silver ores properties of Metal Corporation of India in Zawar, Udaipur district, Rajasthan, together with its Tundoo lead-smelter plant in Bihar were nationalised and taken over by the Hindustan Zinc Ltd. On 9th November, 1967, a separate organisation was established, namely, Hindustan Copper Private Ltd., which took over copper mining projects from the National Mineral Development Corporation. On 4th October, 1967, the Uranium Corporation of India Ltd. was formed which took over uranium mill and Jaduguda mine project in Bihar. By an ordinance issued on 16th October 1971, all the coking coal mines, except those captive mines held by steel plants and National Coal Development Corporation (NCDC), numbering 214 mines and 12 coke oven plants were nationalised. The working of these nationalised including NCDC's mines was handed over to newly created Bharat Coking Coal Private Limited Company (BCCL), which came into existence on 1st January. Non-coking coal mines were taken over on 30th January, 1973 and workings were handed over to the Coal mines Authority Limited (CMAL). The nationalisation of coal industry was ratified in two phases: (i) coking coal mines under Coking Coal Mines (Nationalisation) Act, 1972, and (ii) non-coking coal mines under Coal Mines (Nationalisation) Act, 1973. After nationalisation, the coal mines except the captive mines of TISCO & IISCO and mines managed by the State Governments were placed under the management of three public sector undertakings viz. (1) Coal Mines Authority Ltd. (CMAL), (2) Bharat Coking Coal Ltd. (BCCL), and (3) Singareni Collieries Co. Ltd. (SCOL). Later from 1st November, 1975 the management of CMAL and BCCL was taken over by a holding company Coal India Ltd (CIL). CMAL was abolished and its functions and responsibilities were transferred to 5 subsidiary companies under CIL viz. (1) Eastern Coalfields Ltd. (ECL), (2) Cental Coalfields Ltd. (CCL), (3) Western Coalfields Ltd. (WCL), (4) Bharat Coking Coal Ltd. (BCCL), and (5) Central Mine Planning & Design Institute (CMPDI).

In order to expedite exploration activities, the Government of India created a separate company, namely, Mineral Exploration Corporation Ltd. (MEC) on 2nd October, 1972. As a result the exploratory wing was again separated out from the Geological Survey of India (GSI) and transferred to MEC. Towards the end of 1972 the Ground Water Wing of the GSI was also taken out forming a separate Central Ground Water Board (CGWB) under the Ministry of Agriculture and Irrigation. The merger of Goa with the Indian Union on 19th December, 1961, brought large assets of iron and manganese ores and bauxite closest to the sea coast. There has been phenomenal growth and progress in private sector as well, especially in the expansion of aluminium, cement, steel, ferro alloys, fertilizers and refractory industries. Developments that have taken place in the fields of mining and metallurgy are so numerous and diversified that it would not only be difficult but also unjustifiable to detail them in one chapter. A summarised account of the important developments made in the field is, however, provided here to give a realistic picture of the growth.

COAL

Coal is an important energy mineral. Fortunately India has abundant coal reserves. The Indian reserves of coal and lignite up to 600 m depth are placed at 106,857 million tonnes and 3,024 million tonnes respectively by the Department of Coal, Ministry of Energy. Anticipated world reserves are 6712,501 million tonnes for coal and 2011,401 million tonnes for lignite.

TABLE 20

Break-up Figures of Indian Coal Reserves

in million tonnes

	Total reserves	Measured	Indicated	Inferred
Coking Coal				
Prime coking	5,147	2,693	1,753	701
Medium coking coal	6,005	3,610	2,241	154
Semi to weakly coking coal	6,255	2,610	2,465	180
Total coking coal	16,407	8,913	6,459	1,035
Total non-coking coal	89,564	26,831	35,845	26,888
Total of bituminous coal (Gondwana)	105,971	35,744	42,304	27,923
Tertiary coal	886	82	253	551
Grand Total	106,857	35,826	42,557	28,474

The earliest records of coal mining activities in India is available for during 1774, then during 1815-16 near Chinakuri, Dalamia, etc. in West Bengal. Production of coal caught momentum after laying of railway line connecting Raniganj and Calcutta in 1855. Thereafter extensions of the railway lines gave a great fillip of coal mining industry. In recent times a new orientation to the coal industry was given by setting up of Coal Board (now abolished) during 1952 for planned development, conservation and distribution of coal; function of which has now taken over by a separate committee.

After nationalisation of the coal industry, CIL with headquarters at Calcutta became the major Public Sector Unit looking after the development of coal industry in addition to SCCL and Neyveli Lignite Corporation. CIL has an authorised capital of Rs. 250 crores. Its five subsidiary companies viz., BCCL, ECL, CCL, WCL and CMPDI are having their own authorised capitals. CIL is run by Board of Directors. The activities of CIL are spread over the States of Assam, West Bengal, Orissa, Bihar, Madhya Pradesh, Maharashtra and Uttar Pradesh. The interest of coal in Andhra Pradesh is entirely looked after by SCCL which is a separate company altogether outside the control of CIL having joint interest between the Government of Andhra Pradesh and Government of India. The equity ratio between the two is roughly 54 : 46. Although CIL is mainly responsible for the formulation of programmes and targets, laying down guidelines and monitoring and coordinating the performances, the subsidiary companies are autonomous in character.

CMPDI is largely a research organisation for coordinating science and technology programmes of other institutions like CFRI and CMRI, Dhanbad, RRL, Hyderabad and Jorhat including its own. It also takes up coal exploration work for CIL and engage itself in designing and mine planning.

BCCL with headquarters at Dhanbad is the major coking coal producing company and runs five washeries, three by-product coke oven plants and a number of bee-hive coke plants. It has 84 coal mining units in its control.

ECL with headquarters at Sanctria, West Bengal covers Raniganj coal fields West Bengal, Mugma and Rajmahal coal fields, Bihar and those of Assam. It has 88 mines. The company produces most of the superior grade non-coking coal and soft coke.

CCL with headquarters at Ranchi covers Bokaro, Ramgarh, Giridih and North and South Karanpura coal fields in Bihar, the Talcher coal fields in Orissa and Singrauli coal fields in Madhya Pradesh and Uttar Pradesh. The company has 57 coal mines, five coal washeries and one by-product coal plant.

WCL with headquarters at Nagpur covers all the coal fields of Maharashtra, Madhya Pradesh (except Singrauli) and Ib river of Orissa. There are 65 coal mines and one coal washery under its control. The Parasia coal fields failing under this company also produces semi coking coals. NLC which is outside the control of CIL is an integrated project of Government of India with authorised capital of 220 crores. The project consists of an open cast lignite mine in South Arcot, Tamil Nadu with ultimate capacity of 6.5 million tonnes per year, 600 mw power plant, a fertiliser plant with rated capacity of 152,000 tonnes of urea per year, briquetting and carbonising plant to produce 327,000 tonnes of carbonised briquettes per year and china clay washing plant of 5000 tonnes capacity of washed clay per annum.

The conservation and development of coal mines which was earlier looked after by Coal Board is now looked after by Coal Conservation and Development Advisory Committee set up under the Coal Mines (Conservation and Development) Act, 1974 replacing earlier Coal Mines (Conservation and Development) Act, 1952 under which Coal Board was formed. With the promulgation of 1974 Act the Coal Board was abolished since most of the supervisory and regulatory functions required to be carried out when the coal industry was largely under private sector, became redundant after the nationalisation of coal mines.

A Fuel Policy Committee was set up which in its report in 1974 laid great stress on coal as a primary source of commercial energy and projected requirements by 1990-91 at 339 million tonnes. Later in September, 1977, a Working Group on coal appointed by the Planning Commission for the formulation of policies and programmes during 1982-83 made a little downward demand profile of coal as follows:

	Fuel policy committee in million tonnes	Working group on coal in million tonnes
1978–79	135	112
1979–80	—	119
1980–81	—	130
1981–82	—	139
1982–83	—	153
1983–84	201	—
1990–91	339	—

Nevertheless coal has been regarded as a primary source of commercial energy for the next 15 years in addition to petroleum, hydro electricity and nuclear energy. Wind power, tidal waves and solar energy still remain to be harnessed. India produces now well over 102 million tonnes of coal with targets of further increasing production every year.

Commensurate with the increase in coal production and owing to the fact that Indian coals contain appreciable quantity of ash, a detrimental constituent, there has been good growth in' setting up of coal washeries. Prior to 1952 there was only four coal washeries two of which was owned by TISCO and one each by IISCO and ACC. Ten more washeries were added by the then NCDC, Hindustan Steel Ltd. and Government of West Bengal. All the washeries except those belonging to TISCO IISCO and Government of West Bengal are now held by CIL. Out of the 14 washeries the only one at Nowrozabad working under CIL wash non-coking coal. Details of the washeries till the end of 1977 are given in Table 21.

There are altogether 113 coal fields in India excluding lignite fields covering 89,600 sq km in the States of Assam, Nagaland, West Bengal, Orissa, Bihar, Madhya Pradesh, Maharashtra and Andhra Pradesh. Metallurgical coal is however found in Jharia, Giridih, Karanpura and Bokaro coal fields of Bihar and Raniganj coal fields of West Bengal only. Occurrence of metallurgical grade coal is confined to Barakar formations only. Raniganj formation yield only semi-coking coal. Bihar has the largest number of coal mines.

TABLE 21

Coal Washeries in Operation

(In miilion tonnes)

Sl. No.	Washery	Rated capacity per annum	Clean coal production	
			1980-81	1981-1982
1	2	3	4	5
A	BCCL			
1.	Dugda–I	2.40	0,87	0.89
2.	Dugda–II	2.40	0.92	0.85

TABLE 21   (*Contd.*)

1	2	3	4	5
3.	Bhojudih	2 00	1.39	1.24
4.	Patherdih	2.00	0.95	0.85
5.	Lodna	0.41	0.15	0.17
6.	Sudamdih	N.A.	0.20	0.44
7.	Barora–I	N.A.	—	—
B.	*CCL*			
8.	Kargali	2.74	1.46	1.40
9.	Kathara	3.00	1.07	1.03
10.	Sawang	0.75	0.46	0.42
11.	Gidi	2.84	0.82	0.82
C.	*WCL*			
12.	Nowrozabad	0.47	0.24	0.19
D.	*TISCO*			
13.	Jamadoba	1.34	0.97	0.94
14.	West Bokaro	0.71	0.40	0.31
E.	*SAIL*			
15.	Chasnala (IISCO)	1.20	0.97	0.91
16.	Durgapur (DSP)	1.50	0.69	0.60
17.	Durgapur (DPL)	1.50	N.A.	N.A.

SOURCE: Department of Coal, Annual Report, 1982–83.

Iron and steel: Metallurgy of iron and steel was known to our ancients in medieval time. It is said that the famous 'Damascus' swords were made of iron imported from India and there was a regular trade in iron and steel materials between India on the one hand, and Egypt, East Africa, Arabia, Persia and Near East on the other. In between the medieval time and late nineteenth century, the historical account of iron and steel production is obscure rather practically absent. Account of iron ore production, which is very much linked with iron and steel metallurgy, is somewhat available only from early this century.

The oldest steel plant which has survived through all the vicissitudes till recent times is the Indian Iron and Steel Co. Ltd. (IISCO). This company was reported to be formed sometime between 1830 by one Englishman Joshia Marshall Heath of East India Co. IISCO in 1936 amalgamated the Bengal Iron and Steel Co. Ltd. and in 1937 the Steel Corporation of Bengal forming an integrated Iron and Steel Co. under the managing agency of M/S Martin Burn and Co. IISCO was taken over by the Government on 14th July, 1972. The second oldest integrated company is the Tata Iron and Steel Co. (TISCO) which was established on 26th August, 1907. The first production of pig iron from this company was reported in 1911 and steel in 1912. The third integrated iron and steel plant namely Mysore Iron and Steel Works now renamed as Visvesaraya Iron and Steel Ltd. (VISL) was formed in 1915. Its production began in 1923. The VISL has now been completely geared to produce alloy and special steel. This conversion was completed by February, 1970.

During the end of First Five Year Plan in 1955 a great stress was given for expansion and modernisation of the steel plants and setting up of new units. As a result 3 integrated steel plants viz., Rourkela, Bhilai and Durgapur steel plants were set up under the public sector between 1957 and 1959, initially with a capacity of one million tonnes each, but later on expanded. Execution work of the 3 units viz., Rourkela, Bhilai and Durgapur was organised by Hindustan Steel Ltd. now taken over by Steel Authority of India Ltd. Rourkela was established by technical aid provided by Krupp-Demag of West Germany, Bhilai by Government of the U.S S R. and Durgapur by a consortium of 16 British firms formed in the name of Indian Steel Construction Co. A fourth unit under the public sector was added at Bokaro, Dhanbad district, Bihar with the aid of the U.S.S.R. during 1978. An agreement for setting up Bokaro plant was signed by the Government of India and the U.S.S.R. in January, 1965 for an integrated plant with an initial capacity of 1.7 million tonnes with provision for expansion to 4 million tonnes in its second phase. The foundation of first blast furnace for Bokaro was laid on 6th April, 1968 and completed on 3rd October, 1972.

Bokaro completed the first phase of production of 1.7 million tonnes of steel ingots on 25th February, 1978.

Presently India has got 6 integrated steel plants as per the details given below besides 2 alloy steel plants viz., VISL, Bhadravati and Alloy Steels Plant, Durgapur. The integrated steel plant are:

Name	Capacity
1. Indian Iron and Steel Co. Ltd.	1 million tonnes
2. Tata Iron and Steel Co. Ltd.	2 million tonnes
3. Rourkela Steel Plant	1.8 million tonnes
4. Bhilai Steel Plant	2.5 million tonnes
5. Durgapur Steel Plant	1.6 million tonnes
6. Bokaro Steel Ltd.	1.7 million tonnes

Alloy and Special Steel

1. Visvesvaraya Iron and Steel Ltd.	77,000 tonnes
2. Alloy Steels Plant	1,00,000 tonnes

In addition there are more than a dozen alloy and special steel plants in the country under the private sector viz., Mahindra Ugins Steel Co. Ltd., Guest Keen Williams Ltd., Mukund Iron and Steel Works Ltd., Canara Workshop, Firth India Steel Ltd., JK Iron and Steel Oo., Tex and Tool Co., Upper India Steel, etc. Till the end of 1976, 190 licences for mini-steel plants or letters of intent were issued for a capacity of 4.1 million tonnes of steel. Of these 87 units are reported to have gone in production with a total installed capacity of 2.1 million tonnes.

Of all the 6 integrated steel units Rourkela and Bokaro adopt LD (Linz-Donawitz) process of steel making. LD process is most significant development in the steel making during the last one and half decade. The process was first experimented in November, 1962 by the VOEST of Austria. Since then this process has been fastly replacing the conventional open hearth furnace all the world over.

The process involves blowing of oxygen with jet speed on the top of the convertor. The reaction takes place at a temperature of 2500°C. By this process it is now possible to produce steel at the rate of one tonne per minute which normally takes 6-8 hours by conventional method. The earlier disadvantages faced by this process in producing high carbon steel are got over. It is now possible to produce high carbon steel by controlling the reaction in the convertor suitably. VISL also used L.D. convertors besides electric furnace. Following the description of main plants of the integrated iron and steel units including VISL and ASP.

IISCO has 4 blast furnaces, 2 each with 700 tonnes and 1200 capacity per day, 3 Bessemer Convertors each with 25 tonnes capacity, one O.H. furnace with 300 tonnes capacity (Tilter), five O.H. each with 200 tonnes capacity (Tilter) and one O.H. furnace of 100 tonnes capacity (fixed). It has 5 batteries containing 308 coke ovens with daily throughput of 6235 tonnes.

TISCO follows conventional method of steel production i.e. open hearth. It was expanded to produce 2 million tonnes of steel ingot in 1959. Prior to expansion it had five blast furnaces with daily capacity of 4,200 gross tonnes. During the course of expansion two more blast furnaces F and E were added. Blast furnace F with a rated capacity of 1,650 tonnes per day provides additional hot metal sufficient for the two million tonnes. The blast furnace E is kept as standby status. It is one of the largest and most modern furnaces in the world designed for high top pressure operation and the use of sinter in the burden. The main dimension of furnace F is as follows:

Hearth diameter	28 ft (8.4 metres)
Bosh	31 ft (9.3 metres)
Stock line	22 ft (6.6 metres)
Big Bell	16 ft (4.8 metres)
Height of furnace from centre of the notch to top of the hopper	105 ft (31.5 metres)
Height from the ground	236 ft (70 8 metres)

In VISL, iron ore is smelted by electric furnace because of lack of coking coal in the vicinity. Charcoal is used to serve as a reducing

VISL has one charcoal blast furnace of capacity 80 tonnes a day, four electric pig iron furnaces two with 100 tonnes and 200 tonnes capacities respectively, two 25 tonnes each basic O.H., five electric arc furnaces of varying capacities (20 tonnes; 6 tonnes, $3\frac{1}{2}$ tonnes, $\frac{1}{2}$ tonnes and 56 lbs) and two L.D. converters each of 12 tonnes per blow. In 1962, while VISL was to expand its capacity of mild steel production to 100,000 tonnes from the existing 45,000 tonnes it was decided to convert the mild steel production to that of alloy and special steels by adding ancillary equipments. M/s Bohlets of Austria provided the technical know-how of the production, the technical agreement of which was entered in November 1963. Detailed specification of the plant equipment for the conversion scheme were drawn up by M/s Demag, AEG and OFU of West Germany. Under the conversion scheme the following equipment have been installed.

(1) Facilities for steel melting.
(2) Ingot annealing and conditioning.
(3) Extension to existing blooming and heavy section mill.
(4) Central annealing and heat treatment shop.
(5) Centralised roll turning shop.
(6) Combined section bar and roll mill.
(7) Metallurgical and chemical testing laboratory.

Rourkela unit had initially three blast furnaces, each of 1000 tonnes a day capacity; three L.D. converters each of 40 tonnes capacity, four open hearth furnaces each with 80 tonnes capacity designed for firing with coke oven gas and crude tar and three batteries of 210 ovens each. Under the expansion Rourkela added one blast furnace of 1500 tonnes capacity, two L.D. converters of 60 tonnes each and one battery of 80 ovens. Rourkela after expansion has introduced for the first time machinery for continuous rolling of sheets and two galvanising units. The first blast furnace at Rourkela was inaugurated on 3rd February 1959. The first open hearth furnace was lighted on 16th April and the first tapping of steel was done on 29th April, 1959. The second furnace was commissioned in August, 1959. The plant went into full production in 1961. Attached with this unit is a lime and dolomite calcining plant which operates completely on automatic devices. It was erected by Friedrich Siemens of Dusseldori, West Germany. L.D. process consumes more of basic than acid bricks for lining purpose. Rourkela specialises in production of flat products like plates, sheets, strips, hot and cold rolled tin plates, etc.

The agreement for setting up Bhilai Steel Plant was signed between the Governments of India and the U.S.S.R. on 2nd February, 1955 for one million tonnes of steel plant. The first blast furnace was inaugurated on 4th February, 1959; second on 24th December, 1959 and third on 27th August 1960. An agreement to expand the capacity to 2.5 million tonnes was signed on 12th September 1959. Initially the plant consisted of 3 blast furnaces of 1149 tonnes daily capacity each, 6 open hearth furnaces of 250 tonnes each, three batteries of 195 ovens with 4520 tonnes throughput of coal per day. Under the expansion programme Bhilai has added 3 blast furnaces 1,738 tonnes a day capacity each; four open hearth furnaces of 500 tonnes each and one 250 tonnes open hearth furnace was rebuilt as 500 tonnes furnace and three batteries of 195 coke ovens of same capacity as that with initial stage. Bhilai produces mainly rails and sleepers, bars, beams, channels, angle and other light and heavy structural sections. Bhilai achieved expansion capacity to 2.5 million tonnes of steel ingot capacity in 1966. Another agreement was made in June 1972 to expand its capacity to 4 million tonnes instead of 3.5 million tonnes earlier envisaged.

The Durgapur plant began production on 29th December, 1959 when the first furnace was commissioned. The plant reached the rated capacity of one million tonnes in February, 1963. The main plant initially consisted of three furnaces of 1,270 tonnes a day capacity each, eight open hearth furnaces of 200 tonnes capacity each except one with 100 tonnes capacity only, and three coke oven batteries with 5,280 tonnes a day throughput of coal. It has added during expansion one blast furnace of 1,500 tonnes capacity; one open hearth furnace of 220 tonnes capacity. The capacity of each seven O.H. furnaces installed have been modified to treat 220 tonnes and that of one O.H. to 110 tonnes a day. A skelp mill is another important addition to the rolling mills for making pipes and tubes. Durgapur specialises in the manufacture of wheels, axles, forgings and sleeper bars and blooms.

Bokaro mainly produces flat products. It has three blast furnaces each with daily capacity of 2,640 tonnes of basic iron. The size of blast furnace installed at Bokaro is the largest in India with useful volume of 2,000 cubic metres. The converters of shop No. 1 has been designed to make 2.5 million tonnes ingot per year, which have five 100 tonne converters. Out of these four are installed for the first stage of production of 1.7 million tonnes. Each converter has a capacity to produce 0.85 million tonnes steel per annum with 60 minutes tap to tap time. At a time two converters remain in operation to produce the required quantity and the remaining converters are kept standby or down for repairs and relining. The fifth converter when set up will raise the capacity to 2.5 million tonnes. During the second stage for producing 4 million tonnes of steel, it will add two more blast furnaces of the same capacity and two L.D. converters of 250 tonnes capacity each. Output of one such converter operating continuously with 80 minutes tap to tap time is rated at 1.5 million tonnes per annum. The coke oven built at Bokaro is the tallest built so far in India. During the first stage four batteries each having 69 ovens with facility for coke and coal handling and with complete by-products recovering units have been set up. During the second stage three more batteries with similar facilities will be added. The salient feature of coke oven are:

Length of coke oven	15.040 mm
Height of coke oven	5,000 mm
Width (average) of coke oven	410 mm
on the pusher side	390 mm
on the coke side	430 mm
Useful volume of oven chamber	27.3 m³
Coal charge for one oven	
(on dry basis)	20.7 t
Coking time	16.9 hours

The most added feature of Bokaro steel plant is that its blast furnace has been designed to use more iron ore fines in the form of sinters to the extent of 75% of the burden. It will be operating at a temperature of 1100°C unlike other blast furnaces in the country which are operating normally at a temperature of 900°C.

Alloy Steels Plant, Durgapur has 2 steel melting shops, the shop No. I contains 250 tonnes swing-roof, top charge electric arc furnace and the shop No. II contains one 10 tonnes electric arc furnace together with 2 high frequency induction furnace of 2 tonnes and ½ tonnes capacity. In addition the plant has got a blooming sheet bar, billet and bar mills and a forge shop.

In addition the Government has decided to set up three new steel plants at Vizag, Vijaynagar (Karnataka) and Salem. Per capita consumption of steel in India as estimated in 1968 is very low at 18 kg when compared to several advanced countries viz. Sweden 682 kg; U.S.A. 656 kg; Japan 620 kg; West Germany 540 kg; Canada 526 kg; Czechoslovakia 524 kg; U.K. 424 kg; U.S.S.R. 335 kg; and Belgium-Luxemburg 330 kg. The position of India in 1978 remains the same as that of 1968.

Development of new deposits of iron ore witnessed a great spurt synchronising with the expansion of steel industry and the keen interest show by the Japanese Steel Mills Association for importing iron ore from India in their own country on a long term basis. It marked the development in proving operations of mineral deposits started by the Indian Bureau of Mines towards the end of 1954. Iron ore deposits of Taldih and Dandrahar were blocked out in Sundergarh district, Orissa for feeding Rourkela Steel Plant. A total of 94.28 million tonnes were proved in those two blocks. Barasua mine in Taldih block was opened up by HSL with average annual production capacity of two million tonnes which supplies ore to Rourkela. Thereafter in 1955, proving operations were carried out in Rajhara Pahar in Durg district for feeding Bhilai Steel Plant. In this area, the reserves of 87.6 million tonnes of iron ore were blocked out. Bhilai meets its supply presently from this area. Bolani deposit in Ponposh gorge are, Keonjhar district, Orissa, was developed by Bolani Ores Ltd. for feeding Durgapur Steel Plant. Since then, between July 1958 and 1962, Kiriburu (reserves 176 m tonnes) and Sasangda (133.50 m tonnes) deposits situated bordering Bihar and Orissa, and Daitari (50 m tonnes) in Keonjhar district, Orissa were proved. Kiriburu was developed mainly for export purposes at the rate of nearly two million tonnes of ore per annum through Visakhapatnam port. Additional production from Kiriburu and Sasangda deposits feed Bokaro steel plant. In 1961, Bailadila hill ranges, Bastar district, Madhya Pradesh and in 1964 Chikmagalur and Bellary deposits in Karnataka were taken up for detailed proving. The Bailadila deposits have been developed under Indo-Japanese collaboration for exports of graded iron ore to Japan. The National Coal Development Corporation started regular mining of No. 14 deposit of Bailadila range from September, 1967. Mines has been planned to produce four million tonnes of graded ore per year. Later, the NMDC has also started developing No. 5 deposit for mining additional four million tonnes of graded ore for exports to Japan. Deposits Nos. 4 and 11C are also under development mainly for supplying to proposed steel plant at Vizag.

TABLE 22

Production of Steel Ingots in India During 1947-1982

(In '000 tonnes)

Year	Production
1947	1,276
1948	1,276
1949	1,373
1950	1,459
1951	1,514
1952	1,603
1953	1,532
1954	1,712
1955	1,731
1956	1,765
1957	1,738
1958	1,842
1959	2,468
1960	3,286
1961	4,084
1962	5,158
1963	5,969
1964	6,033
1965	6,413
1966	6,608
1967	6,387
1968	6,448
1969	6,475
1970	6,232
1971	6,322
1972	6,842
1973	6,882
1974	6,820
1975	7,082
1976	9,403

TABLE 22 *(Contd.)*

Year	Production
1977	9,852
1978	8,301
1979	8,162
1980	7,501
1981	8,509
1982	8,670

Kudremukh quartz-magnetic deposit of Chikmagalur and Kumaraswamy and and Donimalai blocks of Bellary-Hospet are under development. In Kudremukh the deposit which is to be mined is the Malleshwara deposit situated in Mudigire Taluk. This deposit is situated about 13 km NNE of Kudremukh, the highest peak in the region which is 1883 metres above msl. The deposit extends for a strike length of 5 km with a reserve of 1000 million tonnes. Kudremukh will meet the export commitment of supplying 150 million tonnes of concentrates to the National Iranian Steel Company over a period of 20 years at the rate of 7.5 million tonnes per year. The Government of Iran is to finance to extent of 630 million dollars for this project.

The shipment of magnetite concentrates to Iran will commence from September, 1980. The construction work of the concentration plant is more or less complete. Laying of the pipeline for carrying magnetite concentrates in the slurry from Mangalore end to the concentration plant has made a good progress and expected to be completed well before September, 1980. The development of Donimalai and Kumaraswamy deposits in Bellary district made a good progress. In fact the Donimalai deposit was opened in October, 1977 and is expected to produce 4 million tonnes of r.o.m. per year for export. The opening of Kumaraswamy deposit is linked with proposed Vijaynagar steel plant. In Donimalai north and south blocks reserves of 95.5 million tonnes containing 65.35% Fe have been proved and in the Kumaraswamy block reserves of the order of 127 million tonnes of grade 64% Fe have been proved. In IISCO property at

Chiria in Singhbhum district of Bihar the Mineral Exploration Corporation has established one of the largest single reserves of the iron ore in the country measuring about 2000 million tonnes. The Geological Survey of India, on the other hand, has made considerable improvement in reserve position of Rowghat iron ore deposit in Bastar district for meeting the expanded need of the Bhilai Steel Plant.

Since 1959, Indian steel mills started making a good progress in the use of sinter prepared in their own plants. Till the end of 1978 about 7.6 million tonnes of sintering capacity was available with the steel mills. TISCO also uses pellets. TISCO has its own pellets plant of one million capacity at Noamundi. Another pelletisation plant, Mandovi Private Ltd., under joint sector located at Shitoda at the bank of river Zuari in Goa with 1.8 million tonnes capacity went into production towards the close of 1978. Voest-Alpine, Austria was the main supplier of the plant and equipment for the project, In addition to these two, there already exists one pelletisation plant owned by Chowgule and Co. having about half a million tonnes capacity per annum. Two more pelletisation plants with 2 million tonnes capacity each one at Donimalai and other at Bailadila are in the offing.

For the purpose of planning, development, regulation and conservation of iron ore reserves, Government of India has constituted an Iron Ore Board which was registered unaer the Societies Registration Act, 1960 on 23rd January, 1973.

Ferro-alloys: Prior to Second World War the manufacture of ferro-alloys and special steel was practically unknown in India. During the war time, efforts were made by the TISCO and Ordnance factories to manufacture special steels which had to be discontinued after the war was over as the production cost was too high and even substantial tariff protection could not be of much help. The production of ferro-manganese, the chief ingredient for the production of steel, was first started during 1917 to 1919 by the Bengal Iron and Steel Co. Ltd., now amalgamated with the Iron and Steel Co. Ltd. Production was stopped by this firm but resumed in 1928. Tata started production in 1919 and continued intermittently in small quantity limited to its own use. Later IISCO and VISL also produced intermittently small quantities of ferro-manganese for their own use. Till then, ferro-manganese was produced in the blast furnace which was not of standard quality and contained high phosphorus about 0.6%. Urgent need was felt for the production of standard grade ferro-manganese in electric furnaces during the Second Plan period. Consequently seven units were set up between 1956 and 1961, with total installed capacity of 183,600 tonnes; two each in Maharashtra, Karnataka and Orissa, and one in Andhra Pradesh. The eighth unit was set up again in Maharashtra in 1977 with 50,000 tonnes annual capacity. The ferro-manganese produced in the electric furnaces contains on an average, Mn, 75% P, 0.35% max.; S, 0.05% max.; Si, 0.15%; C, 7% max. and remaining Fe. Silica percentage is however quite variable which may vary from 0.15% to 2%. It has not been possible to produce ferro-manganese with higher Mn content because our manganese ore does not contain Mn: Fe ratio exceeding 6.8 : 1.

Manufacture of ferro-silicon which is next in importance to ferro-manganese was produced in 1942 by Visvesvaraya Iron and Steel Ltd. (formerly Mysore Iron and Steel Ltd.). Prior to 1942, requirements were met from imports chiefly from Norway and Sweden. The VISL started production of ferro-silicon in two electric furnaces (each 1500 KVA) with combined capacity of 1828 tonnes of 70–75% grade. A third electric furnace (9000 KVA) capable of producing 3830 tonnes was added in 1951. The capacity was raised to 7500 tonnes in 1959 and to 20,000 tonnes in 1965. The same furnaces are used by the VISL for the production of ferro-chrome and also ferro-manganese depending upon the requirements. On 3rd May, 1967, another firm, Indian Metals and Ferro-Alloys Ltd., P.O. Therubali,

TABLE 23

Name of Ferro-manganese Plants and Their Annual Capacities

Name of unit	Year of commence- ment	Capacity (in tonnes)
1. The Visvesvaraya Iron and Steel Ltd. Bhadravati, Shimoga Dist., Karnataka.	1956	1,600

TABLE 23   (*Contd.*)

2.	The Dandeli Ferro-Alloys Pvt. Ltd., (formerly Electro-Metallurgical Works Dandeli, North Kanara Dist.) Karnataka.	1957	12,000
3.	The Tata Iron & Steel Co. Ltd., Joda, Keonjhar District, Orissa.	1958	30,000
4.	Ferro-Alloys Corporation, Garividih, Srikakulam Dist., Andhra Pradesh.	1958	50,000
5.	The Jeypore Sugar Co. Ltd., Rayagada, Koraput Dist., Orissa.	1958	24,000
6.	Universal Ferro and Allied Chemicals Ltd., Tumsar Road, Bhandara Dist., Maharashtra.	1959	30,000
7.	Khandelwal Ferro-Alloys Ltd., Kanhan, Nagpur District, Maharashtra.	1961	36,000
8.	Maharashtra Electro Smelt Ltd., Chandrapur.	1977	50,000
9.	Uniferro International (A Division of Universal Ferro & Allied Chemicals Ltd.)	1981	45,000
10.	Utkal Ferro Alloys Pvt. Ltd., Mirza Ghalib Street, Calcutta.	—	300
11.	The India Thermit Corporation Ltd., Fazalganj, Kanpur.	—	430

Production of Ferro-alloys in India During 1960–1982

(Qty. in tonnes)

Year	Ferro-boron	Ferro-chrome	Ferro-manganese	Ferro-molybdenum	Ferro-silicon	Ferro-titanium	Ferro-tungsten	Ferro-vanadium	Ferro-silicon zirconium
1960	—	—	85,677	—	7,118	—	—	—	—
1965	—	N.A.	149,331	—	20,539	0.504	—	—	—
1966	—	641	137,482	—	19,394	1.441	—	—	—
1967	—	138	130,467	2.010	22,187	8.297	2.750	—	—
1968	—	1,229	145,767	19.988	24,160	11.882	6.750	—	—
1969	2.810	5,188	155,489	34.064	27,186	35.829	36.985	24.007	0.495
1970	0.302	15,364	175,612	145.001	27,590	41.862	7.088	9.184	—
1971	0.580	12,314	162,342	250.675	29,472	94.845	19.056	51.612	—
1972	0.250	1,525	160,020	237.584	31,481	142.994	4.223	60.060	3.716
1973	0.400	6,556	139,650	92.885	21,526	43.291	6.963	⁻66.728	0.500
1974	1.025	15,300	146,015	209.072	29,682	7.160	49.580	65.309	—
1975	0.802	10,128	142,398	196.627	39,972	56.356	34.947	50.090	0.175

1976	0.72	17,005	175,760	181,765	83,947	32,023	18.35	69.90	0.590
1977	—	18,068	196,798	314,791	44,745	74,866	4.45	128.93	0.295
1978	.140	21,545	219,087	327,726	52,366	75,714	12.903	236.980	.850
1979	1.336	22,249	186,942	302,157	53,087	128,540	22.861	190.608	—
1980	.460	16,494	158,303	162,291	54,319	321,233	21.206	92.262	6.400
1981	.200	31,903	205,571	172,696	60,354	192,270	12.147	117.023	6.578
1982	.434	41,625	157,884	118,745	40,253	68,952	3.269	89.447	—

Koraput; Orissa commenced regular production of ferro-silicon. Two more ferro-silicon plants were added namely Nava Bharat Ferro-Alloys at Kottagudem, A.P. with capacity of 10,000 tonnes per annum and Sandur Manganese & Iron Ore Ltd., Yeshwantnagar, Karnataka with capacity of 24,000 tonnes per annum in 1975 and 1977 respectively. This company has a licensed capacity of 7200 tonnes. During the years 1966 and 1967 a number of firms were set up and commenced production of various alloys which were hitherto not produced in the country. M/s Electric Control Gear (P) Ltd., Ahmedabad, Gujarat, started production of ferro-chrome, ferro-titanium, ferro-tungsten and ferromolybdenum in 1967. M/s Mehra Ferro-Alloys, Amritsar commenced production of ferro-chrome from December 1966, the grade being 65–67% Cr. M/s Ferro-Alloys Corporation, Garividih also commenced production of ferro-chrome and silico-manganese. The India Thermit Corporation Ltd., Kanpur is one of the oldest applied ferro alloys manufacturers producing ferro-titanium, ferro-molybdenum, ferro-vanadium, ferro-tungsten. ferro-boron, etc. The Government of Orissa has also set up a ferro-chrome plant at Jaipur Road, Cuttack, having 10,000 tonnes capacity. This plant also produces silico-chrome and ferro-silicon according to requirements. These firms along with many others have been licensed to manufacture several other ferro alloys like ferro boron, ferro-vanadium, ferro-columbium, etc. A breakthrough in the manufacture of these strategic alloys has been achieved.

Fertiliser: Much stress on the use of chemical fertiliser and its production has been given only during the recent years.

By the end of 1978, there were 28 units for nitrogenous fertiliser, and 37 units for phosphate fertiliser dispersed in various part of the country.

Nitrogen, phosphorus and potassium are the three main plant nutrients and introduced to the soil in the form of soluble chemical compounds, the other elements present therein merely act as a carrier. Use of fertiliser has been found to be one of the readiest means of increasing agricultural output. Our average rate of consumption of fertiliser ($N + P_2O_5 + K_2O$) during 1968 was 8 kg per hectare against 84 kg in European countries, 48 kg in Japan and 390 kg in the U.S.A. The scope of augmenting production of fertiliser is, thus, unlimited. However, there was an increase in consumption of fertiliser in India from 8 kg to 15 kg per hectare in 1974 and to 25 kg in 1978. It is estimated that 44.7% of our land area is under

TABLE 24

Production of Nitrogenous and Phosphatic Fertilisers

(In '000 metric tonnes)

Year Fertiliser	1969	1970	1971	1972	1973	1974	1975	1976	1977	1978	1979	1980	1981	1982
Nitrogenous 'N' content	716	820	906	1,070	1,050	1,185	1,432	1,800	2,000	2,035	2,243	2,054	2,987	3,359
Phosphatic 'P$_2$O$_5$' content	222	229	278	325	315	327	325	418	597	710	752	834	916	995

cultivation, which is quite high compared to many other countries. It is only 30% in the U.K., 20% in the U.S.A., 16% in Japan, 10% in the U.S.S.R. and only 4% each in Australia and Canada which are called granary of wheat. Survey of Indian soil has indicated that 85% of its arable land are low and medium in available nitrogen and phosphorus and that 65% are low or medium in potash.

Of all the three nutrients, N is very important and responds to all type of soil. Phosphatic and potassic fertilisers are added to soil to meet its respective deficiencies. However the response of soil to potassic fertiliser is least, because all soil contains potash in some soluble form. Trial production has shown that a 20 kg application of N per hectare can increase yield of rice by 259 kg and that of wheat by 350 kg. What types of fertiliser and in what proportions they should be used per unit of area in agricultural economics? We shall confine our discussion mainly on resources of fertiliser raw materials and economics of their production and manufacture.

Nitrogenous fertiliser is chiefly composed of ammonia. Air and coal are the two main sources of nitrogen. Coal contain 1 to 2% nitrogenous matters. Only about 15 to 20% of total nitrogen present in coal is recoverable, and thus 5.5 to 6.5 lbs of ammonia per tonne of coal coked is obtained. The coke ovens with the steel mills or elsewhere like that of Sindri Fertiliser Unit form a useful source for the production of ammonia. Three coal based urea fertiliser plants each with a capacity of 228,000 tonnes in term of N content are being set up at Talchir in Orissa, Ramagundam in Andhra Pradesh and Korba in Madhya Pradesh. Water, crude oil, naphtha, natural gas, refinery gas, etc. form the source of hydrogen required for ammonia synthesis. For the present water is the major source from which hydrogen is obtained but it requires lots of energy. Naphtha is cheaper source of hydrogen if it is available in sufficient quantity, same is the potition with natural gas. When naphtha or gas is heated, they give $CO + H_2$. The mixture is reacted with steam to give more hydrogen. Hydrogen is reacted with nitrogen obtained from the liquefaction of air and converted into ammonia (NH_3). Nitrogenous fertilisers are prepared into many forms like ammonium sulphate, ammonium nitrate, calcium ammonium nitrate, ammonium phosphate, nitrophosphate, ammonium chloride and urea. Ammonia shares the major cost in the production of N fertiliser. For example, ammonia shares 50%, 65% and 80% of the cost of ammonium sulphate, urea and ammonium chloride. For countries having no resources of phosphorus and sulphur, the manufacture of urea (NH_2—CO—NH_2) is of a great significance. It does not involve use of sulphur. It is produced by reaction of ammonia with CO_2. The advantage of urea over other ammonium salts is in the nitrogen content. Nitrogen present in urea is 46%, which is more than double of ammonium sulphate and 12% more than the ammonium nitrate. The nitrolime manufactured by Rourkela and Nangal units contains only 25%N. For the manufacture of phosphatic fertilisers, especially of superphosphate, the use of phosphatic minerals and sulphur is essential. However, in the manufacture of triple superphosphate, ammonium phosphate, the use of sulphur as sulphuric acid can be avoided, provided the phosphoric acid is prepared by furnace process instead of reacting with sulphuric acid. Gypsum forms useful source of sulphate radical. Sindri Fertiliser Unit utilises gypsum for the manufacture of ammonium sulphate. By-product calcium sulphate is utilised by the Fertiliser and Chemical Travancore Ltd. for the same purpose.

Potassic fertiliser is generally used in the form of nitrate, chloride, chlorate and sulphate. Reh matti, lake and sea bitterns are the main sources of potash in the country. Bitterns contain 2 to 2.5% KCl. Felspar is another useful source but it does not occur in soluble form till it is converted into soluble compounds by the geological processes.

Cement: Cement, although by its name, might not have been known earlier. However, evidence of its use is believed to have been found in the Indus Valley civilisation. The use of lime, in the last century began in 1870 when the blue clays found locally around Calcutta was mixed with chalk imported from England for the preparation of cement. Shortly after that date a small company was formed in Calcutta using kankar and imported limestone from England. The company did not last long. It seems nothing was done to establish this industry until 1904 when a small factory was set up in Tamil Nadu (Madras) by the South India Industrial Ltd. This venture also failed. Afterwards during 1913, the Indian Cement Company Ltd. established a cement plant at Porbandar in Gujarat under the Managing Agency of Tata Sons & Co. and by 1914 this company was able to produce 14,000 tonnes a year. Within next three years two new factories were set up at Kymore near Katni in Madhya Pradesh and at Lakheri, Bundi district in Rajasthan. By the end of World War I in 1918, the three companies together were producing 85,000 tonnes a year. Between 1919 and 1924 four factories (including a factory at Wah now gone to Pakistan) were installed at Japla, Dwarka and Banmore and capacities of three old plants expanded. The combined installed capacity in 1924 was 559,000 tonnes. By 1934, many factories were expanded and two new uni.s set up at Shahabad and Maddukarai raising the installed capacity to 1089,000 tonnes. In 1936 old units with exception of Sone Valley Portland Cement Co. Ltd., Japla of Dalmia Jain (now Sahu-Jain Group) united to form the Associated Cement Companies Ltd. A new unit of Jaipur Udyog Ltd. was set up during the same year at Sawai Madhopur, Rajasthan. In 1937 the Kalyanpur Lime and Cement Works at Banjari in Bihar was established by Dalmia-Jain Group. In 1938, the first State enterprise was established at Bhadravati in Mysore. During 1938–39 the Dalmia-Jain Group established five cement factories (including two of Dandot and Shantinagar which now fall in Pakistan) located at Dalmianagar in Bihar, Dalmiapuram in Madras, and Dalmia Dadri in Haryana. The ACC also expanded and added four more factories including one at Rohri (now in Pakistan) located at Khalari, Bihar; at Surajpur, Haryana and at Kistna in Andhra Pradesh. The Vijaywada Plant of Andhra Pradesh Cement Co. Ltd. went into production in 1939. At the eve of World War II in 1939, further expansion of old and establishment of new units were halted. After the partition in 1947, India was left with 18 cement plants. Four cement plants located at Wah, Dandot, Shantinagar and Rohri went to the then West Pakistan and the one at Chatak of Assam Bengal Cement Co. Ltd. set up in 1937 to East Pakistan, now Bangladesh.

In post-Independence era, a total of 37 new plants were added within 1948–78 and many existing units expanded. This period witnessed not only the setting in of new units but also diversifying the types of cement products. White cement is now produced by Kottayam, Kymore and Porbandar plants. The Associated Cement Co. has started producing oil-well cement at Shahabad factory in A.P., portland blast furnace slag cement at Jamul and Chaibasa factories, and hydrophobic cement at Shahabad and Khalari factories. Orissa Cements Ltd., Jaipur Udyog Ltd., Madras Cements Ltd. and many others have started manufacturing pozzuolana cement. Saurashtra Cement and Chemical Industries Ltd. has been licensed to manufacture pozzuolana cement.

The first cement plant commissioned after independence was by

Bagalkot Cement Ltd., Bagalkot in Bijapur district, Karnataka, on 21st December, 1948. During the following year three more units were set up. The India Cements Ltd. established a plant at Sankarnagar (Talaiyathu) in Tirunelveli district, Tamil Nadu; Travancore Cements Ltd. (a Government of Kerala Undertaking) at Nattakam in Kottayam district, Kerala and Shri Digvijaya Cement Co. Ltd. at Sikka in Jamnagar district, Gujarat. Again in 1951, two plants were added by Orissa Cement Ltd. at Rajgangpur, Sundergarh district, Orissa and at Sevalia, Kaira district, Gujarat. In 1953, a plant at Jhinkpani near Chaibasa, Singhbhum district, Bihar, was set up by A.C.C. to manufacture portland slag-cement. In 1954, a second State-owned factory was set up at Churk in Mirzapur district, Uttar Pradesh. For the first time a cement plant based on calcium carbonate sludge obtained from the Sindri Fertiliser Plant was set up near Sindri, Dhanbad, Bihar by Associated Cement Co. which came into production in September, 1955. During 1957 and 1958 three more units came under production namely, the Ashoka Cement Ltd., at Dalmianagar, Shahabad district, Bihar, A.C.C's at Mancherial, Adilabad district, Andhra Pradesh and K.C.P. Ltd. at Macherla, Guntur district, Andhra Pradesh. In 1959, two cement plants at Bagganipalli, Kurnool district, Andhra Pradesh, belonging to Panyam Cement and Mineral Industries Ltd. at Satna, Satna district, Madhya Pradesh by Satna Cement Works of Birlas came under production. Madras Cement Ltd. commissioned its plant at Talukkapatti, Ramnad district, Tamil Nadu in 1960. The Ranavav plant in Junagarh district, Gujarat, belonging to Saurashtra Cements and Chemical Industries Ltd. went into production in 1961. Again in 1963 two more plants at Ammasandra Tumkur district, Karnataka and at Sankaridrug in Salem district, Tamil Nadu went into production, owned by Mysore Cements Ltd. and India Cements Ltd. respectively. In 1964, a third State owned cement plant was set up at Wuyan in Anantnag district, J and K State under the management of J and K Minerals Ltd. In 1965, a second slag cement plant at Jamul, Durg district Madhya Pradesh owned by A.C.C. went into production. In 1966, Mawmluh-Cherra Cements Ltd. (formerly Assam Cements Ltd.), a Government of Meghalaya undertaking plant at Cherrapunji, Khasi, and Jaintia hills district, commenced production and the A.C.C. established another unit at Porbandar in Junagarh district, Gujarat to manufacture special cements. In 1967, three new plants, at Bargarh in Sambalpur district, Orissa, by Orissa

TABLE 25

Yearwise Progress in Setting up of Cement Units in India, 1914-1982

Year	No. of units	Year	No. of units
1914	1	1954	26
1915	1	1955	27
1916	2	1956	27
1917	3	1957	28
1920	3	1958	30
1921	5	1959	32
1922	6	1960	33
1924	6	1961	34
1925	7	1962	34
1933	7	1963	36
1934	8	1964	37
1935	8	1965	38
1936	9	1966	40
1937	10	1967	43
1938	13	1968	45
1939	16	1969	46
1940	17	1970	49
1945	17	1971	50
1946	18	1972	51
1947	18	1973	51
1948	19	1974	51
1949	22	1975	54
1950	22	1976	55
1951	24	1977	55
1952	24	1978	55
1953	25	1979	57
		1980	63
		1981	68
		1982	69

Industrial Development Corporation, a fifth State enterprise under cement; and at Chittorgarh, Rajasthan by Birla Cement Works and at Puliyur, Tiruchirapally, Tamil Nadu owned by Chettinad Cement Corporation came into production. In 1968, again two plants, one at Palakurthy, Karimnagar district, Andhra Pradesh belonging to M/s Kesoram Cements having 200,000 tonnes a year capacity and another at Wadi, Karnataka with an installed capacity of 400,000 tonnes belonging to A.C.C. went into production. The first plant of Cement Corporation of India, a Government of India undertaking, went into trial production in November, 1969. The plant is located at Mandhar, near Raipur, Madhya Pradesh. Its annual capacity is 200,000 tonnes of cement.

The regular production of Mandhar plant, however, commenced only from the beginning of 1970. During the same year 3 more plants located at Ghughus in Chandrapur district, Maharashtra; Bajajnagar in Udaipur district, Rajasthan and Alangulam in Ramnathapuram district, Tamil Nadu came into existence. The Ghu-

TABLE 26

Installed Capacity and Production of Cement in India During
1974–1982

Year	Installed capacity in tonnes	Productions in tonnes
1	2	3
1947	2,149,450	1,470,355
1948	2,149,450	1,577,645
1949	2,860,243	2,136,038
1950	3,174,797	2,654,198
1951	3,612,490	3,246,730
1952	3,856,330	3,594,202
1953	4,255,008	3,840,480
1954	4,514,698	4,468,368
1955	4,819,498	4,558 589
1956	5,794,858	5,007,254

TABLE 26 (*Contd.*)

1	2	3
1957	6,638,544	5,691,226
1958	7,172,554	6,165,454
1959	8,456,351	6,936,029
1960	8,710,300	7,844,400
1961	9,474,000	8,245,200
1962	9,728,400	8,586,000
1963	10,285,200	9,355,200
1964	10,725,600	9,714,500
1965	11,694,000	10,577,612
1966	12,570,800	11,052,980
1967	12,781,200	11,292,242
1968	15,154,060	11,929,872
1969	15,743,430	13,624,206
1970	17,360,000	13,956,037
1971	19,390,000	14,931,946
1972	19,764,000	15,801,005
1973	19,764,600	15,006,241
1974	19,815,000	14,264,597
1975	20,826,000	16,176,000
1976	21,670,000	18,700,000
1977	21,870,000	19,190,000
1978	21,870,000	19,508,000
1979	23,538,000	18,338,000
1980	25,647,000	17,894,000
1981	29,222,000	20,874,000
1982	31,729,000	22,637,000

ghus plant is owned by A.C.C. having an installed capacity of four lakh tonnes a year. The Alangulam plant is owned by Tamil Nadu Cement Ltd., a Tamil Nadu Industrial Development Corporation Undertaking, and its capacity is also 4 lakh tonnes a year. This plant went into production on 5th February, 1970. The Bajajnagar cement plant is owned by Udaipur Cement Ltd. having an installed capacity of two lakh tonnes a year only and it went into production on 26th March, 1970. During 1971 a cement plant owned by U.P. Government went into production. The plant is located at Dalla in Mirzapur district having an installed capacity of 4.20 lakh tonnes a year. Again in 1972, one more plant owned by Cement Corporation of India located at Kurkunta in Gulbarga district, Karnataka having an installed capacity of 2 lakh tonnes a year went into production. Thereafter only in 1975, three new plants were commissioned. These are: Durgapur Cement Works, P.O. Durgapur, West Bengal having four lakh tonnes installed capacity; Century Cement Works, R.S. Tilda, Raipur, M.P. having six lakh tonnes installed capacity and J.K. Cement Works, Nimbhara, Chittorgarh having three lakh tonnes installed capacity. The first two plants were commissioned in January, 1975 and the third one in March, 1975. One more plant owned by CCI located at Bokajan in Assam was commissioned in 1976.

ALUMINIUM

Amongst various industries aluminium industry is the one which has made phenomenal growth in the country. The first production was started in March, 1943 by the Aluminium Production Company of India incorporated on 17th December, 1938. Later on 21st June, 1944 its name was changed to Indian Aluminium Company Ltd. The smelter was set up at Alupuram, Alwaye in Kerala. The main consideration in locating the plant in Kerala was availability of cheap hydro-electricity, which is a must in aluminium industry. The initial capacity of the smelter was 2,000 tonnes a year. It was based on imported alumina till the company commenced its own production at Muri in Ranchi district, Bihar. In 1955, the smelter capacity was expanded to 5,500 tonnes and further raised to 15,850 tonnes in 1963. Another pioneering effort was made by the Aluminium Corporation of India which although registered in 1937, started production in 1944 in its integrated plant at Jaykaynagar, Burdwan, West Bengal. Its initial capacity was 2,500 tonnes. The Indian Aluminium Company Ltd. set up another smelter of 10,000 tonnes capacity at Hirakud in Sambalpur district, Orissa in 1959 when power was made available from Hirakud project. Later its capacity was increased to 20,320 tonnes in November, 1961. Consequently the capacity of alumina plant at Muri was raised to 75,000 tonnes. The company has fabricating facilities at Kalwa in

Maharashtra, Belur in West Bengal and Alupuram, Kerala, where metal is processed into sheet, extrusions, foil, powder and paste. The Aluminium Corporation of India Ltd. also expanded the capacity of its Jaykaynagar plant from 2,500 tonnes to 7,500 tonnes and 9,000 tonnes in 1967. A new company Hindustan Aluminium Corporation was formed in 1959 sponsored by M/s Birla Brothers in collaboration with Kaiser Aluminium and Chemical Corporation Ltd., of U.S.A. with an authorised capital of Rs. 200 million. The latter firm hold 25% of the capital. The plant was erected at Renukut in Mirzapur district, Uttar Pradesh, with an initial capacity of 20,000 tonnes of metal. The alumina reduction plant was commissioned in May, 1962 and the alumina plant in November, 1962. The capacity was doubled in 1965 and trebled to 60,000 tonnes in July, 1967, raised to 90,000 tonnes in 1973 and further expanded to 95,000 in 1975. The company has been licensed to enlarge its capacity to 120,000 tonnes a year. In May, 1965, the Madras Aluminium Co's integrated plant of 10,000 tonnes metal went into production. The plant is located at Mettur Dam, Salem district, Tamil Nadu. By 1967 its capacity was increased to 12,500 tonnes. It was further licensed to increase its capacity to 25,000 tonnes metal a year which was duly achieved by the end of 1975.

During 1969, the Indian Aluminium Co. Ltd. set up an integrated plant in Belgaum, Karnataka with an installed capacity of 30,000 tonnes of metal and 75,000 tonnes of alumina. It was further licensed to expand capacities to one lakh tonnes of alumina and 60,000 tonnes of metal per year. The production capacity of alumina plant of one lakh tonnes was duly achieved during 1973. The smelter capacity has now reached to 60,000 tonnes during 1978. The Belgaum plant gets supply of electricity from the Sharavathy hydro-electric installations. The first integrated plant of Bharat Aluminium Co., a Government of India undertaking is located at Korba, Madhya Pradesh. It has a capacity to produce 2 lakh tonnes of alumina and one lakh tonnes of metal per year. The alumina plant was commissioned in April, 1973 with 50% of its installed capacity. The smelter plant was commissioned in May, 1975 with only 25,000 tonnes capacity and the second pot of same capacity in 1977 thus making the total smelter capacity of 50,000 tonnes.

The aluminium industry could have made more impressive progress; but for the shortage of adequate supply of electricity. By the end of 1978, a total of 275,170 tonnes of installed capacity of alu-

TABLE 27

Locations and Capacities of Aluminium Smelters During 1978

Name of the company		Locations of plants	Installed capacity
1. Indian Aluminium Co. Ltd.	(i)	Alupuram (Kerala)	20,000
	(ii)	Hirakud (Orissa)	20,000
	(iii)	Belgaum (Karnataka)	73,000
*2. Aluminium Corporation of India Ltd.		Jaykaynagar (W. Bengal)	9,000
3. Hindustan Aluminium Corporation Ltd.		Renukut (U.P.)	100,000
4. Madras Aluminium Co. Ltd.		Mettur Dam (Tamil Nadu)	25,000
5. Bharat Aluminium Co.		Kobra (Madhya Pradesh)	100,000
			275,170

*Under lockout since September 1973. BALCO took over the management of Aluminium Corporation of India Ltd. on 2nd May, 1978. The new company has been named as Alucoin Jaykaynagar Industrial Undertaking.

minium metal was available in the country. The break-up of the licensed and installed capacity as on 1st January 1979 is tabulated on the previous page.

TABLE 28

Production of Aluminium Metal in India During 1948–82

Year	Qty. in tonnes	Year	Qty. in tonnes
1948	3,476	1964	56,667
1949	3,548	1965	67,169
1950	3,652	1966	83,761
1951	3,911	1967	96,546
1952	3,623	1968	120,100
1953	3,818	1969	132,544
1954	4,934	1970	161,083
1955	7,340	1971	178,181
1956	6,605	1972	179,103
1957	7,909	1973	154,336
1958	8,316	1974	1,28,917
1959	17,355	1975	1,59,678
1960	18,255	1976	2,09,549
1961	18,382	1977	1,83,909
1962	35,209	1978	2,05,355
1963	55,230	1979	2,11,400
		1980	1,84,700
		1981	2,12,800
		1982	2,16,700

COPPER

Available records indicate that the copper lodes were prospected as early as 1857 in Dalbhum sub-division of Singhbhum district. A Hindustan Copper Company is reported to have worked in 1862 copper ore between Rajdah and Badia. The ore was carried to Calcutta for smelting but the heavy preliminary expenditure restricted the venture. Further work was done by the Rajdoha Mining Co. in 1891. During 1906–1908 the G.S.I. carried out a series of diamond drilling operations The encouraging results drew the attention of the Cape Copper Company under the management of John Taylor and Sons which took over the Rakha Mines property of Rajdoha Mining Co. The Rakha mine was developed and the concentration and refining plants were erected between 1919 and 1920. The total copper metal produced between 1919 and 1923 was 3549.76 tonnes. The mine closed down in 1923 after working down to a depth of 240 metres. In 1920, the Cordoba Copper Company also under the management of Messers John Taylor and Sons prospected the Mosabani area on an option from the Cape Copper Company and met with promising results. During the same time few other companies prospected the areas in Singhbhum district but without much success.

Later the Indian Copper Corporation Ltd.,[1] was formed by re-organisation of Cordoba Copper Company in 1924. This company started production in December, 1928 when the copper smelter set up by the Corporation at Maubhandar near Mosabani went into operation. Since then this area has been under regular production although the smelting capacity has remained the same at about 10,000 tonnes a year of fire-refined copper till the middle of 1965. A new flash smelter to 16,560 tonnes capacity of blister copper has been added out of which 8,500 tonnes is for electrolytically refining and remaining quantity is for fire-refining. The company has also added a nickel sulphate plant of 260 tonnes per year capacity for extraction of nickel, a sulphuric acid plant of 54,000 tonnes capacity, based on sulphurous gases from flash smelter which is to be used in the production of superphosphate. The company operates four mines namely Mosabani, Surda, Kendadih and Pathargora. The

---

[1]ICC was nationalised and taken over by Hindustan Copper Ltd. on 10th March, 1972.

Badia mine which lies on the same lode of Mosabani has been amalgamated with the latter. The underground workings of Mosabani and Badia extend for a horizontal distance of 4 km and it is possible to walk from one end to another. The present workings in the Mosabani mines have gone down to the depth of more than 660 metres. Presently ore worked contains on an average 1.5% Cu. Prior to Mosabani mines, Rakha mine in the same district was established in 1919 by the Cape Copper Company. This mine was worked probably down to a depth of 240 metres and then closed. It is now being thoroughly investigated for reopening.

The shortage of copper metal in the country led to concerted search on selected prospects to establish once for all whether a particular prospect was worth exploitation or not. Exploration for copper was started in full swing in early 1957 first in Rajasthan and later the work was extended to Bihar, Andhra Pradesh, Gujarat, Jammu and Kashmir, Karnataka and in recent years in Madhya Pradesh. About forty prospects have been subjected to detailed studies so far. The study has resulted in discovering four major fields viz., (i) Khetri copper belt in Jhunjhunu district, Rajasthan, (ii) Malanjkhand deposit in Balaghat district, M.P., (iii) Agnigundala lead-copper belt in Guntur district, Andhra Pradesh, and (iv) Rakha mine area in Singhbhum copper belt in Bihar. Several small yet significant deposits, e.g., Chitradurga in Karnataka, Ambamata in Gujarat, Dariba and Chandmari, etc. in Rajasthan have been established. In Khetri copper belt four mineralised sectors have been proved viz., Madan-Kudan, Kolihan, Akwali and Satkui. Mines have been developed in Madan-Kudan and Kolihan sectors. These mines were developed for production of about 10,000 tonnes of ore per day. An output rate of about 2,500 tonnes a day has been achieved so far. The output from these mines feeds the newly commissioned electrolytic flash smelter plant which went on stream on 10th November, 1974 with an installed capacity of 31,000 tonnes of metal per annum with provision of 198,000 tonnes of by-product sulphuric acid, 8,000 ounces of gold and 64,000 ounces of silver. Since the Khetri and Kolihan mines are presently not capable of raising the required quantity of ore, a part of the requirements of copper concentrate for the smelter is met by imports. Dariba copper mine in Alwar district has been developed with a production capacity of 100 tonnes a day. The run-of-mine is concentrated at the mine site and the concentrate is transported to the Khetri smelter.

Chandmari deposit was opened up for production of 500 tonnes of ore per day. It is being expanded to produce 2,000 tonnes per day.

In Rakha area, two blocks viz., Mine block and Roam-Sidheswar block have been subjected to detailed exploration. These two blocks are being developed to produce 1,000 tonnes of ore a day under phase I programme. A matching concentration plant has been set up to treat run-of-mine for feeding Ghatsila smelter. Under phase II, the mine output will be doubled with matching increase in copper concentration facility.

Exploration in Agnigundala belt was started in 1959 and practically completed in 1968 by the Geological Survey of India. Since then, the Hindustan Copper Limited is developing the properties. In this belt, three important blocks have been established viz., Bandalamottu, Nallakonda and Dhokunda. Results of the exploratory underground development activities of Nallakonda block which was regarded chiefly copper rich have, however, not been encouraging.

Comparable to Khetri copper belt and that of Singhbhum copper belt, Malanjkhand copper deposit of Balaghat district, Madhya Pradesh holds equally important position so far as copper resource is concerned. The deposit was initially explored by the GSI then for some time by the Mineral Exploration Corporation on behalf of the Hindustan Copper Limited. The HCL has entered into an agreement with a Soviet agency for preparation of feasibility report for mine development. The report has since been received. Malanjkhand deposit would be developed by open-pit with ultimate capacity of two million tonnes of ore per day.

The Chitradurga Copper Co. Ltd., a Government of Karnataka undertaking, has also started mining operation on a small scale in its Ingaldhal copper deposit. The concentrate is sold to HCL. In addition, the HCL and HZL buy all the concentrates of copper, lead and zinc from Sikkim Mining Corporation in which Government of India holds equal share. The Corporation operates Bhotang mine at Rangpo, the average mine ore contains 1.3% copper, 1.2% lead and 3.2% zinc.

Lead-Zinc-Silver: The discovery of famous lead-zinc-silver deposits of Zawar in Udaipur district, Rajasthan was made during the time of Rana Lakha Singh (1382–97 A.D.) and probably worked until the time of the great famine in 1812–13 for a period of more than 400 years. There is little information after that period about the working of Zawar mine or any other lead-zinc deposit. In 1942, a deposit of lead ore at Banjari about 100 km from Jaipur, Rajasthan, was acquired by the Eastern Smelting and Refining Co. Ltd. This company set up a plant in 1974 at Tundu in Dhanbad district, Bihar with the help of Burmese engineers who had fled to India during the Japanese invasion. About the same time a prospecting licence was granted to the Mewar Mineral Company of Udaipur for prospecting of Zawar area but the company did not show much interest. The Utilisation Branch of the Geological Survey of India then carried out exploratory operations for two and half years from August 1942 to March 1945 in the Mochia Magra Hill of the Zawar area which lead to indication of promising zones of mineralisation. When the Geological Survey of India ceased operation, it was acquired by the Metal Corporation of India promoted by Eastern Smelting and Refining Co. Ltd. The Metal Corporation of India set up an ore-dressing plant in June, 1950 to treat run-of-mines ore for preparing zinc and lead concentrates. Lead concentrate is treated at Tundu. They started recovery of silver in 1954. As there was no zinc smelter in the country, the zinc concentrate was used to be sent to Japan for the treatment on toll basis. The first zinc smelter plant based on imported concentrate was set up by the Cominco-Binani Zinc Ltd. It was commissioned on 22nd July, 1967. It is a joint company formed in 1962 by Binani and Consolidated Mining and Smelting Co. of Canada. The plant is situated on the bank of Periyar river 19 km from Cochin in Kerala. It has a rated capacity of 20,000 tonnes of zinc and 36,000 tonnes of sulphuric acid per annum. By-product cadmium at the rate of 35 to 40 tonnes annually is also obtained. The second smelter of 18,000 tonnes capacity went into production on 18th March 1968. The plant is located at Debari, near Udaipur in Rajasthan. Its construction was started in 1962 by the Metal Corporation of India, later taken over by the Hindustan Zinc Ltd., a Government of India undertaking, in 1965. This plant is attached with ancilliary units for the recovery of about 75 tonnes of cadmium and by-product sulphuric acid at the rate of 29,500 tonnes per year which is converted into about 85,000 tonnes of superphosphate.

Both the plants are under expansion the capacity of the former is being doubled and the latter to 45,000 tonnes a year. Another zinc plant based on imported concentrates having 30,000 tonnes of zinc metal and 10,000 tonnes of lead metal capacity was commissioned towards the close of 1977 at Vizag. The capacity of Tundu lead smelter is expanded to 8,000 tonnes per year. Much needed zinc and lead concentrate for the expanded capacities of Debari and Tundu plants will come from increased mine production from Zawar properties and adjoining new area being developed at Dariba-Rajpura. Other smaller prospects like Bondumottu in Andhra Pradesh; Sargipalli in Sundergarh district, Orissa are to supplement the requirements. It is expected that the country will have 115,000 tonnes of zinc metal capacity within the period of three years.

Oil: The history of Indian oil industry is fairly old. It was in 1825 that an army officer Lt. R. Wilcox while surveying the inhospitable jungles of Assam recorded several gas and oil seepages from a bed in the Burhi Dihing river. He recorded in his memoirs, "the jungles are full of odour of petroleum". On the recommendation of H.B. Medlicott of the Geological Survey of India, six wells were drilled during 1866-67 by a Calcutta firm, two in Makum near Margherita and four in Jaipur near Nahorkatiya. Two more holes were put by the same firm. Oil was struck at a depth of 35.4 metres (118 ft). But on account of extremely poor transport facility the venture failed. Again in 1889, Assam Railways and Trading Company Ltd. got interested in oil and completed a hole in 1890 which struck oil at the depth of 198.6 metres (662 ft) near Digboi. It confirmed the possibility of Digboi developing into a commercial field. In April, 1899, the Assam Oil Company was formed to work Digboi oil field. Till then 14 wells were drilled. The company maintained a precarious existence for over 20 years, but was able to drill 80 wells ranging up to 600 metres. In January, 1921, the Burmah Oil Company took over the commercial and technical management, giving it much needed financial assistance and required facilities of equipment, transport and refining. In about ten years' time the production stepped up to 180,000 gallons per day compared with about 12,000-14,000 gallons per day before 1921. During 1922 to 1925, the Assam Oil Company carried out thoroughly geophysical prospecting over a large tract of Assam and indicated presence of oil at great depth But on account of non-availability of deep hole oil rigs, followed by moratorium on drilling during the war, the first well in Nahorkatiya, Sibsagar district could not be spudded in until 1952 which heralded an epoch making discovery oil in the area other than Digboi in Assam. The well struck oil at a depth of 3,000 metres. Till 1953, Digboi oil field remained the only field of oil production and also the only refinery in India. The production and refining capacity ways barely half a million tonnes. The other refineries of Burmah Shell, ESSO (renamed as Hindustan Petroleum Corporation Ltd. on July 1974 with 74% of ownership with the Government of India), Caltex, etc., came into existence only from 1954.

Further drilling work in Hugrijan-Nahorkatiya and Moran areas of Sibsagar district proved these to be highly potential of oil. At Moran oil was struck at 4,122 metres (13,739 ft). It is the deepest well in Asia. It is pertinent to mention that over 200 tests and geological interpretation bore holes were drilled before it was possible to locate oil in these areas. The government's anxiety for speedily raising the internal output of crude oil led to the formation of Oil India Ltd., on a 50:50 partnership basis in the properties of the Assam Oil Co. The agreement was signed on 14th January, 1958. Since then Oil India Ltd. has required more areas. Two refineries were set up by the Indian Refineries Ltd. now incorporated into the Indian Oil Corporation Ltd.; one at Noonmati near Gauhati, Assam and the other at Barauni. Monghyr district, Bihar, with the help of Rumanian and Russian Governments respectively. These two refineries were connected by pipeline for supply of crude from Nahorkatiya and Moran areas. Gauhati refinery went into production on 1st January, 1962 and Barauni in July, 1964. Another refinery at Bongaigaon is being set up with initial throughput capacity of one million tonnes.

The Oil and Natural Gas Directorate set up on 14 August, 1956 later converted into Commission on 15th October, 1959, started search in favourable horizons for possible oil reserves. Their several efforts in Jawalamukhi area in Hoshiarpur district, Punjab, Purnea, Raxaul in Bihar, Jaisalmer in Rajasthan and J & K have so far proved abortive. However, their efforts in Lunej in Cambay area, Khaira district, Ankleshwar in Broach district, Kalol and Navagaon in Mehasana district, Vadsar in Baroda district, Gujarat brought fruitful results. At Lunej oil was struck on 8th September, 1958 and at Hazal in Ankleshwar on 14th May, 1960. The Cambay area has proved to be mostly gas bearing and while Ankleshwar is rich in oil. Trial production from Ankleshwar field commenced in August 1961 with 100 tonnes a day which was stepped up to 600 tonnes in February, 1962, 1,200 tonnes in June, 1962 and 1,500 tonnes in December, 1962. The crude oil was initially despatched to Burmah Shell and Hindustan Petroleum Corporation Ltd. (formally Esso) refineries in Bombay till ONGC's own refinery at Koyali was commissioned in 1965 with Russian aid. Initially the capacity of Koyali refinery was 3 million tonnes a year which has now been raised to 7.3 million tonnes a year. Gas from Gujarat field is sold to Dhuvaran and Uttaran thermal stations, Gujarat State Fertiliser Corporation and several other industries in Baroda district. Further prospecting work by ONGC has proved distinct oil fields in Lakwa and Rudrasagar in Upper Assam Valley and gas field in Jaisalmer area, Rajasthan. Rudrasagar has been in regular production since April, 1966. The production from Lakwa field has also been commenced thereafter. The crude from Rudrasagar and Lakwa is transported to Nahorkatiya and Barauni refineries.

The ONGC and OIL are the two principal organisations exploring for oil. The Government of India in order to expedite exploration in offshore areas had granted exploration and mining rights to two American firms, namely Reading and Bates Groups and Carlsberg Group. The former Company was given mining and exploration right in the Gulf of Kutch and the latter in the Bay of Bengal. The period of agreement varied between 24 and 27 years from 1st August, 1974. In October, 1975, the government awarded the Asamera group of Canada the contract for off-shore oil exploration and production in the Cauvery basin.

The presence of oil in Bengal basin was for the first time reported on 11th October, 1975 by Natomas Company, contractor of Carlsberg India in the very first well which was spudded on 21st September, 1975. The well is situated 160 km south of Sunderbans and about 136 km from the Orissa coast. Drilling for oil in Kutch started on 28th October, 1975. All these foreign companies after making some initial efforts abandoned the projects. The ONGC has launched an all-India programme for systematic geological and geophysical prospecting to decipher all possible petroliferous structures. The commission has established several structures in Cambay basin namely, Kadi, Sobhashan, Sanad, Bakarol, Dholka, Kathana structures, Broach Synclines, Kosamba and Navagaon field. In the offshore areas, two petroliferous structures namely Bombay High and Bassein structure of considerable signifinance have been located. The Bassein structure and Bombay High are 80 km and 190 km NW of Bombay, the latter covers an area of 2,500 sq km. Depths of sea bottom at these two points vary between 75 to 90 metres. Production from Bombay High has started from 21st May, 1976. The present rated production is 4 million tonnes and within next four years it is planned to raise production by 10 million tonnes, by adding 2 million tonnes every year. Production from Bassein structure is expected soon initially with 2 million tonnes then raising up to 10 million tonnes per year.

Another important structure in the off-shore has been located in the Arabian Sea known as Tarapore structure. The first well at Tarapore structure in the Arabian Sea adjoining Gulf of Cambay was spudded in on 11th October, 1973 by self-elevating jack up type drilling platform named Sagar Samrat. The well had to be abandoned in January, 1974 at the depth of 2781.5 metres due to severe downhole complication. Thereafter, the Sagar Samrat was shifted to another location in Bombay High. The first well was spudded in on 3rd February, 1974 and the oil was struck in the last week of February, 1974 at a depth of 962 metres which showed a pressure of 500 lbs/sq inch. In this area possibility of striking petroleum reservoir down to 2,000 metres depth is expected, as two more oil bearing limestone horizons are found to be present at a depth of 1,330 metres and 1,612 metres respectively. Another major oil field discovery was made in 'Ratnagiri structure', in the Arabian sea, 80 km south-west of Bombay where drill well on October 6, 1979 struck an oil column at a depth of 1,900 m below sea level. Above oil zone is a gas column of 15 m. A significant feature of the discovery is that the oil has been found here in limestone of Eacene age compared to younger formation of Miocene age in Bombay High.

Four more off-shore drilling rigs have been added namely, Haakon Magnus, the Dalmehoy, Shenandoah, and Gellysburg for vigorous search. The geological sequence leading to the formation of crude oil in Bombay High can be compared to that one found in North Sea where also petroleum reservoirs are found to be in limestone formation. In order to survey the vast coastal shelf areas the commission have acquired a survey ship "Anweshak" equipped to carry out seismic and magnetic surveys and have also chartered a few for deploying in the Arabian Sea and on the coast of Andamans & Nicobar islands. OIL is also going in for drilling in the Bay of Bengal.

Deeper Continental Shelf Project covering 51,200 sq km north and north-west of Bombay has been formed to be operated by ONGC. On the land, in the eastern region, ONGC struck oil for the first time during September 1975 in Arunachal. Oil horizon was struck between 2,842 to 2,865 metres depth. Another horizon is expected at 4,500 metres depth. Hydrocarbons India Private Ltd. a a wholly owned subsidiary of ONGC has been formed for exploring and mining oil in Persian Gulf jointly with Iranian Oil Company of Iran, Philip Petroleum Company of the U.S A. and AGIP of Italy. The last firm holds 50% of the share and the remaining three hold equal shares in remainder 50%. Oil is produced by these firms from Rastum and Raksh structures. ONGC also signed a contract on 20th November, 1973 with the Government of Iraq for exploration and development of areas in Iraq for petroleum.

The country has to move a long way in attaining self-sufficiency in petroleum by exploring vast off-shore areas and also onshore areas especially in the North-east, North and North-west on the foot hills of Himalayas. Imports of crude and petroleum products alone eat away about fifty per cent of the total foreign exchange earnings of the country. On geological grounds, there is every possibility of finding oil along the coast. The submerged basins of the continental margins are regarded areas of high potentiality for oil. Location of new oil fields in the offshore areas have brightened the prospects many times. Government of India took effective control over oil industry both in refining and marketing and in exploration and production. The first step was acquiring 74% share of ESSO in March, 1974. The ESSO was then renamed to Hindustan Petroleum Corporation on 15th July, 1974. It became 100% government company when its balance 26% share was acquired on 1st October, 1976. Burmah-Shell was taken over completely in January, 1976 and renamed as Bharat Petroleum Corporation CALTEX was completely taken over by the Government of India on 9th May, 1978 and was merged with Hindustan Petroleum Corporation Ltd.

Road, Rail and Port : A well developed road, rail and port links are the prerequisite infrastructure for the development of industries including mineral-based industries. Added to this should there be adequate supply of electricity and water which are other infrastructure necessary for the development of industry. Considerable development have taken place in these directions as well.

Prior to 1958, the volume of export trade in minerals and ores was comparatively small and none of our ports were capable to accommodate vessels of capacity more than 30,000 DWT. It was only after 1958 due to growing interests shown by the Japanese for

importing iron ore in increasing quantities that a great need was
felt of bringing efficiency in ore loading capacites at ports by improved mechanical system, and also to have good rail connections for
hauling ores from hinterland areas. These aspects were given special
impetus during the Second and Third Plan periods and caught the
same momento subsequently. By and large developments of ports
may be attributed largely due to tremendous increase in export of
iron ore which rose from less than 12 million tonnes in 1965 to
over 23 million tonnes in 1977. Iron ore export is expected to rise
over 35 million tonnes within a period of four to five years. Developments followed immediately with the signing of the first agreement with the Japanese in 1958 for export of iron ore from Kiriburu
area and then by Bailadila agreement in 1960. As a result two new
ports at Haldia and Paradeep on the east coast were constructed
and available facilities at Visakhapatnam port were considerably
improved. New rail links connecting Sambalpur-Titagarh for moving
Kiriburu ore and Kirindul-Kottavalsa-Visakhapatnam for transporting Bailadila ore to Visakhapatnam port were laid. The Bailadila

TABLE 29

Position of Refineries with Their Capacities

(as on 1.5.1985)

Name & Location	Year of commencement	Capacity in million tonnes
Indian Oil Corporation Ltd.		
Digboi	1901	0.50
Gauhati	1962	0.85
Barauni	1964	3.30
Koyali	1965	7.30
Haldia	1974	2.50

Mathura	1982	6.00
Hindustan Petroleum Corporation Ltd. Bombay	1954	3.50
Vizag	1957	4.50
Bharat Petroleum Corporation Ltd. Bombay	1955	6.00
Cochin Refineries Ltd. Cochin	1966	4.50
Madras Refineries Ltd. Madras	1969	5.60
Bongaigaon Refinery & Petrochemicals Ltd. Assam	1979	1.00
	Total	45.55

(Kirundul Railway Station)—Visakhapatnam rail link involved construction of 46 tunnels with total length of 14 km, 15 covered ways,
87 major bridges, some with piers over 40 metres in height and
1,200 minor bridges. The line crosses the difficult terrain of Eastern
Ghats and by attaining altitude of about 1,000 metres could be
considered one of the highest broad gauge rail links in the world.
Haldia Paradeep provide outlet for Barajamda sector (Bihar-Orissa)
and Daitari (Orissa) ores. An express highway has been constructed
linking Daitari with Paradeep with movement of ores by road. It
is now connected with rail link joining Banspani-Navagarh and
Cuttack to Paradeep for which construction work has started. A
new line has been laid for connecting Haldia. Simultaneously improve
ments have been made at Madras and Marmugao ports to handle
increasing supplies from Bellary-Hospet and Goan ores respectively.
The narrow gauge section between Bellary and Guntakal was converted into broad gauge to link it directly to Madras port.

One very significant phase has emerged in our port development
programmes. At the time of Kiriburu/Bailadila agreements the
Japanese buyers visualised the use of carriers up to 35,000 DWT and

had, therefore, desired provision of adequate facilities of that size at Visakhapatnam port, through which major quantities of ore were to be shipped to Japan. But well before this facility could be completed in 1965, ore carriers of larger size up to 60,000 DWT had already come on the high seas. Since then over 100,000 DWT vessels have also come into operation and possibility of using 200,000 DWT carriers is already on the way. Viewed with this context, the entire port facilities are required to be further improved. The main features of reconstruction and development programmes are as under:

HALDIA

This port envisages the construction of an enclosed dock basin consisting of six berths: a coal berth, an ore berth, a fertiliser berth, general cargo berth, containers berth, and oil jetty. When in full operation, it will have an annual capacity of three million tonnes and a draft capable of accommodating 45,000 DWT vessels. This facility is in the final phase of completion.

MANGALORE

This port, on the west coast, is receiving major uplift for shipping 7.5 million tonnes of iron ore concentrates per annum from Kudremukh. For the present it has four berths, out of which one berth is earmarked for export of ores with handling capacity of 5 lakh tonnes of iron ore and 2 lakh tonnes of manganese ore annually. The deepest draft is 8.5 metres. Under the expansion programme which is under execution, the draft is being deepened to 12.4 metres for receiving ships up to 60,000 DWT. Simultaneously mechanical ore handling plants comprising of stackers, reclaimers and two ship loaders of capacity of 3,000 tonnes/hour each are being installed.

PARADEEP

The port has been suitably modified to receive vessels of up to 100,000 DWT and will have annual capacity to ship 5 million tonnes of iron ore mainly originating from Orissa and Bihar. Bucket wheel reclaimer has been provided to improve mechanical loading capacity for loading at the rate of 6,000 tonnes an hour.

VISAKHAPATNAM

The existing inner harbour can accommodate vessels up to 35,000 DWT and load 6 million tonnes ore at the rate of 6,000 tonnes an hour. With a view to increasing its capacity, an outer harbour capable of accommodating vessels of 100,000 DWT initially and 200,000 DWT or more at a later stage with ore loading facility at a rate of 8,000 tonnes per hour is completed. Ultimate capacity of this port is aimed at loading 12 million tonnes of iron ore annually. The Visakhapatnam outer harbour is one of the deepest artificial draft harbours in the world.

MADRAS

The present facility at the inner harbour can handle about 2.5 million tonnes of ore. Loading operations are semi-mechanised. The outer harbour has been developed for receiving 8,000 tonnes oil tankers and provided with an ore berth with adequate mechanical loading facilities for receiving 70,000 DWT vessels initially with provision to increase the capacity up to 100,000 DWT vessels. The port is designed to handle 8 million tonnes of ore per year with loading rate of 8,000 tonnes per hour.

MARMUGAO

At present this port handles about 14.5 million tonnes of ore per annum. Berthing capacity being inadequate, loading is done mostly in mid-stream through barges. A channel is being deepened to receive vessels of 60,000 DWT initially and 100,000 DWT at a later stage with facility of mechanised loading at the rate of 8,000 tonnes an hour.

Besides the development work undertaken as enumerated above, other ports through which mineral traffic takes place also received due attention. At Kandla an oil berth was provided. At Bombay port three deep-water berths capable of berthing the largest oil tankers with connecting submarine pipelines to the mainland were completed. At Cochin, a new coal berth and oil jetty have been completed. There are 150 minor ports of which 20 are considered important. Important ports under the 'minor port' category handling mineral traffic, namely, Kakinada, Machlipatnam, and Cuddalore on the east coast, and Coondapur, Honawar, Belikeri, Karwar, Vengurla, Bhavnagar, Porbander, Bedi and Okha on the west coast including Mangalore, have been modified to a large extent.

The transport capacity of the railways improved considerably under their massive development programmes through successive plan periods by way of dieselisation, electrification, remodelling of several yards and loading stations, introduction of larger capacity (55 tonnes) wagons, and converting some narrow/metre gauge rail routes to broad gauge. Railway carries about 250 million tonnes of mineral traffic annually, accounting over 65% of the total goods carried by it. New lines laid specially for mineral traffic, other than the two mentioned earlier, are: (1) Champa to Kobra for coal, (2) Bhilai to Dhalli-Rajhara for iron ore, (3) Rourkela to Barasua for iron ore, (4) Garhwa Road to Robertsganj via Renukut mainly for bauxite, (5) Billi to Katni via Singrauli coal field, and (6) Bijuri to Bishrampur via Karonji for coal. A new line linking Panskura with Haldia port for a distance of 70 km for movement of iron ore from Barajamda sector and coal from Bihar and West Bengal has been opened up. Presently Paradeep is not linked with railway line. Movement of iron ore to this port is done by road only. This port is also being connected by a rail link with important iron ore mining areas via Barajamda, Daitari, Gandhamanden, Malangtoli and Banspani. A 40 km Tornagallu-Mudukalaparta rail line for movement of iron ore to Madras port from Donimalai has been opened up. Construction of metre gauge connecting Hassan with Mangalore port for a distance of 189 km is in progress. In the meantime Hassan to Sakleshpur (43 km) and Mangalore to Subramaniya Road (92 km) were completed and opened during 1976 and 1977 respectively.

TABLE 30

Mineral Production in India, 1948—1982

Year	Agate (tonnes)	Apatite (tonnes)	Asbestos (tonnes)	Ball-clay (tonnes)	Barytes (tonnes)	Bauxite ('000 tonnes)
1948	N.A.	1,132	83	—	23,515	32
1949	N.A.	598	148	—	21,456	43
1950	N.A.	3,074	212	—	12,155	65
1951	N.A.	423	527	—	10,639	68
1952	N.A.	452	879	—	10,191	65
1953	N.A.	4,429	730	—	9,551	72
1954	N.A.	2,329	395	—	19,094	76
1955	N.A.	5,652	1,418	—	7,746	82
1956	N.A.	8,926	1,250	—	6,416	101
1957	N.A.	9,325	1,746	—	13,121	110
1958	N.A.	14,806	1,181	—	15,908	169
1959	N.A.	16,350	1,366	—	13,552	218
1960	N.A.	14,921	1,711	—	16,906	387
1961	N.A.	20,140	1,473	—	24,246	476
1962	397	29,018	1,692	—	32,662	587
1963	731	13,127	2,790	3,302	38,063	567
1964	262	4,049	3,375	9,597	47,205	593
1965	423	7,076	4,775	8,122	48,458	707
1966	493	16,275	6,979	6,801	52,608	750
1967	457	11,631	7,911	7,777	53,016	801
1968	630	6,695	9,187	11,080	57,747	961
1969	504	9,316	9,876	7,110	65,478	1,085
1970	743	15,997	10,128	12,865	78,634	1,374
1971	728	11,307	11,139	12,620	58,695	1,517
1972	798	11,613	12,359	17,491	48,348	1,684
1973	1,104	9,981	12,460	17,483	120,054	1,297
1974	988	12,034	23,685	22,396	146,490	1,114
1975	1,662	30,338	20,586	25,941	226,099	1,274
1976	3,128	38,280	24,119	33,322	236,240	1,449
1977	1,768	35,361	22,177	48,369	330,989	1,519
1978	2,031	29,460	19,280	35,562	350,280	1,653
1979	2,164	20,548	32,334	128,090	490,699	1,952
1980	1,549	20,082	33,716	126,988	442,326	1,785
1981	1,326	16,862	27,521	123,964	405,138	1,955
1982	1,186	18,319	26,949	121,424	376,845	1,920

TABLE 30 (*Contd.*)

Year	Calcite (tonnes)	Chalk (tonnes)	China clay Processed (tonnes)	China clay Natural (tonnes)	Chromite (tonnes)	Coal ('000 tonnes)
1948	249	—	N.A.	N.A.	22,918	30,534
1949	2,193	—	N.A.	N.A.	19,728	32,135
1950	N.A.	—	N.A.	N.A.	16,998	32,805
1951	1,403	—	N.A.	N.A.	16,970	34,952
1952	3,640	—	N.A.	N.A.	35,752	36,840
1953	2,629	—	N.A.	N.A.	65,809	36,522
1954	5,110	—	N.A.	N.A.	46,237	37,441
1955	2,641	—	N.A.	N.A.	90,783	38,810
1956	5,763	—	N.A.	N.A.	53,532	40,037
1957	4,742	—	N.A.	N.A.	79,802	44,186
1958	2,917	—	N.A.	N.A.	63,957	46,056
1959	7,679	—	61,641	N.A.	98,695	47.800
1960	9,320	—	82,916	71,129	106,896	52,593
1961	11,118	--	79,175	N.A.	48,785	56,065
1962	13,907	—	81,640	N.A.	66,648	61,370
1963	14,776	150	83,447	N.A.	69,013	65,956
1964	13,906	1,835	87,641	N.A.	34,969	62,440
1965	20,481	8,046	93,115	202,326	59,685	67,162
1966	18,039	22,811	98,088	204,635	77,770	67,974
1967	16,973	35,521	102,403	190,021	113,868	68,223
1968	15,445	48,930	97,029	180,974	205,675	70,813
1969	17,339	52,220	97,793	185,259	226,568	75,411
1970	17,644	47,749	101,128	193,403	273,679	73,698
1971	20,425	48,840	106,303	223,653	275,405	71,824
1972	28,633	60,053	117,211	296,939	294,599	75,658
1973	23,867	65,652	100,073	274,171	288,814	77,870
1974	25,336	54,336	110,905	316,555	394,913	84,102
1975	13,614	45,339	99,034	271,734	500,294	95,911
1976	21,567	61,349	103,036	334,637	157,656	100,876
1977	27,445	61,414	95,684	349,143	352,535	100,247
1978	28,753	72,907	101,141	322,935	265,907	101,549
1979	30,079	79,786	116,320	372,831	309,893	103,364
1980	26,118	87,245	105,193	354,396	321,318	109,152
1981	27,403	86,017	118,840	389,380	343,601	123,104
1982	21,854	90,473	125,058	429,598	364,204	128,504

TABLE 30 (*Contd.*)

Year	Lignite ('000 tonnes)	Copper ore ('000 tonnes)	Corundum (tonnes)	Diamond (carats)	Diaspore (tonnes)	Dolomite ('000 tonnes)
1948	73	327	284	2,426	—	83
1949	68	335	197	1,632	—	49
1950	21	366	303	2,769	—	54
1951	34	375	562	1,674	—	14
1952	45	330	647	2,054	—	4
1953	35	242	329	2,207	—	17
1954	30	348	4,777	1,955	—	140
1955	29	359	244	1,787	—	92
1956	25	392	357	1,499	—	100
1957	22	410	451	790	—	143
1958	19	411	395	1,540	—	177
1959	33	404	214	682	—	341
1960	47	448	250	1,159	—	650
1961	64	423	329	1,313	—	723
1962	211	492	301	1,141	—	914
1963	999	474	658	1,432	160	1,093
1964	1,569	473	540	2,260	685	521
1965	2,300	468	481	4,460	531	977
1966	2,568	481	385	2,113	1.511	1,054
1967	2,929	459	326	7,626	1,489	1,167
1968	4.126	484	326	8,643	902	1,284
1969	4,188	511	452	11,794	3,453	1,283
1970	3,545	518	411	20.325	6,172	1,148
1971	3,660	666	318	19,383	4,831	1,320
1972	3,067	873	391	20,099	5.198	1,348
1973	3,320	1,102	266	21,427	8,451	1,449
1974	3,044	1,429	337	20 975	3,221	1,195
1975	2.822	1,838	313	19,994	2,734	1,457
1976	3,895	2,395	526	20.487	10,090	1,886
1977	3,632	2,551	1,306	18,297	7,900	2,152
1978	3,613	2 130	1,076	16.665	4,909	1,942
1979	3,264	2,156	909	15,229	6,437	2.179
1980	4,549	2,005	1,454	14,432	5,928	2,031
1981	5,966	2,109	1,289	14.834	6,099	2 068
1982	6,673	2,478	1,471	12.022	6 595	2,186

TABLE 30 (*Contd.*)

Year	Emerald ('000 carats)	Felspar (tonnes)	Fire-clay ('000 tonnes)	Fluorspar Graded (tonnes)	Concen- trates (tonnes)	Garnet Abrasive (tonnes)	Gem variety (kg)
1948	N.A.	1,002	124	—	—	—	—
1949	751	862	108	—	—	—	—
1950	421	1,800	125	—	—	—	—
1951	253	3,436	114	—	—	—	—
1952	462	2,052	121	—	—	—	—
1953	551	3,942	85	—	—	—	—
1954	509	6,580	93	—	--	—	—
1955	192	5,314	89	—	—	150	—
1956	474	3,972	142	—	—	181	—
1957	338	7,998	166	—	—	—	—
1958	80	8,567	209	—	—	—	—
1959	230	9,896	234	—	—	367	—
1960	322	10,613	272	—	—	469	—
1961	295	10,629	299	332	—	241	—
1962	352	21,744	390	657	—	426	1,985
1963	386	24,002	441	708	—	401	3,408
1964	53	25,249	426	389	—	295	2,554
1965	65	27,114	460	651	—	237	3,651
1966	54	27,972	470	2,089	—	218	5,170
1967	38	30,654	426	1,603	—	592	6,068
1968	23	35,261	487	1,315	—	1,988	4,962
1969	9	32,219	532	1,880	—	1,637	3,619
1970	12	34,580	584	4,979	1,528	988	5,268
1971	23	43,954	623	3,363	9,831	1,391	2,962
1972	21	52,801	732	3,301	12,715	3,235	1,883
1973	3	42,802	718	2,951	6,479	2,741	814
1974	3	56,093	803	3,749	8,879	3,702	632
1975	38	42,572	672	3,067	11,598	4,432	420
1976	1	55,094	682	3,643	13,980	2,075	3,673
1977	1	54,710	726	3,586	15,209	1,825	5,529
1978	21	47,783	682	3,432	13,901	3,024	4,411
1979	4	54,264	794,847	4,135	17,360	6,820	5,035
1980	7	60,190	761,877	4,236	17,101	3,748	4,559
1981	—	52,290	847,506	4,274	18,720	3,289	1,335
1982	8	46,968	877,848	7,028	18,203	5,396	3,220

TABLE 30 (Contd.)

Year	Gold (kilogrammes)	Gypsum ('000 tonnes)	Ilmenite ('000 tonnes)	Iron ore ('000 tonnes)	Kyanite (tonnes)	Lead concentrates (tonnes)
1948	5,612	80	233	2,321	12,806	1,346
1949	5,107	142	313	2,854	20,265	1,316
1950	6,125	210	216	3,013	36,057	2,008
1951	7,041	207	228	4,152	43,182	1,812
1952	7,877	418	229	4,475	27,314	2,033
1953	6,948	595	219	5,010	15,619	2,807
1954	7,439	622	244	5,758	43,009	2,894
1955	6,559	701	255	6,964	11,930	3,112
1956	6,508	867	341	7,521	20,459	3,972
1957	5,573	937	301	8,115	23,881	4,928
1958	5,291	794	314	9,065	26,026	5,341
1959	5,144	860	303	10,768	16,499	6,488
1960	4,995	997	250	16,609	20,156	6,245
1961	4,868	866	174	18,705	27,155	5,532
1962	5,080	1,122	138	19,674	49,712	6,384
1963	4,305	1,191	26	20,602	32,085	5,920
1964	4,619	882	12	21,388	34,173	6,130
1965	4,063	1,160	30	23,738	37,481	5,496
1966	3,736	1,294	30	26,783	63,820	5,151
1967	3,161	1,034	42	25,699	50,374	3,995
1968	3,583	1,338	59	27,961	64,361	3,556
1969	3,062	1,391	48	29,567	84,172	3,300
1970	3,241	926	**	31,366	120,923	3,880
1971	3,656	1,089	—	34,311	63,482	4,262
1972	3,290	1,105	—	35,391	67,897	5,005
1973	3,278	887	—	35,563	58,215	7,672
1974	3,145	1,073	—	35,545	42,217	11,035
1975	2,825	816	—	41,794	52,673	15,117
1976	3,152	727	—	43,740	47,172	15,856
1977	3,014	778	—	42,598	42,123	16,744
1978	2,724	854	—	38,156	28,214	16,840
1979	2,637	877	—	39,859	40,709	20,965
1980	2,452	866	—	41,936	49,141	16,763
1981	2,495	957	—	41,589	36,583	19,951
1982	2,244	922	—	42,721	37,721	21,747

TABLE 30 (*Contd.*)

Year	Limestone ('000 tonnes)	Lime kankar ('000 tonnes)	Calcareous sand ('000 tonnes)	Magnesite (tonnes)	Manganese ore ('000 tonnes)	Mica-Crude (tonnes)
1948	1,539	N.A.	N.A.	49,102	534	N.A.
1949	2,011	N.A.	N.A.	92,018	656	N.A.
1950	2,304	N.A.	N.A.	53,708	897	N.A.
1951	2,965	N.A.	N.A.	119,351	1,398	N.A.
1952	2,832	N.A.	N.A.	90,470	1,597	N.A.
1953	4,137	N.A.	N.A.	94,237	2,121	N.A.
1954	6,263	N.A.	N.A.	71,639	1,542	N.A.
1955	7,485	N.A.	N.A.	58,432	1,745	23,625
1956	8,385	N.A.	N.A.	93,183	1,999	28,484
1957	9,571	N.A.	N.A.	90,312	1,877	30,943
1958	10,533	N.A.	N.A.	104,236	1,379	31,942
1959	10,847	N.A.	N.A.	157,967	1,698	28,846
1960	12,935	N.A.	N.A.	156,331	1,452	29,226
1961	14,755	N.A.	N.A.	209,744	1,401	28,347
1962	16,939	413	668	212,888	1,636	28,481
1963	17,347	321	699	235,066	1,316	25,429
1964	17,017	275	846	207,936	1,408	22 810
1965	19,957	393	857	238,905	1,647	23,840
1966	19,830	449	56	232,053	1,707	22,915
1967	19,571	359	872	246,448	1,580	18,152
1968	21,030	284	822	253,073	1,610	18,265
1969	22,517	306	889	297,893	1,486	18,508
1970	23,843	381	996	354,291	1,702	16,581
1971	25,079	440	1,023	295,604	1,841	15,095
1972	26,053	338	960	250,931	1,643	14,114
1973	25,490	340	1,004	193,266	1,492	13,830
1974	25,948	295	729	275,532	1,504	13,804
1975	26,519	281	902	313,453	1,605	11,501
1976	29,987	85	1,074	329,698	1,835	9,494
1977	30,380	265	898	402,007	1,865	9,352
1978	30,169	223	932	419,641	1,567	9,246
1979	31,481	,102	772	396,211	1,771	9,073
1980	29,183	9	771	380,113	1,695	7,934
1981	32,437	18	685	462,534	1,532	8,534
1982	34,274	90	660	418,909	1,490	8,776

TABLE 30 (Contd.)

Year	Ochre (tonnes)	Pyrophyl- lite (tonnes)	Pyrites (tonnes)	Quartz & silica sand (tonnes)	Rutile (tonnes)	Salt		Sapphire (crude) (kg)
						Others (000' tonnes)	Rock salt (tonnes)	
1948	10,725	N.A.S.	—	4,650	129	2,297	4,123	—
1949	6,067	N.A.S.	—	4,218	—	2,018	4,297	—
1950	12,178	N.A.S.	—	4,445	36	2,609	5,130	—
1951	11,474	N.A.S.	538	14,840	46	2,678	5,530	—
1952	17,882	N.A.S.	2,203	4,695	149	2,715	6,088	—
1953	60,587	N.A.S.	281	15,855	106	3,210	5,865	—
1954	76,716	N.A.S.	—	17,454	106	2,553	4,071	—
1955	16,480	N.A.S.	813	12,848	151	2,928	5,222	—
1956	12,814	N.A.S.	15	23,326	550	3,238	3,494	—
1957	15,816	N.A.S.	—	26,193	481	3,666	4,405	—
1958	20,458	N.A.S.	—	41,725	456	4,230	5,197	—
1959	28,414	N.A.S.	—	63,168	389	3,174	3,695	—
1960	20,675	N.A.S.	—	123,825	982	3,431	4,311	—
1961	18,804	N.A.S.	—	160,196	815	3,476	4,300	—
1962	22,890	2,157	—	217,624	1,616	3,882	4,401	—
1963	27,908	4,485	—	249,966	1,871	4,541	3,359	N.A.
1964	32,523	6,071	—	247,662	1,871	4,644	2,829	203
1965	34,845	11,828	—	297,805	1,317	4,700	2,877	225
1966	32,033	8,399	—	280,454	1,816	4,504	4,000	185

TABLE 30 (*Contd.*)

Year	Ochre (tonnes)	Pyrophyllite (tonnes)	Pyrites (tonnes)	Quartz & silica sand (tonnes)	Rattle (tonnes)	Salt		Sapphire (crude) (kg)
						Others ('000 tonnes)	Rock salt (tonnes)	
1967	37,854	4,934	—	271,953	2,534	4,466	3,233	189
1968	39,555	10,676	13,914	348,278	2,686	4,027	2,882	145
1969	38,704	10,919	38,686	456,072	2,500	5,173	3,444	140
1970	59,504	13,922	25,643	579,542	**	5,588	4,017	163
1971	53,018	11,780	40,886	671,778	—	5,426	3,845	—
1972	67,409	15,086	30,723	720,533	—	6,517	4,252	85
1973	53,560	15,027	41,507	548,971	—	6,860	3,599	—
1974	80,500	15,611	35,660	579,518	—	5,912	5,273	—
1975	98,321	15,102	50,633	596,496	—	5,843	3,330	—
1976	92,053	34,080	47,531	684,881	—	4,076	4,438	—
1977	75,935	34,619	31,085	749,227	—	5,328	3,759	—
1978	62,016	23,665	54,163	770,652	—	—	4,380	—
1979	107,431	34,842	67,172	871,921	—	—	4,363	—
1980	92,938	35,077	82,905	819,673	—	—	4,683	—
1981	90,436	41,681	72,278	865,339	—	—	4,326	—
1982	100,987	43,794	56,438	868,037	—	—	4,123	—

TABLE 30 (Contd.)

Year	Sillimanite (tonnes)	Silver (kilogrammes)	Slate (tonnes)	Steatite (tonnes)	Vermiculite (tonnes)	Tungsten concentrates (tonnes)	Zinc concentrates (tonnes)	Petroleum (crude) ('000 tonnes)
1948	216	398	N.A.	18,384	—	—	—	450
1949	990	351	N.A.	21,535	—	—	—	450
1950	1,499	488	N.A.	25,895	—	—	666	450
1951	4,114	454	N.A.	35,186	—	—	2,144	450
1952	5,158	550	N.A.	21,105	53	14	3,929	450
1953	5,579	455	N.A.	29,603	236	10	4,371	450
1954	3,115	5,013	N.A.	43,000	22	15	4,038	450
1955	2,462	4,788	N.A.	43,069	3	1	4,943	450
1956	4,712	3,253	N.A.	47,604	125	1	6,940	450
1957	7,536	3,915	N.A.	44,681	942	2	7,589	450
1958	14,067	3,416	N.A.	46,181	—	—	7,391	450
1959	7,696	3,881	N.A.	64,485	2	1	9,978	450
1960	8,483	4,128	N.A.	93,392	213	3	9,787	454
1961	8,113	5,941	N.A.	98,147	632	9	9,254	513
1962	8,328	4,315	1,117	108,292	433	10	9,837	1,078
1963	11,493	3,991	1,104	116,495	680	4	10,627	1,652
1964	12,362	4,735	1,142	133,840	429	8	10,744	2,212
1965	11,276	5,235	1,408	156,411	732	14	9,641	3,022
1966	10,286	1,220	1,065	149,051	511	25	8,900	4,647

TABLE 30 (Contd.)

Year	Sillimanite (tonnes)	Silver (kilogrammes)	Slate (tonnes)	Steatite (tonnes)	Vermiculite (tonnes)	Tungsten concentrates (tonnes)	Zinc concentrates (tonnes)	Petroleum (crude) ('000 tonnes)
1967	5,797	3,471	1,309	135,310	263	28	10,029	5,667
1968	4,657	2,926	859	173,629	2,352	39	12,839	5,853
1969	3,946	3,278	1,074	176,580	3,981	41	13,781	6,723
1970	4,565	1,540	1,361	159,314	1,530	35	15,888	6,809
1971	4,326	3,773	1,549	178,877	538	30	15,855	7,185
1972	4,046	4,427	2,071	210,619	1,541	32	17,055	7,373
1973	3,138	4,258	9,325	209,732	2,723	26	23,913	7,198
1974	2,950	4,581	10,524	292 896	2,895	24	29,117	7,490
1975	8,278	2,582	15,395	217,353	2,245	38	39,150	8,283
1976	14,896	3,183	21,129	220,461	3,435	44	45,322	8,659
1977	15,023	13,228	21,826	246,601	2,878	43	46,113	10,185
1978	13,295	12,084	12,878	276,439	1,806	41	65,867	11,251
1979	16,108	11,515	19,399	372,500	3,109	38	71,847	12,839
1980	14,488	11,377	15,078	359,567	3,678	44	46,489	9,399
1981	12,273	17,298	12,969	367,325	3,477	48	52,925	14,925
1982	12,512	14,403	6,785	341,145	2,196	53	52,839	19,734

TABLE 31

Export of Principal Ores and Minerals from India During 1960—1982-83

(In '000 tonnes)

Year	Bauxite	Chromite	Coal	Ilmenite	Iron ore	Kyanite	Magnesite	Manganese ore @	Mica	Sillimanite	Steatite blocks and powder	Limestone	Barytes	Diamond (cut) Rs. million
1960	76	41	1,322	229	9,080	26	28	1,318	30	8	7	87	25	—
1961	99	42	942	125	9,994	32	32	1,123	26	5	10	105	7	—
1962	249	17	1,082	101	8,680	37	32	903	31	5	16	98	6	—
1963	135	10	874	76	9,379	28	33	946	34	5	14	110	9	17
1964	94	31	1,245	27	10,482	29	36	1,598	30	9	12	113	8	26
1965	63	33	925	15	11,264	31	53	1,147	38	6	11	77	5	41
1966	78	44	361	32	13,658	36	34	1,249	31	6	10	—	3	91
1967	61	77	250	21	13,574	41	20	1,108	21	2	10	1	7	160
1968	99	109	469	49	15,635	50	22	1,189	22	3	14	—	3	242
1969	126	112	280	74	15,118	46	26	1,208	21	2	25	—	14	296
1970	135	153	393	54	21,205	69	34	1,614	28	2	18	N.A.	20	257
1971	54	130	266	50	19,355	43	24	1,243	25	2	10	2	37	332
1972	28	110	267	78	21,864	29	16	861	25	1	7	72	41	548
1973	28	234	493	62	21,285	25	14	692	28	1	10	72	34	711
1974	18	301	389	119	21,900	25	16	1,035	36	1	12	61	145	870

TABLE 31 (Contd.)

Year	Bauxite	Chromite	Coal	Ilmenite	Iron ore	Kyanite	Mange-site	Manga-nese ore	Mica @	Sillima-nite	Steatite blocks and powder	Lime-stone	Barytes	Diamond (cut) Rs. million
1975	14	371	425	69	22,796	24	10	793	35	**	7	121	174	856
1976	21	278	478	121	23,403	11	8	714	22	**	9	219	151	1.876
1977	48	167	623	77	23,191	9	13	554	23	1	18	139	177	3,754
1977–78	42	86	640	131	21,561	8	12	443	23	**	18	145	207	5166.8
1978–79	20	114	265	120	21,225	1	12	628	22	**	11	171	285	6686.2
1979–80	46	213	93	121	24,768	1	6	627	29	**	14	221	342	4677.0
1980–81	84	130	108	112	22,408	1	4	609	33	**	9	200	300	5598.7
1981–82	132	148	118	76	23,676	—	1	552	26	**	10	209	190	7322.3
1982–83	187	157	181	194	21,260	1	1	422	22	**	11	148	247	8708.2

TABLE 32

Import of Principal Ores and Minerals into India, 1960–1977

Year	Antimony ore and concentrates (tonnes)	Apatite and rock phosphate (tonnes)	Asbestos (tonnes)	Borax (tonnes)	Cryolite (tonnes)	Fluorspar (tonnes)	Graphite (tonnes)	Manganese ore (tonnes)	Sulphur (tonnes)	Crude Petroleum Quantity ('000 tonnes)	Crude Petroleum Value (Rs. million)	Petroleum products Value (Rs. million)	Zinc ore & concentrates ('000 tonnes)
1960	1,214	240,071	22,064	5,993	1,447	4,559	1,581	6,879	178,904	5,723	227	529	—
1961	1,059	231,051	19,672	6,317	2,811	5,891	1,752	490	191,433	5,968	287	516	—
1962	1,398	287,609	22,378	4,884	3,230	9,375	2,242	20,687	248,031	6,022	402	540	—
1963	1,920	297,683	36,960	6,091	2,746	8,690	1,084	9,320	261,018	6,519	478	548	—
1964	1,344	449,407	35,787	8,822	1,916	2,144	2,016	7,838	242,144	6,791	426	538	—
1965	1,702	494,746	38,396	9,150	4,767	5,004	2,490	1,064	283,893	6,811	404	447	—
1966	2,110	849,774	29,955	2,173	4,880	5,908	2,186	6,382	287,956	7,457	5C1	513	—
1967	1,528	598,009	29,785	5,573	4,814	7,069	1,248	5,603	592,978	8,704	796	390	21
1968	1,733	860,625	25,924	5,769	3,001	4,288	972	7,263	390,390	10,450	939	407	19
1969	1,636	661,747	33,609	10,257	4,948	13,042	1,088	7,992	305 010	10,702	940	382	30
1970	699	667,657	39,766	18,654	2 744	9,058	1,244	5,368	521,423	11,665	1,024	305	49
1971	1,640	813,261	45,715	9,568	2,529	16,869	855	7,285	420,607	12,688	1,399	435	46
1972	1,334	809,476	46,379	5,105	1,896	9,956	1,703	5,298	559,484	12,310	1,442	567	38
1973	1,241	874,746	43,643	13,965	—	677	1,074	4,123	551 476	13,443	2,446	1,009	35

TABLE 32 (Contd.)

Year	Antimony ore and concentrates (tonnes)	Apatite and rock phosphate (tonnes)	Asbestos (tonnes)	Borax (tonnes)	Cryolite (tonnes)	Fluorspar (tonnes)	Graphite (tonnes)	Manganese ore (tonnes)	Sulphur (tonnes)	Crude Petroleum Quantity ('000 tonnes)	Crude Petroleum Value (Rs. million)	Petroleum products Value (Rs. million)	Zinc ore & concentrates ('000 tonnes)
1974	717	1,106,161	60,176	5,343	—	16,857	780	2,776	588,113	13,972	9,005	2,203	10
1975	724	567,314	41,514	18,996	—	4,206	404	5,430	617,115	13,669	9,792	1,995	23
1976	1,523	471,521	47,167	11,504	1,961	4,062	647	190	588,763	14,032	11,437	1,566	38
1977	1,453	951,032	65,968	10,446	2,571	4,752	698	5,048	767,346	14,850	12,845	2,521	19
1977–78	811	911,693	61,690	12,385	1,406	4,733	632	5,963	—	833,229	14,507	12,462	32,737
1978–79	91	980,575	60,573	11,305	257	5,003	505	1,858	—	722,478	14,657	12,511	81,911
1979–80	1	1,162,690	58,689	4,533	20	5,024	896	8,047	185	930,577	16,121	21,875	48,162
1980–81	—	1,201,538	63,248	12,552	30	5,264	169	12,173	226	766,241	16,247	33,489	11,344
1981–82	—	1,218,349	72,127	10,525	4	277	428	3,807	621	682,029	15,298	37,363	35
1982–83	—	913,391	48,643	5,927	50	5,024	129	2,640	592	700,001	16,949	40,437	20,511

TABLE 33

Total Value of External Trade in Mineral, Ores, Metals and
Alloys Excluding Petroleum in India During 1960–1977

(Value in Rs. '000)

Year	Export		Import	
	Minerals & Ores	Metals & Alloys	Minerals & Ores	Metals & Alloys
1960	673,546	142,717	111,767	1,584,612
1961	666,828	194,116	116,369	1,504,347
1962	606,742	50,478	153,276	1,412,967
1963	693,034	86,492	158,449	1,411,775
1964	808,528	170,121	166,154	1,624,872
1965	812,170	200,129	204,823	1,836,815
1966	1,135,588	284,007	343,265	1,412,124
1967	1,389,670	644,089	661,542	2,275,729
1968	1,620,671	977,543	655,490	1,709,018
1969	1,714,424	1,036,613	649,015	1,565,248
1970	2,032,857	1,200,090	681,083	2,419,912
1971	1,898,296	676,968	704,117	3,461,086
1972	2,368,137	499,929	791,751	3,153,514
1973	2,487,774	649,574	1,158,737	3,468,818
1974	3,128,803	1,481,321	1,613,598	5,626,294
1975	3,985,526	2,716,315	1,835,498	4,786,764
1976	5,472,590	5,334,979	2,428,035	4,000,404
1977	7,435,664	4,315,440	4,288,987	4,397,976
1977–78	8,711,000	3,845,300	17,092,000	4,832,600
1978–79	10,503,000	3,307,300	18,634,900	7,705,100
1979–80	9,215,200	1,223,600	28,053,900	12,278,000
1980–81	10,240,000	847,400	40,272,200	14,110,000
1981–82	14,279,400	911,900	44,478,100	17,786,100
1982–83	23,967,900	796,700	50,709,400	16,381,300

19

Quantitative Techniques for Estimating Mineral Demand

INTRODUCTION : THE GENERAL PROBLEM OF ESTIMATION

It will be a truism to assert that we can only observe and measure phenomena that have occurred till today. Future, definitionally, cannot be measured or quantified in advance. However often, for planning and policy-making and other purposes, we are interested in knowing how certain things like mineral demand and its output will behave in the future: their values in the coming years, the increment in values each year.

Estimation refers to the process of arriving at such future values of variables. Typically data is available on a variable, say like demand for talc/steatite for a number of years up to the present year. Such a series of observations made overtime is called a time-series. The general problem of estimation is *how to use* this time-series to arrive at a *reasonable* estimate for the future.

The problem thus resolves itself into two parts: one, finding a suitable methodology of how to use the time-series to make predictions and two, how to ensure that the prediction is reasonable. The first broadly is the domain of statistical techniques and the second of judgemental analysis. However, it should be noted that both these tools are complementary in nature and cannot be considered in isolation from one another.

A. Statistical Techniques

Statistical techniques are concerned exclusively with organising data available to us and providing a method or formula to estimate future values. Different statistical techniques have been developed.

But in estimating mineral demand there have developed primarily two different methods of employing these statistical techniques. These are: trend analysis and end-use method.

a) *Trend Analysis*

The basic idea behind trend analysis may be expressed as follows: Observation in the real world rarely correspond to a fixed pattern. For instance demand for talc/steatite does not exactly double each year. Or the output of iron ore does not precisely increase by 5 million tonnes per year. There are variations, fluctuations, ups and downs. Still if we observed the values of a particular item over a sufficiently long period of time we would find that there is an underlying pattern or tendency in these figures. The series is either secularly growing or diminishing or moving cyclically over time.

This underlying movement or rate of change of a series is called its trend. The task of trend analysis is to identify the trend for a particular series and carry it forward to make estimates for the future.

i) *The graphical method*

The main principles involved in trend analysis may easily be elucidated through a discussion of the graphical method: the simplest technique of trend analysis.

For example, consider the data on consumption of talc/steatite in India for the period of 1965 to 1972:

TABLE 34

Consumption of Talc/Steatite in India
(In thousand tonnes)

Year	Consumption	Moving-average (3 yr.) estimates
1	2	3
1966	76	—
1967	83	80.7
1968	83	84.7
1969	88	88.0
1970	93	93.0
1971	98	100.3

TABLE 34   (*Contd.*)

1	2	3
1972	110	109.7
1973	121	121.3
1974	133	133.7
1975	147	

This data has been graphed in Fig. 40. For reasons that should be obvious, it is called a scatter diagram. The various points do not fall in a fixed pattern. The movement from year to year varies very much. Such a movement cannot obviously be used to obtain estimates for the future because it is very sensitive to individual values and not to the whole series. For example, with present data, simple projection would predict a consumption demand of 160.5 thousand tonnes in 1976. However, if consumption in 1975 were, for some reasons, 130 thousand tonnes, the projected demand would be 127 thousand tonnes. If, however, consumption were 150 thousand tonnes in 1975 the estimated demand for 1976 would be 167 thousand tonnes (See Fig. 40).

A better idea in this case is to draw a line indicating the general movement or trend of the series and use it to make future predictions. By consention the trend line is a smooth, free hand curve (or line) drawn so as to best fit the data. In Fig. 40 the unbroken line represents the trend line and gives an estimate of 144.5 thousand tonnes for 1976.

The drawbacks of the graphical method are obvious. The drawing of the curve is partly a subjective process depending upon the skill of the statistician and the relative weight he attaches to different points.

ii) *The moving-averages method*

A different way to elicit the trend is to use the method of moving-averages. Moving averages are a series of arithmetic means calculated from overlapping groups of a time-series data. Each average is based on values covering a fixed time interval called the "period of moving averages" and is shown against the centre of the time interval. In calculating successive averages the period is adjusted by replacing the first observation of the previously averaged group by the next observation below the relevant group.

If the period of moving averages is odd, e.g. 3 years, 5 years, 7 years, the mean is naturally associated with the middle-year of the time-interval.

In an even period moving average, the calculated value falls between two middle-years. In this case successive moving-averages are

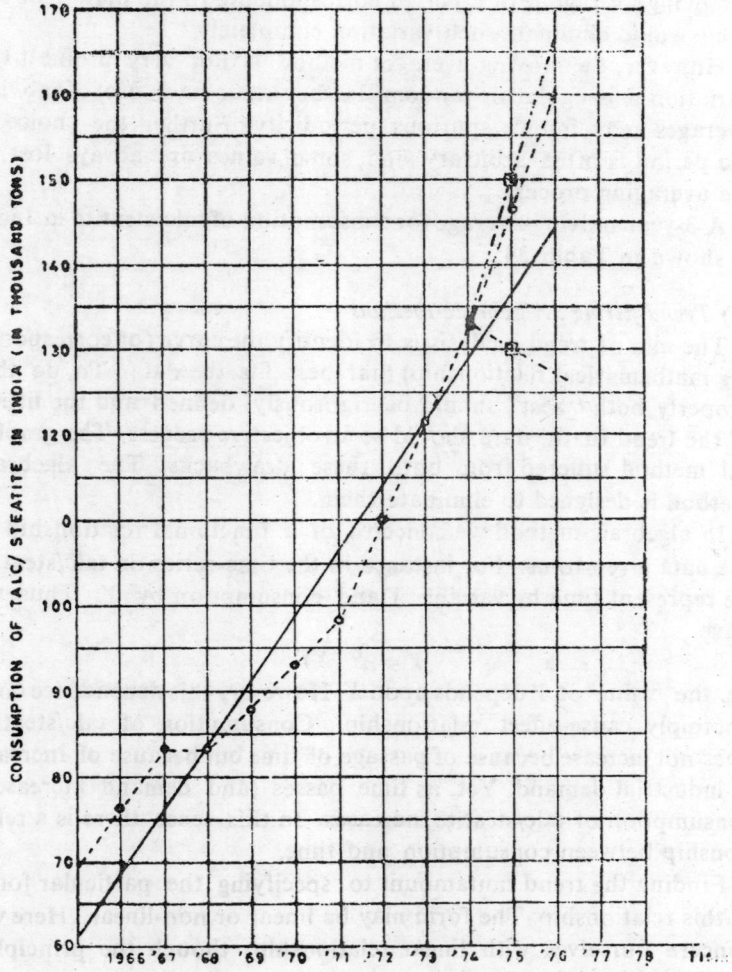

Fig. 40

themselves averaged and this new average is put against the centre
of the two previous moving averages. Thus in a 4-year period
the first value gets centred at the 3rd year, the next at the 4th and
so on.

The theory of moving averages is simple. By stabilising values
it irons out variations in the series and reduces the intensity of
fluctuations. In fact if a time series data exhibits cyclical fluctuations
a moving-average with a period corresponding to the period of the
cycle would eliminate such variation completely.

However, the moving averages method is not very useful if the
variation is irregular or random. In the latter case use of moving
averages may induce spurious periodicity. Further the choice of
the period is often arbitrary and some values are always lost in
the averaging process.

A 3-year moving-average for consumption of talc/steatite in India
is shown in Table 34.

iii) Trend fitting : Algebraic method

The aim of trend analysis is to identify the curve (or corresponding mathematical relationship) that best fits the data. To do this
properly both "best" should be rigorously defined and the fitting
of the trend to the data should be an objective process. The graphical method suffered from both these drawbacks. The algebraic
method is designed to eliminate them.

In algebraic method we conceive of a functional relationship in
the data given to us. For instance in the time-series on talc/steatite
we represent time by variable X and consumption by Y. Thus we
have

$$Y = F(X)$$

i.e. the value of Y depends upon X. However, this dependence does
not imply cause-effect relationship. Consumption of talc/steatite
does not increase because of passage of time but because of increase
in industrial demand. Yet, as time passes (and demand increases)
consumption of talc/steatite increases. In this sense there is a relationship between consumption and time.

Finding the trend tantamount to specifying the particular form
of this relationship. The form may be linear or non-linear. Here we
concern ourselves with linear relationships though the principles
elaborated hold for non-linear relations as well.

The best straight line to be fitted through the scatter of points should have the following characteristics. The sum of distances of points above the line should equal the sum of distances of those below. The sum of square of these distances should be the minimum.

Mathematically, if we denote the estimate obtained from the trend line as $\hat{Y}$, the conditions are:

$$\Sigma(Y - \hat{Y}) = 0$$

$$\Sigma(Y - \hat{Y})^2 \text{ is the minimum.}$$

In finding the best fitting line we first postulate the relationship between X and Y as

$$Y = a + bX \tag{1}$$

We have to determine the values of a and b such that the conditions of "bestness" above are satisfied. These values of a and b can be obtained by the formula:

$$a = \frac{\Sigma Y}{n}$$

$$b = \frac{\Sigma XY - n\bar{X}\bar{Y}}{\Sigma X^2 - n\bar{X}^2}$$

The essential steps involved in fitting the trend line are:

(i)   Number the year, start with 1. These represent the X values.
(ii) Prepare columns of XY and X^2. Sum the columns to compute ΣX, ΣY, ΣXY and ΣX^2.
(iii) Compute a and b, remembering that

$$\bar{X} = \frac{\Sigma X}{n}$$

Substituting these values of a and b we obtain the trend equation. This equation can now be used for forecasting.

An illustrative exercise for talc/steatite consumption in India has been carried out in Table 35.

---

(1) This is the general equation of the straight line. a refers to the distance on the Y-axis above the origin. (the y-intercept b represents the slope of the line: the change in value of Y for a unit change in X. In our context b would represent the increase in consumption per year.

TABLE 35

Years	X	Consumption Y	XY	X²	Estimated Consumption (Y')
1966	1	76348	76348	1	69404
1967	2	83173	166346	4	76951
1968	3	83346	250038	9	84498
1969	4	88275	353100	16	92044
1970	5	92759	463795	25	99591
1971	6	98108	588648	36	107138
1972	7	110392	772744	49	114685
1973	8	121356	970856	64	122232
1974	9	133093	1197837	81	129779
1975	10	146797	1467970	100	137326

From the above:

$\Sigma X = 55$ $\Sigma Y = 1033648$ $\Sigma XY = 6307682$

$\Sigma X^2 = 385$ $n = 140$

and

$$a = \frac{\Sigma Y}{n} = \frac{1033648}{10}$$

$$= 103364.8$$

$$b = \frac{\Sigma XY - n\overline{X}\,\overline{Y}}{\Sigma X^2 - n\overline{X}^2}$$

Now $Y' = a + b(X - \overline{X})$

$$= 103364.8 + 7546.8(X - 5.5)$$

$$= 61857 + 7457X$$

Using this equation the estimated values for the various years have been calculated.

For the year 1978, putting $X = 13$ we obtain the following estimate: 159968 tonnes.

The above represent the main statistical tools used to forecast mineral demand. Actual statistical estimation is a much more complex process involving exercises in regression, correlation and the estimate of standard error. Further, non-linear curves are more appropriate in describing real-life situations.

b) *The End-use Method*

In this method the main user-industries of the particular mineral are identified. For example talc/steatite in India is used by paper, insecticides, cosmetic, textile, rubber, ceramic, paint and other industries. In the next step the production and demand trend for these industries are analysed and estimates for future obtained by using the techniques described above. In these forecasts care is exercised to take into account the various economic and non-economic developments (like change in composition in inputs, in tastes and therefore demand, in investment patterns) that may modify the observed trends.

After obtaining estimates of future demand for the product of these industries, input-output exercises are carried out to determine how much of the mineral in question is required to produce one unit of output in that particular industry. Based on this input coefficient the demand for the mineral in that particular industry is worked out. Similarly demand for the mineral in other industries is worked out and these are summed up to obtain total estimated mineral demand.

The demand for talc/steatite in India by this method comes out to be 190,000 tonnes in 1978–79.

B. Judgemental Analysis

The techniques described above are all quantitative in nature. An alternative means of forecasting a given quantity or some future event is to rely on the knowledge, judgement and intuition of an expert or group of experts. Apart from simplicity, the advantage in such a procedure lies in the fact that this judgement can attempt to take into account and evaluate the importance of certain factors whose nature precludes quantitative analysis.

A recent example of this method is the Delphi technique, propounded by Olaf Helmer and T.J. Gordon. Basically the Delphi technique consists in obtaining the views from a group of experts through questionaires which are revised and adjusted by feedbacks based on the analysis of earlier responses. The ultimate objective is to reach a consensus.

Another new method of forecasting is called 'multiple contingency forecasting.' It consists in predicting and simulating alternative futures or "scenarios" based on contingencies assumed for technological, economic, social, environmental, and other relevant influences. Contingencies and assumptions for each contingency are identified, quantified and analysed. This method is preferred for long range predictions and is being used by the United States Bureau of Mines.

Typically, forecasts are rarely based on an application of only one technique, but rather on elements of several. In particular, some degree of subjective judgement is desirable in any forecasting situation.

Selected Bibliography

Mineral Economics, Political and General

Peter T. Flawn, *Mineral Resources—Geology, Engineering, Politics, Law*, Rand McNally and Co., 1966, New York.

B.C. Roy, *Indian Mineral Resources*, Industries and Economics, M.M.G.I., 1974, Calcutta.

Steve H. Hawke, *Mineral Economics—Its Past and Future*, Colorado School of Mines Magazine, pp. 22-23, April 1969.

W. Sealey, Mudd Series, *Economics of the Mineral Industries*, A.I. M.E., 3rd Edition, 1974, New York.

W.K. Buck, *Mineral Economics—Its Definition and Application*, M.R. 127, Department of Energy, Mines and Resources, Canada.

H.F. Bain, C.K. Leith, and others, *Mineral Economics*, A.I.M.E. Series, McGraw-Hill Book Company Inc. 1932.

Walter H. Voskuil, *Minerals in World Industry*, McGraw-Hill, 1955.

Leith, Furness and Lewis, *World Minerals and World Peace*, Brookings Institution, 1943.

C.K. Leith, *World Minerals and World Politics*, Whittlesey House, McGraw-Hill, 1931.

T.S. Lovering, *Minerals in World Affairs*, Prentice Hall, 1943.

C.K. Leith, and others, *Elements of a National Mineral Policy*, Prepared by the Chairman, Mineral Enquiry Committee, New York, 1933.

John B. DeMills, *Strategic Minerals*, McGraw-Hill, 1947.

C.A. Roush, *Strategic Minerals Supplies*, McGraw-Hill, 1939.

A.K. Madan, *Economic Survey of Minerals in India,* Economics, Industrial Publications, New Delhi.

Dennis L. Meadows, *The Limits to Growth,* The New American Library, Inc. N.Y. 1972.

D. Cartisle, *Natural Resources,* McGraw-Hill.

Paul M. Tyler, *From the Ground Up,* McGraw-Hill, 1948.

R.K. Sinha, *A Treatise on Industrial Minerals of India,* Allied Publishers Pvt. Ltd., Bombay, 1967.

A.K. Dey, "Role of Geology in Defence", *Metals and Minerals Review,* January, 1965.

N.V. Sovani, *The European Economic Community,* J.S.S. Institute of Economic Research, Dharwar.

K.W. Clarfield and others, *Eight Mineral Cartels,* McGraw-Hill, 1975.

Mineral Legislation and Taxation

B. Rao, Inner Space: A natural claim or a U.N. takeover, *The Indian Year Book of International Affairs,* 1968.

C. Richard Tinsley, Mining of Manganese Nodules: An Intriguing Legal Problem, E/MJ, October, 1973.

S.N. Bose, *Indian Labour Code* Eastern Law House (Pvt.) Ltd., Calcutta.

Northcutt Ely, *Summary of Mining and Petroleum Laws of the World,* U.S. Bureau of Mines, Inf. Cir, 8544, 1972.

James Boyed, *The Future of Exploration and the Mineral Industries,* A lecture presented at Stanford University, January, 1963.

E.C. Hodgson, *Digest of Mineral Laws of Canada,* Mineral Resources Division, Department of Energy, Mines and Resources, Ottawa, 1966.

E.C. Hodgson and W.J. Beard, *Summary Review Federal Taxation and Legislation Affecting the Canadian Mineral Industry,* Mineral Inf. Bull., M.R. 82, 1966, Mineral Resources Div., Ottawa.

T.L. Gibbs, *The Part Played by the Government in the South African Mining Industry* (Trans. of the Seventh Commonwealth Mining and Metallurgical Congress) South African Institute of Mining and Metallurgy, 1961.

R.B. Tombs, *A Survey of the Mineral Industry of Southern Africa,*

Mineral Inf. Bull., M.R. 58, Mineral Resources Div., Dept. of Mines and Technical Surveys, Ottawa.

T. Gonsalves and B. Banerjee, *Mineral and Mining Laws of India*, Manager of Publications, Government of India.

M.J. Pandya, *Mineral Economics of Taxation in India*, Mining Geological and Metallurgical Institute of India.

S.K. Borooah, *The Mineral Laws in India*, Indian Mining Journal, March, 1954.

IBM, Digest of Minor Mineral Laws of India, published by Indian Bureau of Mines, 1974.

ECAFE, *Survey of Mining Legislation with Special Reference to Asia and Far East*, 1957.

N.L. Sharma and R.K. Sinha, *Mines and Minerals Legislation in India*, Dhanbad Publications.

U.N.O.—*U.N. Conference on the Law of the Sea*, A/Conf/13158, Pub. No. 58, V4, Vol. II.

P.L. Charles, "Fiscal Policies and Development", *Eastern Economist*, January 1967.

S.L. Rai, Mining Laws of the World, 1987.

Mineral Resources

Peach, *Zimmermann's World Resources and Industries*, 3rd Edition, 1972.

S. Krishnaswamy, *India's Mineral Resources*, Oxford & IBH Publishing Co. Pvt. Ltd., New Delhi, 1979.

J.C. Brown and A K. Dey, *India's Mineral Resources*, Oxford, 1955.

Indian Bureau of Mines, *Indian Minerals Year Book*—Annual Numbers.

Indian Bureau of Mines, *Bulletin of Mineral Statistics and Information*.

M.L. Sethi, *The Mineral Resources of Rajasthan*, Government of Rajasthan.

B.C. Roy, *The Economic Geology and Mineral Resources of Rajasthan*, G.S.I. Mem. 86.

M.S. Krishnan, *Mineral Resources of Madras*, G.S.I. Mem. 80.

J.A. Dunn, *The Economic Geology and Mineral Resources of Bihar Province*, Vol. 78, G.S.I.

World Mining, *Japanese Supplement*, April 1967.

World Mining, *Nepal—Mineral Discoveries Offer New Mineral Opportunity*, p. 35, August 1967.

U.S. Bureau of Mines, *Mineral Facts and Problems*.

United Nations, *World Iron Ore Resources (E 2655 ST/ECA/27)* 1955.

Dept. of Mines (Ottawa), *Canada and the World*, Rept. 860, 1957.

D.B. Simkin, *Minerals—A Key to Soviet Power*, 1953.

Mineral Resources of World (Atlas), Van Royen and Bowles, 1952.

N.L. Sharma, K.S.V. Ram, *India's Economic Minerals*, Dhanbad Publications, Dhanbad.

R.K. Sinha and N.L. Sharma, *The Mineral Riches of West Asia*, Ecohomic Times, November 6, 1967.

World Mining, *African Mining Today*, July 1966.

Harold A. Quinn, *Geology and Mining in Ethiopia*, World Mining February and March 1964.

World Mining—*Recent Mineral Discoveries in Tibet and Sinkiang*, November 1966.

S.K. Borooah, *Minerals in Madhya Pradesh*.

R.K. Sinha and N.L. Sharma, *World Resources of Minerals and their Strategic Importance*, Metals and Minerals, Review, Oct., Nov., Dec., 1967.

Mining Magazine, *Mauritania Vast Mineral Wealth*, April 1968.

Historical

Capt. Munn, *Gold Mining in South India*, Tran. Min. Geol. Institute of India, Vol. XXX.

R.C. Mazumdar, *Cultural Heritage of India*, Vol. III.

Tagore, *Manimala*.

P.C. Ray, *History of Chemistry in Ancient and Medieval India Incorporating History of Hindu Chemistry*, Indian Chemical Society, Calcutta, 1966.

Cultural India's Heritage, *Ramkrishna Centenary Volume*.

S. Piggot, *Prehistoric India,* 1950.

P. Neogi, *Copper in Ancient India*, Calcutta.

P. Neogi, *Iron in Ancient India*, Calcutta.

R. Shamasastry, *Kautilyas Arthasastra*, Mysore, 1929.

Specific Minerals and Metals

Geological Survey of India, Various Economic Bulletins, Records and Memoirs.

N.L. Sharma and K.S.V. Ram—*Introduction to Geology of Coal and Indian Coal Fields*—Dhanbad Publications, Dhanbad.

Stanley Abkowitz and others, *Titanium in Industry*, Technology of Structural Titanium, D. Van Nostrand Co., Inc., New York, 1955.

A. Nelson, *Beach Sands of Australia*, Mining and Minerals Engineering, Sept. 1965.

D.R. Horn et. al., Ocean Manganese Nodules, Metal Values and Mining Sites, *Technical Report No. 4*, NSF—GX 33616, 1973, Washington.

IBM, *Monograph on Bauxite*, 1977.

IBM, *Manganese ore, Facts and Problems*, 1974.

A.A. Linari, *Occurrence, Mining and Recovery of Diamonds*, Mining and Minerals Engineering, August and Sept. 1968 London.

Anon., *Digboi Completes 75 Years*, Digboi Batori, November 1965.

W.H. Emmons, *Gold Deposits of the World*, 1937.

H.C. Jone, *The Iron Ore Deposits of Bihar and Orissa*, G.S.I., Mem. LXIII, Pt. 2.

M.S. Krishnan, *Iron Ore, Iron and Steel*, Bull. G.S.I. Series A, No. 9.

L.L. Fermor, *Manganese Ore Deposits of India*, G.S.I., Mem. XXVII.

D.N. Vishnoi, *Oil from Coal, a Necessity*, Metals and Minerals Review, July 1960.

World Mining, *Russian Manganese Ore Deposits*, November 1967.

Field Mining and Economic Geology

R.N.P. Arogyaswamy, *Courses in Mining in Geology*, Oxford & IBH Publishing Co. Pvt. Ltd., New Delhi, 1980.

V.M. Kreiter, *Geological Prospecting and Exploration*, Mir Publications, Moscow, 1968.

A.A. Archer, *Progress and Prospects of Marine Mining*, Mining Magazine, March 1974.

G.S.I., *A Manual on Mineral Exploration and Borehole Deviation*, Sept. 1970 (unpublished).

U.S.B.M., *Model as a Basis for Making Decisions during Mineral Deposit Evaluation*, Rept. of Inv. No. 6778.

B.A Kennedy and E.J. Wade, *Feasibility Studies for large open pits mines*, World Mining, 1972.

H.E. McKinstry, *Mining Geology*, Asia Publishing House, 1960.

G.J. Young, *Elements of Mining*, McGraw-Hill, 1951, New York.

R.D. Parks and C.H. Baxter, *Examination and Valuation of Mineral Property*, 1957.

Robert Peele and John A. Church, *Mining Engineer Handbook*, John Wiley and Sons, New York.

M.R. Huberty and W.L. Flock, *Natural Resources*, McGraw-Hill.

USGS, Bull. 1450—A, 1976 Principles of the Mineral Resource Classification System of the USBM and USGS.

GSI, *Report of the Committee on Standardisation of Terminology and Classification of Ore and Mineral Reserves*, Mis. Pub. No. 58, June 1981.

C.C. Popoff, *Computation of Reserves of Mineral Deposits, Principles and Conventional Methods*, USBM, Inf. 8283.

J.D. Forrester, *Principles of Field and Mining Geology*, John Wiley and Sons, New York, 1946.

A.M. Bateman, *Economic Mineral Deposits* (Indian Edition), Asia Publishing House, 1959, Bombay.

W. Lindgern, *Mineral Deposits*, McGraw-Hill, 1931, New York.

Shevyakor, *Mining of Mineral Deposits*, Foreign Language Publishing House, Moscow.

S.K. Borooah, *Economic Mineral Deposits of India*, Sewati Prakash Bhawan, 1963, Nowgong, Assam.

D.N. Wadia, *Geology of India*, Macmillan and Co., 1953, London.

M.S. Krishnan, *Geology of India and Burma*, Higginbothams (Private) Ltd., Mount Road, Madras-2.

Mehdiratta, *Geology of India*, Burma and Pakistan.

F.H. Lahee, *Field Geology*, McGraw-Hill, New York.

F.D. Adams, *The Birth and Development of the Geological Sciences*, Dover Publication, 1954, New York.

P.C. Pande, *Economic Minerals of India*, Dattsons Press, Nagpur.

S J. Truscott, *Mine Economics*, Mining Publications Ltd., London, 1962

Krafts and Hawkins, *Applied Reservoir Engineering*.

Mineral Dressing

E.J. Pryor, *Mineral Dressing*. Mining Publication Ltd., London.

A.F. Taggart, *Handbook of Mineral Dressing*, Chapman and Hall Ltd., London, 1945.

A.M. Gaudin, *Principles of Mineral Dressing*, McGraw-Hill, 1939.

F.H. Micheli, *The Practice of Mineral Dressing*, Mine and Quarry Engineering, London.

S.J. Truscot, *Textbook on Ore-Dressing*, Macmillan and Co. Ltd., London, 1923.

A.F. Taggart, *Handbook of Mineral Dressing in Ores and Industrial Minerals*, John Wiley, New York.

Industrial Growth and Statistical

Planning Commission, *Programme of Industrial Development during 1956–61*, Manager of Publications, Delhi, 1956.

Indian Bureau of Mines, *Indian Minerals Year Book*, Annual Numbers.

Indian Bureau of Mines, *Monthly Bulletin of Mineral Statistics and Information*.

C.S.I.R , *The Wealth of India*, Industrial Products, New Delhi.

United States, *Mineral Year Book*, Annual Numbers

Bureau of Mines, *Mineral Trade Notes*, Monthly, Washington.

Bureau of Mineral Resources, Geology and Geophysics, *The Australian Mining Industry*, Canberra, Annual Numbers.

Canadian Mineral Year Book, Mineral Resources, Deptt. of Energy, Mines and Resources, Ottawa.

Director General of Commercial Intelligence, *Monthly Statistics of Foreign Trade of India*.

Director General of Mines Safety, *Quarterly Bulletin of Metalliferous Mines in India,* Dhanbad.

World Mining, Annual Numbers.

The Mining Journal, Annual Numbers.

G.S.I. *Quinquennial Review.*

G.S.I., *Thirteen Year Review of Mineral Production in India,* Record 80 (1933–46).

G. D. Kalra, *Calcium Carbide Industry of India,* Metals and Minerals Review, Oct. 1967.

Cement Manufacturers Association, 50 years of the Cement Industry in India, 1914–1964, Bombay.

Mining Geological and Metallurgical Institute of India, Golden Jubilee Commemorative Volume—Progress of the Mineral Industry of India, 1906–1955.

Asoka Mehta, *Fertiliser: Reassuring Prospects,* Commerce, Annual Number, 1967.

C R. Ranganathan, *Meeting Farmer's Need: Reassuring Prospects,* Commerce, Annual Number, 1967.

Anon., *Fertiliser Requirements during 2nd Plan, Technology,* Vol. No. 2, Sindri.

ECAFE, *Chemical Age of India,* Proceedings of a ECAFE Conference on the Development of the Fertiliser Industry in Asia and the Far East, March, 1964.

Iron and Steel Review, Worldwide distribution of L.D. Steel Plants, Nov. 1964.

R.K. Sinha, Cement Industry, *The Economic Times,* 7th Sept. 1970.

Chemical Age of India, *Indian Cement Industry, 1982,* Special issue, July 1982.